DETAILED INTERPRETATION OF CONTEMPORARY ARCHITECTURAL DESIGN

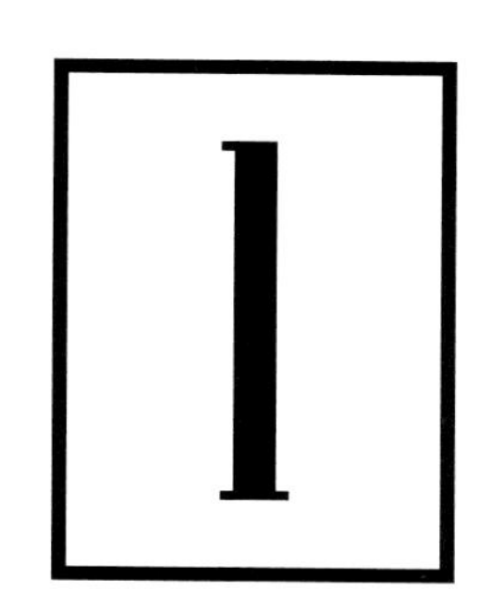

当代建筑设计详解

佳图文化 主编

中国林业出版社

图书在版编目（CIP）数据

当代建筑设计详解 . 1 / 佳图文化主编 . -- 北京 : 中国林业出版社 , 2018.10

ISBN 978-7-5038-9791-7

Ⅰ . ①当… Ⅱ . ①佳… Ⅲ . ①建筑设计－作品集－中国－现代 Ⅳ . ① TU206

中国版本图书馆 CIP 数据核字 (2018) 第 239859 号

中国林业出版社
责任编辑：李 顺 薛瑞琦
出版咨询：（010）83143569

出 版：中国林业出版社（100009 北京西城区德内大街刘海胡同 7 号）
网 站：http://lycb.forestry.gov.cn/
印 刷：固安县京平诚乾印刷有限公司
发 行：中国林业出版社
电 话：（010）83143500
版 次：2019 年 9 月第 1 版
印 次：2019 年 9 月第 1 次
开 本：889mm×1194mm 1 ／ 16
印 张：20
字 数：200 千字
定 价：298.00 元

Preface 前言

"Modern Chinese Architecture" series is a set of professional books that introduce and present modern Chinese architecture comprehensibly. It selects the excellent works with different design concepts of different types that could reflect the level and development trend of contemporary Chinese architecture the most to represent the style and features. It revolves around the architectural works and analyzes further into overall planning, design concept, challenge, architectural layout, detailed design, relationship between the old and new in reconstruction, relationship between the building and the city, etc, thus to interpret the design essence from various angles. The selected works include office buildings, education buildings, commercial buildings, cultural buildings, hotel buildings, sports buildings, transportation buildings, healthcare buildings and so on. All the works are introduced with plans, elevations, sections, construction drawings, joint details and photos, presenting diverse architectural forms and details in a comprehensive and visual way for dear readers.

《当代中国建筑》系列图书为一套全面介绍和展示当代中国建筑的专业建筑类书籍。全书精选最能反映当代中国建筑水平和发展趋势的优秀作品，通过对这些不同类型、不同设计理念的建筑作品的展示和解读，反映当代中国建筑的风貌。全书内容围绕建筑设计案例展开，深入分析建筑案例的整体规划、设计构思、设计难点、建筑布局、细部设计、改扩建中新与老的联系、建筑与城市的关系等众多方面的问题，以期从各个角度展现建筑案例的设计精髓。所选案例囊括了办公建筑、教育建筑、商业建筑、文化建筑、酒店建筑、体育建筑、交通建筑、医疗建筑等众多建筑类型。案例中展示的平面图、立面图、剖面图、施工图、节点详图，以及建成实景图等，将不同的建筑形式以及建筑细节更为详尽、直观地呈现出来，以供关注当代中国建筑状况的读者借鉴、参考。

Contents

目录

Office Building 办公建筑

Office Building
办公建筑

Hotel Building
酒店建筑

Commercial Building
商业建筑

Contents

目录

Commercial Building 商业建筑

Hotel Building 酒店建筑

Office Building
办公建筑

Huge Volume
体量宏大

Elegant Facade
立面简洁

Highly Intelligent
高智能化

Ecological and Energy-saving
生态节能

KEY WORDS
关键词

L-shaped Terrain “L”形地形

Spatial Level 空间层次

Reasonable & Convenient 合理便捷

Anhui Province Administrative Center

安徽省政务服务中心大厦及综合办公楼

FEATURES
项目亮点

The project is divided into three parts, such as government affairs service center, comprehensive office building, talent exchange center & convention center, according to functionality and modular design principle.

根据功能划分和模块化的设计原则，设计将建筑主要分为三部分，分别为政务服务中心、综合办公楼、人才交流中心及会议中心。

Location: Hefei, Anhui, China
Architectural Design: Tongji University Architectural Design (Group) Co., Ltd.
Total Floor Area: 67,000 m^2

项目地点：中国安徽省合肥市
设计单位：同济大学建筑设计研究院（集团）有限公司
总建筑面积：67 000 m^2

Overview

Located in the northwest corner of the intersection of Maanshan Road and Taihu Road in Hefei, this highly intelligent administrative center with advanced equipment boasts a total floor area of 67,000 m^2.

项目概况

安徽省政务服务中心大厦及综合办公楼位于合肥马鞍山路与太湖路交口西北角，总建筑面积 6.7 万 m^2。大厦智能化程度高，设备先进。

Site Plan 总平面图

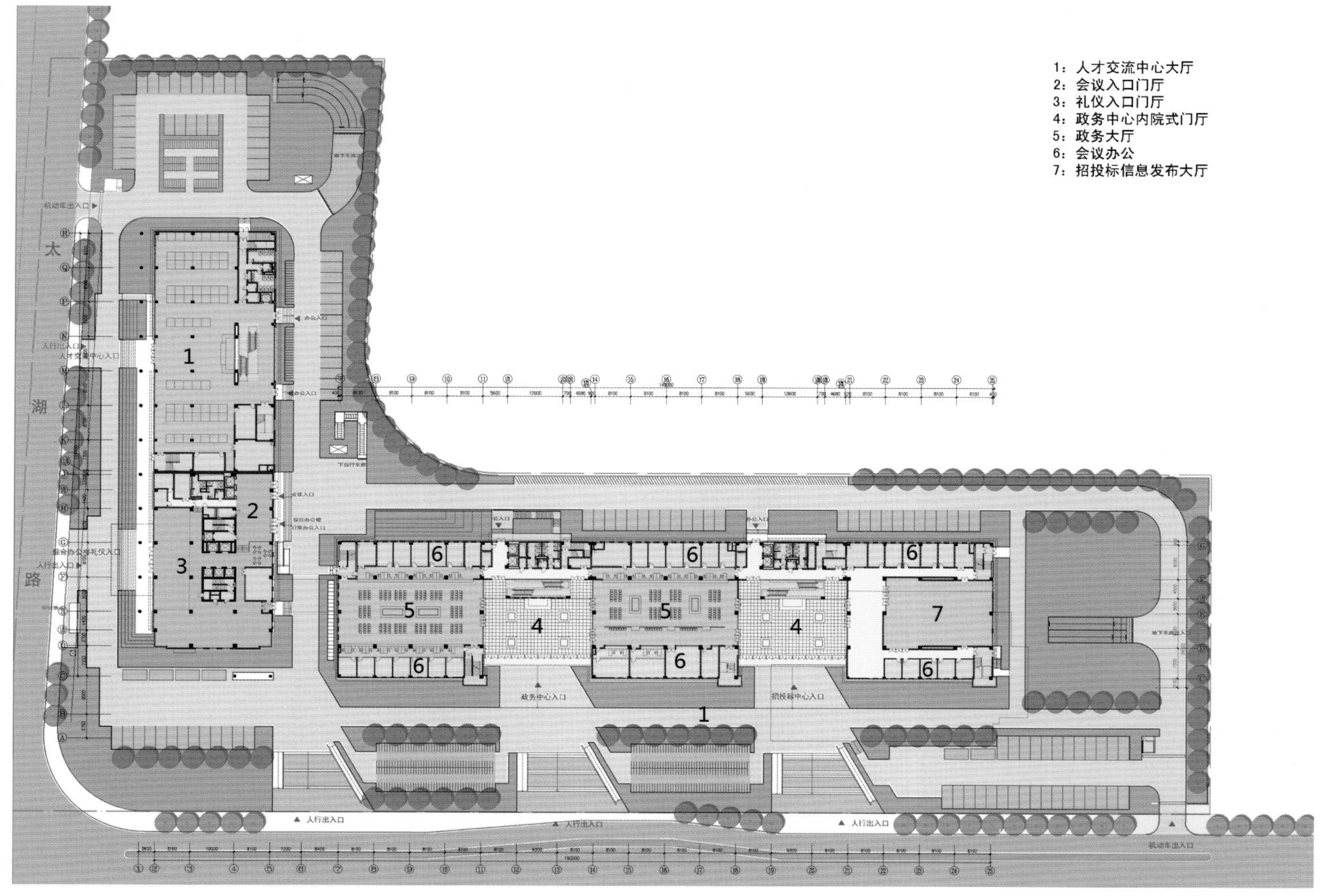

First Floor Plan　一层平面图

Overall Layout

The project is divided into three parts, such as government affairs service center, comprehensive office building, talent exchange center & convention center, according to functionality and modular design principle. Since the entire site is L-shaped and in consideration of the orientation of the main building, the amount of people flow, city landscape and the impact on the northwest residential land, three building volumes are arranged as follows:

Comprehensive office building (high-rise tower) is arranged in the corner of the L shape, facing south to demonstrate its distinct iconic image, which is conducive to facilitate the relation between the main offices and the offices in the other two parts, and has the minimal impact on the sunlight for the northwest residential land.

Government affairs service center is arranged along Maanshan Road and completely isolated from the high-rise tower, which can form a complete urban interface and ease the flow of people. It will function reasonably and conveniently in this way.

Talent exchange center is arranged along the urban interface along Taihu Road. Thanks to the nearby vehicle entrance and ground parking lot, it allows people to move quickly. In addition, the convention center serves both the government affairs center and the society, which improves efficiency.

总体布局

根据功能划分和模块化的设计原则，设计将建筑主要分为三部分，分别为政务服务中心、综合办公楼、人才交流中心及会议中心。由于整个基地呈L形，考虑到主楼的朝向、人流量的大小、城市景观以及对西北面居住用地的影响，三个建筑体量做如下布置：

综合办公楼部分（即高层塔楼）布置在L形转角处，长边朝南，便于展示其标志性，在道路转角处塑造鲜明的建筑形象，同时也有利于建立主要办公机构与其余两部分办事机构之间的联系，并且对西北面居住用地的日照影响最小。

政务服务中心部分沿马鞍山路布置，与塔楼基本分开，可形成完整的城市界面，同时方便大量人流的进出，功能安排上合理且便捷。

人才交流中心部分沿太湖路的城市界面布置，以靠近基地车行出入口和地面停车场，有利于人流快速集散，同时会议中心既可对内服务于政务中心内部，又可对外服务于社会，有利于提高效率。

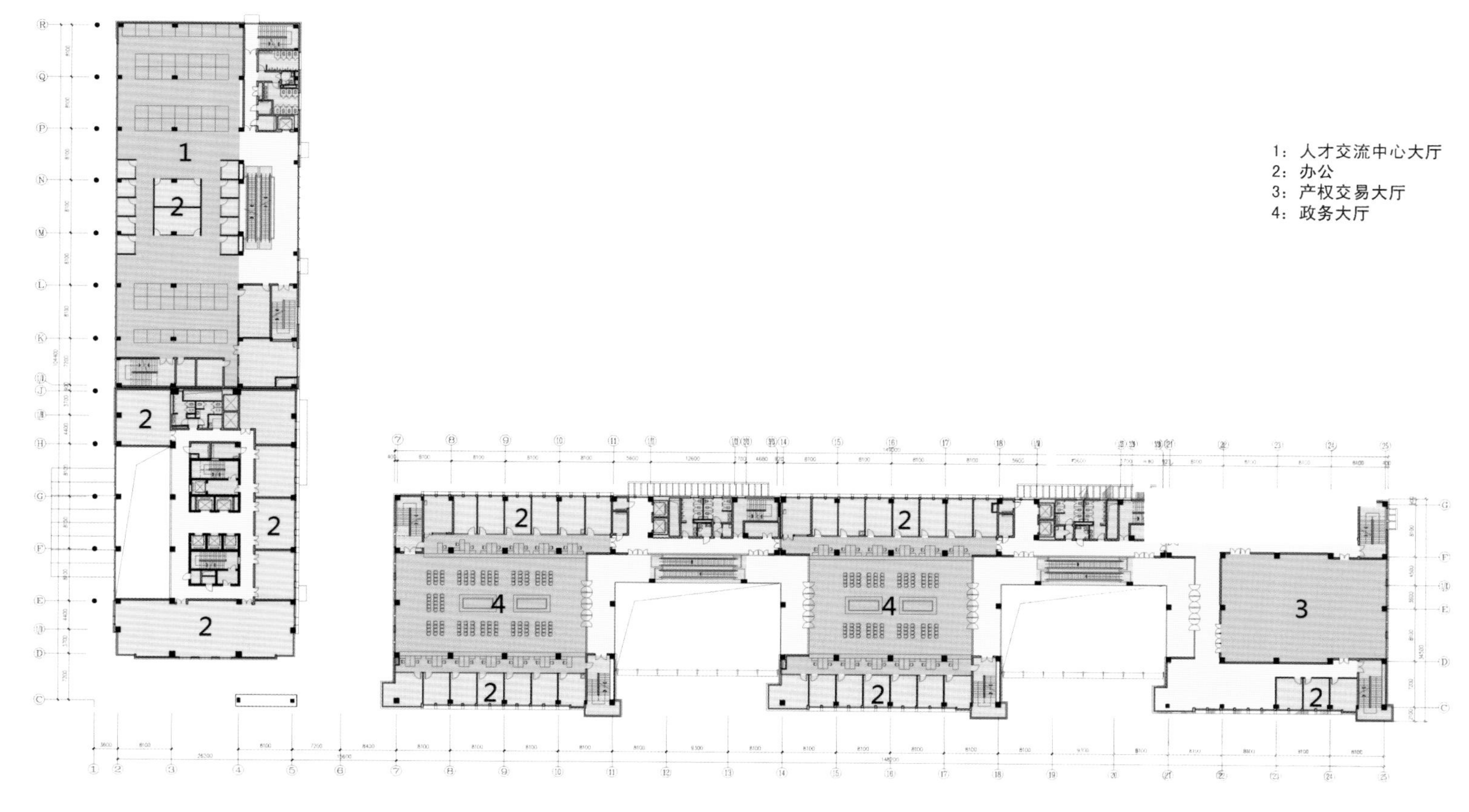

1：人才交流中心大厅
2：办公
3：产权交易大厅
4：政务大厅

Second Floor Plan 二层平面图

Landscaping

Green can be viewed around the site and along the street. A greenbelt, isolation belt, is created between the building and the urban space to prevent the external interference. At the bottom of the office building, a large area of water and falling water landscape are developed around the street corner, which provides a favorable transition space for the interior and the exterior on one hand and enriches the city landscape on the other hand. Green grass slope is shaped at the bottom at the talent exchange center with several prominent glass objects embedded, which presents an interesting environment and solves the lighting problem for the network computer room under the slope.

景观绿化

在基地周边界面以及沿街面设计周边绿化，在城市空间与建筑之间形成一条绿化缓冲隔离带，以屏蔽外界环境对基地内建筑的干扰。在综合办公楼底部，街道转角处打造大片水面以及叠水景观，一方面有利于办公楼营造良好的室内外过渡空间，另一方面有利于形成城市转角处丰富的空间层次，为城市景观做出贡献。在人才交流中心及报告厅部分底层设计绿化草坡，几个突出的玻璃体嵌入其中，既形成了丰富有趣的环境空间，又解决了草坡下网络机房等的采光问题。

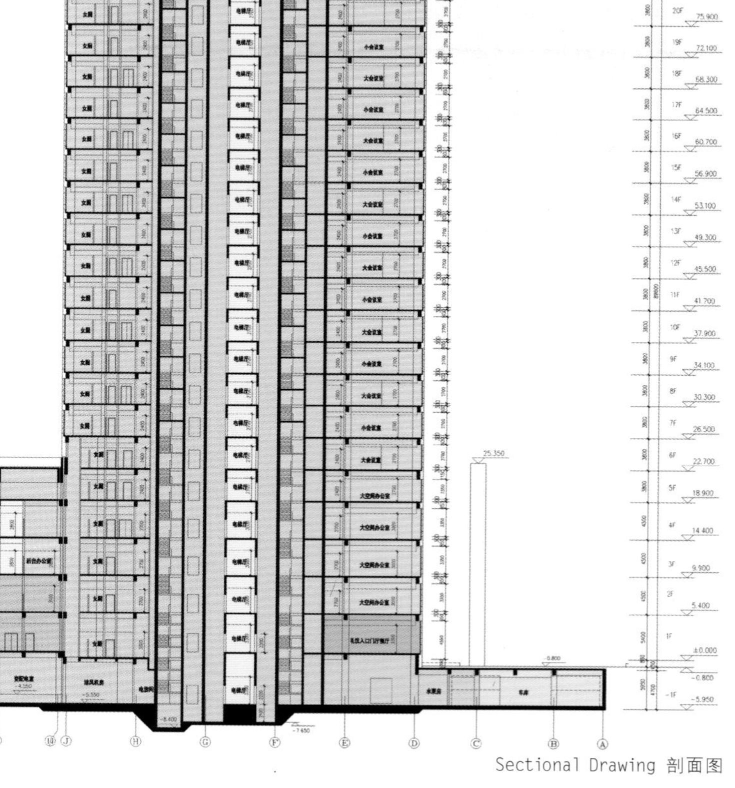

Sectional Drawing 剖面图

Traffic Flow

Two entrances for vehicles are set on both ends (the north and the west) of the site and are connected by dual carriage way. And a single carriageway is arranged around the office building and talent exchange center & convention center only for office staff, right side in and left side out. A loop comes into being and separates the office staff with outside people.

交通流线

基地最北端马鞍山路以及基地最西端分别设置两个车行出入口，两出口之间以双行车道联结，沿综合办公楼和人才交流中心及会议中心东南两侧布置一圈单行车道，右进左出，设置门卡，仅供内部办公人员使用。这样既使内部交通流线形成环路，又将办公人员与外来人流分离。

KEY WORDS
关键词

Modern & Generous **现代大气**

Succinct Shape **形体简洁**

Rich Facade **立面丰富**

Dalian International Shipping Building
大连国际航运大厦

FEATURES
项目亮点

The facade is comprised of stone curtain wall and glass curtain wall. Space variation and contrast of straight lines and arcs reflect the subtle relationship beneath the overall form. Detailed concave-convex changes emphasize the light and shade contrast and change. In general, the overall style is simple, modest, modern and generous.

立面主要为石材幕墙与玻璃幕墙相结合。通过直与弧的空间变化和对比体现整体形体关系。细部突出局部的凹凸变化，强调立面的光影明暗对比与变化。整体风格简洁稳重、现代大气。

Location: Dalian, Liaoning, China
Architectural Design: Dalian Architectural Design & Research Institute Co., Ltd., Mo Atelier Szeto

项目地点：中国辽宁省大连市
设计单位：大连市建筑设计研究院有限公司、莫平建筑事务所

Overview

This project is an office building located in a flat plot in Dalian Bonded Area. The site area is 25,610 m^2 and the plot ratio is 2.36. It's a comprehensive office building.

项目概况

本工程位于大连市保税区，地势平坦。基地用地面积为 25 610 m^2, 容积率为 2.36。建筑性质为综合性办公建筑。

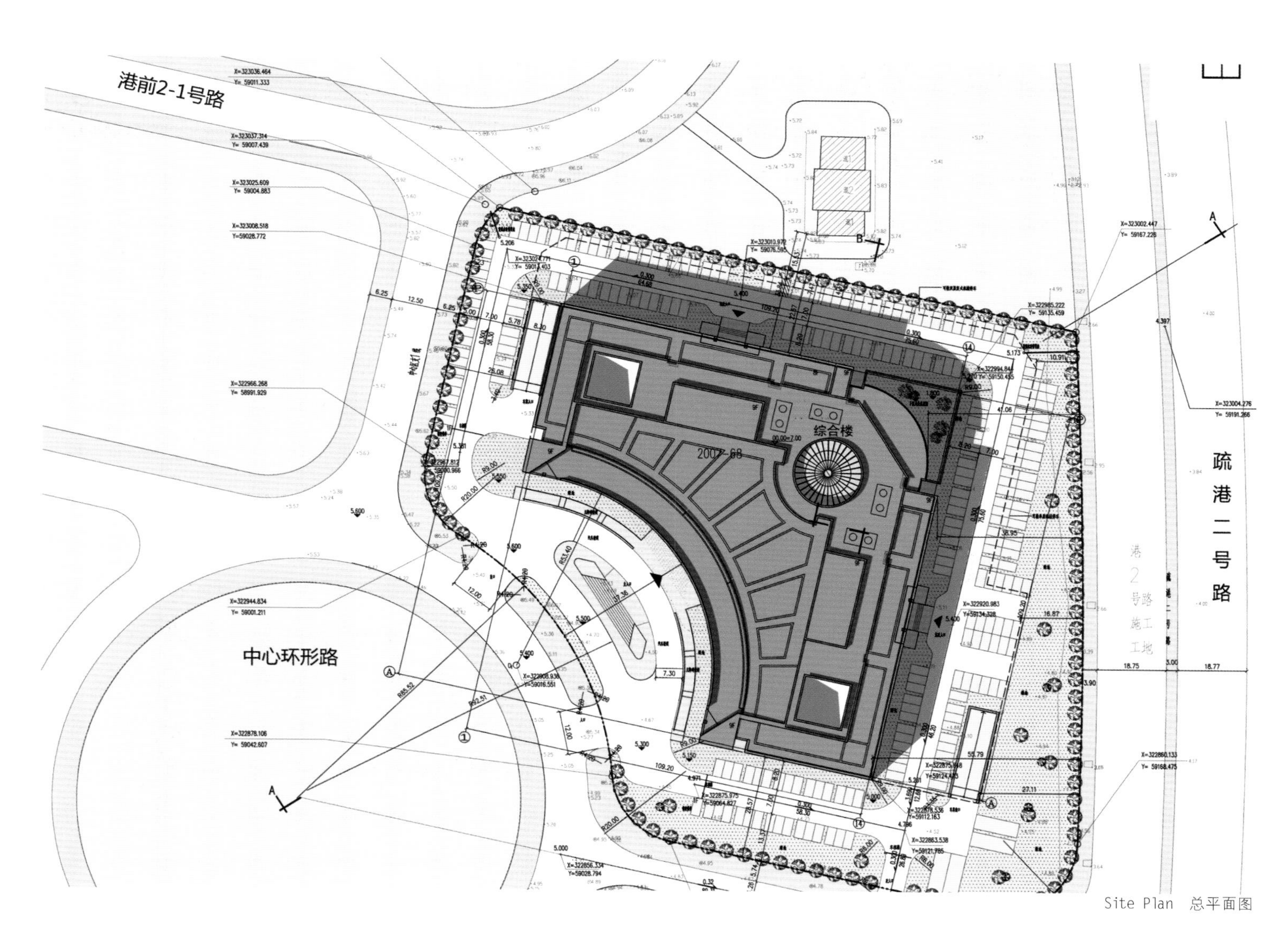

Site Plan 总平面图

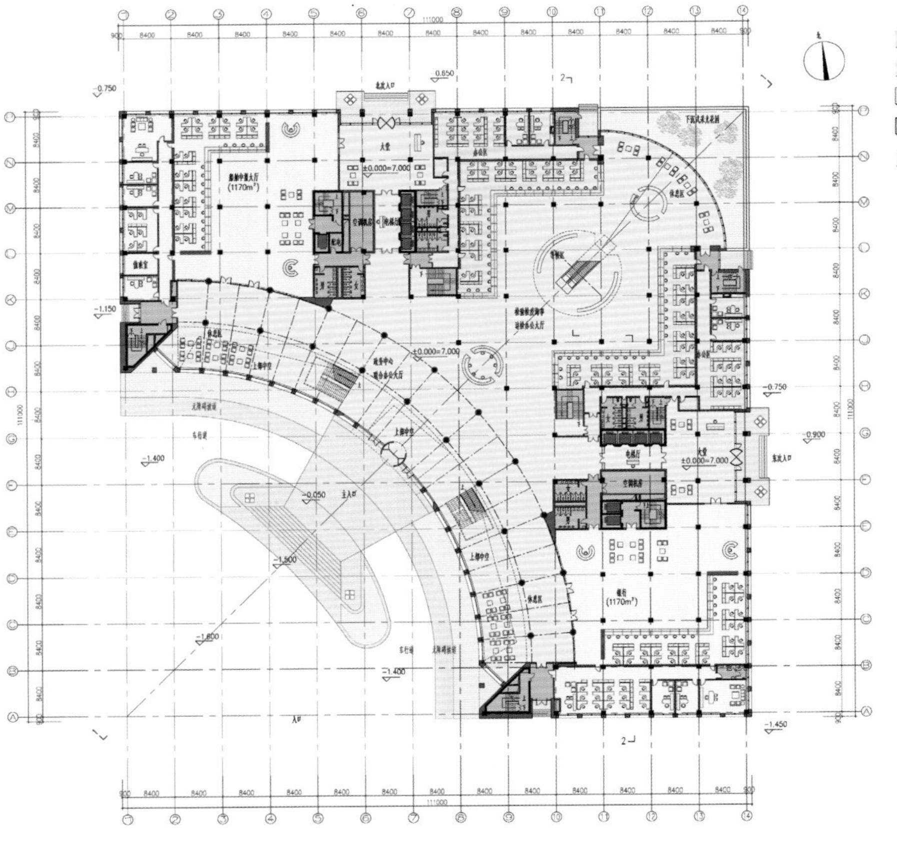

办公区域
交通区域
商业区域
服务,设备区域
绿化区域

Plan 平面图

Architectural Design

The gross floor area is 74,252 m^2 and the building height is 40.50 m. The 1st to 8th floor is comprehensive office area and the 9th floor is the plant room. The layout is appropriate and the horizontal circulation is concise and clear. Different interlaminations and varied atrium & outdoor courtyard link up each floor, creating a changeful integral space. The facade is comprised of stone curtain wall and glass curtain wall. Space variation and contrast of straight lines and arcs reflect the subtle relationship beneath the overall form. Detailed concave-convex changes emphasize the light and shade contrast and change. In general, the overall style is simple, modest, modern and generous.

建筑设计

建筑总面积为 74 252 m^2，建筑总高度为 40.50 m。地上八层为综合办公功能区；九层屋面为机房层。本工程方案平面布局合理，层间横向流线简洁、明晰；竖向通过不同层间、不同形式的中庭、室外庭院将各层有机联系成富于变化的整体空间。立面主要为石材幕墙与玻璃幕墙相结合。通过直与弧的空间变化和对比体现整体形体关系。细部突出局部的凹凸变化，强调立面的光影明暗对比与变化。整体风格简洁稳重、现代大气。

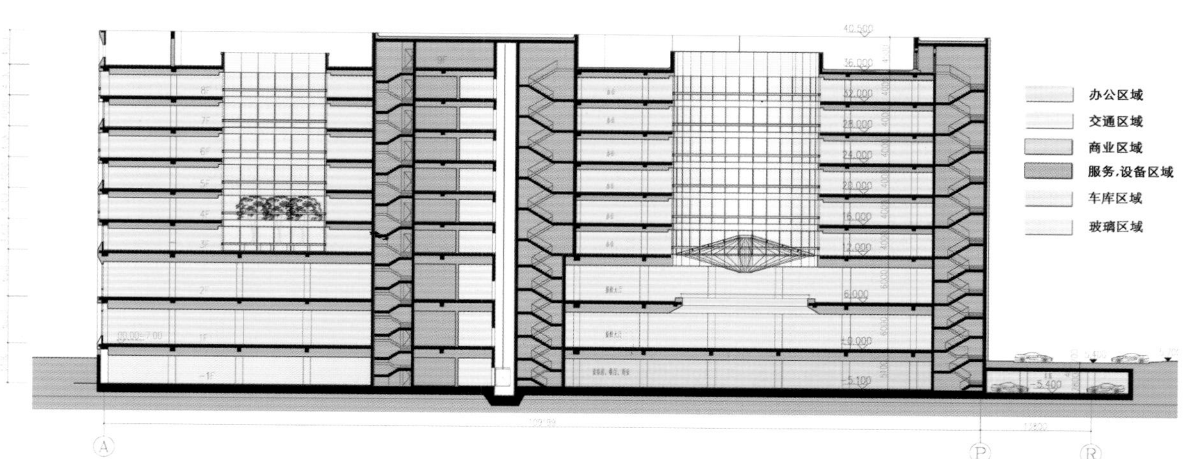

Elevation 立面图

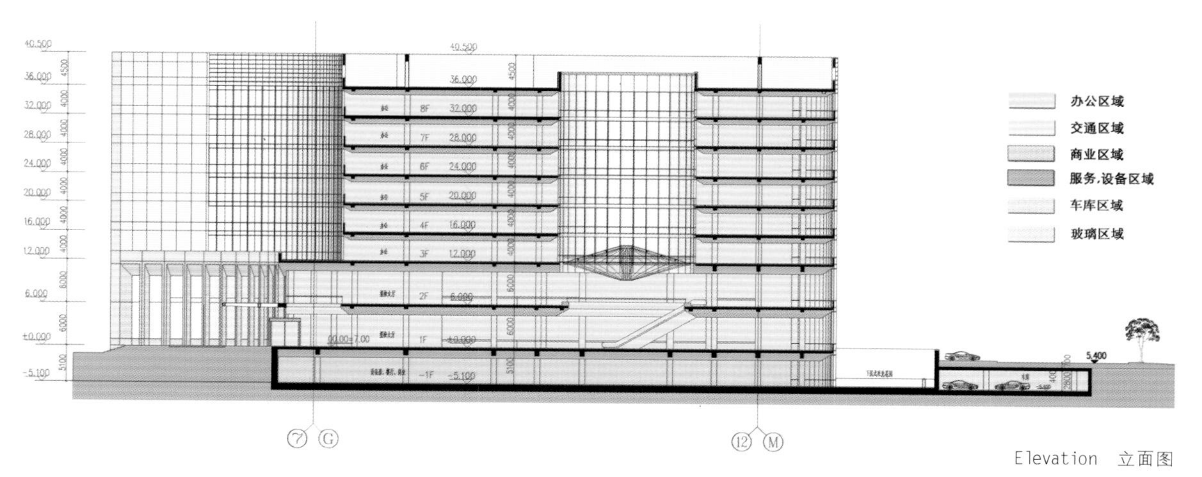

Elevation 立面图

电梯→

KEY WORDS 关键词

Beautiful Environment 环境优美

Core Building 核心建筑

Into the Environment 融入环境

Fujian Public Security Intelligence Control Room

福建公安指挥情报中心

FEATURES 项目亮点

This project emphasizes an axial sequence, and the main part is located at the end of the axis as the most important core building in the site.

项目设计强调一种轴向的序列，建筑主体位于大院轴线的收官之处，成为统领院区最重要的核心建筑。

Location: Fuzhou, Fujian, China
Architectural Design: Beijing Institute of Architectural Design, Fujian Architectural Design Institute
Land Area: 2,407.49 m^2
Floor Area: 16,086.37 m^2
Overground Floor Area: 9,706.67 m^2
Underground Floor Area: 6, 379.7 m^2
Building Height: 21.15 m

项目地点：中国福建省福州市
设计单位：北京市建筑设计研究院有限公司、福建省建筑设计研究院
建筑用地面积：2 407.49 m^2
建筑面积：16 086.37 m^2
地上建筑面积：9 706.67 m^2
地下建筑面积：6 379.7 m^2
建筑高度：21.15 m

Architectural Design

This project emphasizes an axial sequence, and the main part is located at the end of the axis as the most important core building in the site. Landscape environment is taken into account and a number of lush trees are reserved as much as possible to add a green view for the building, which not only upgrades the landscape of the courtyard, but forms a lee and sunny beautiful environment with flourishing trees & lawns and fragrant flowers.

建筑设计

项目设计强调一种轴向的序列，建筑主体位于大院轴线的收官之处，成为统领院区最重要的核心建筑。建筑形态从景观环境出发，通过庭院和形体上的组织，将基地上若干棵郁郁葱葱的大树尽可能多地保留了下来，使建筑在建成之际就掩映在一片绿意盎然之中，在提升大院整体景观的同时，形成避风向阳、草木欣欣、鸟语花香的优美环境。

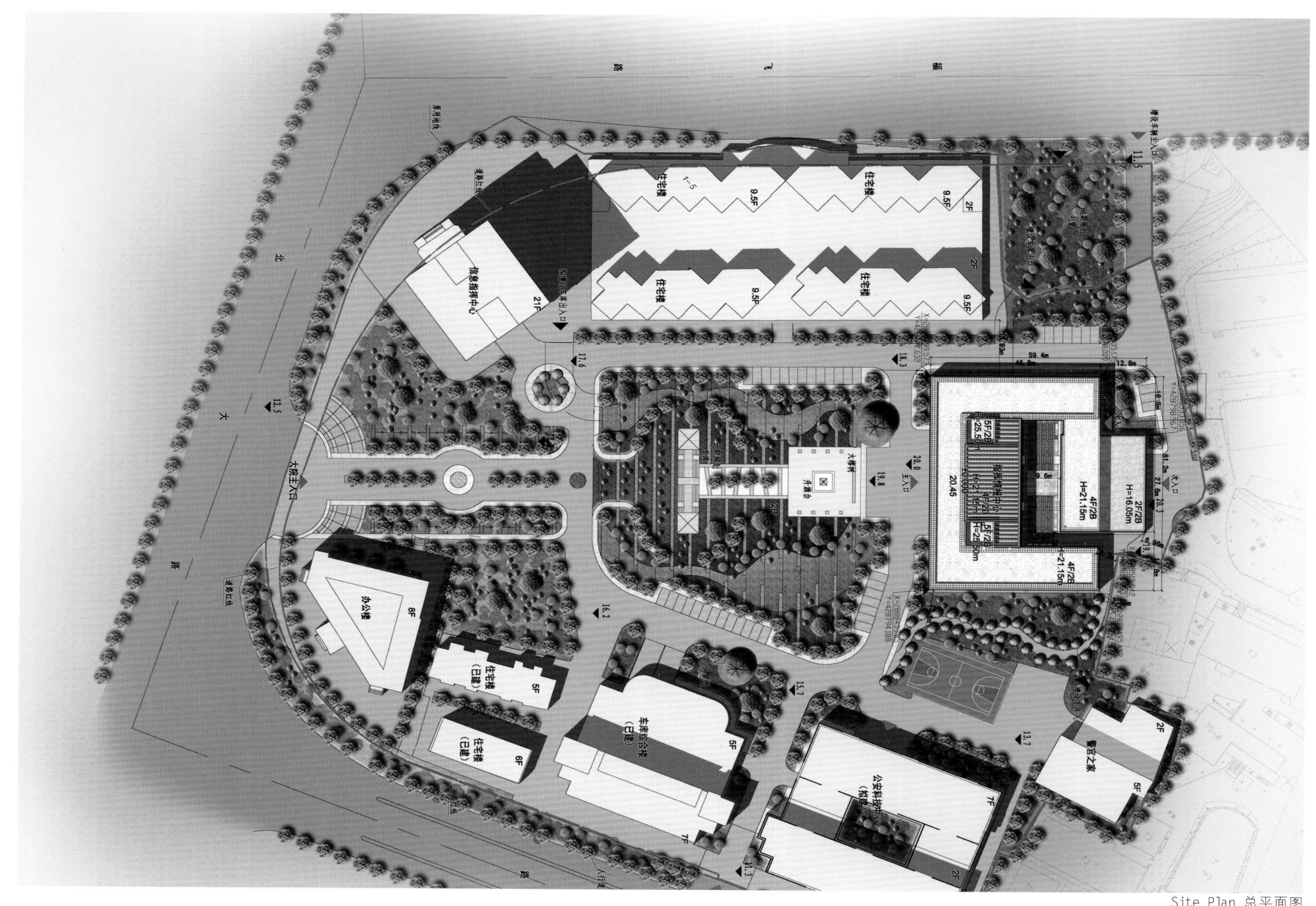

Site Plan 总平面图

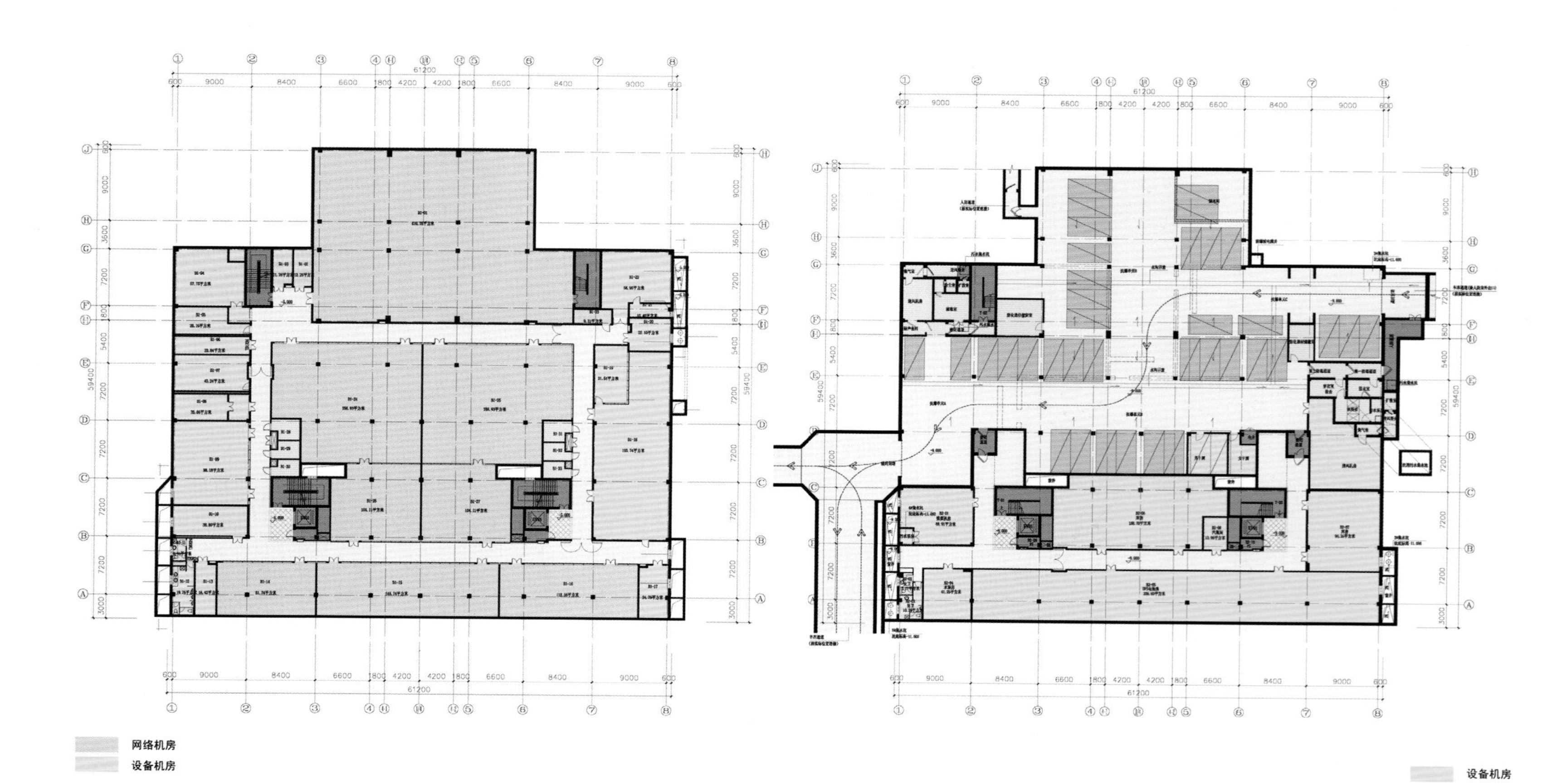

Plan for Basement First Floor 地下一层平面图

Plan for Basement Second Floor 地下二层平面图

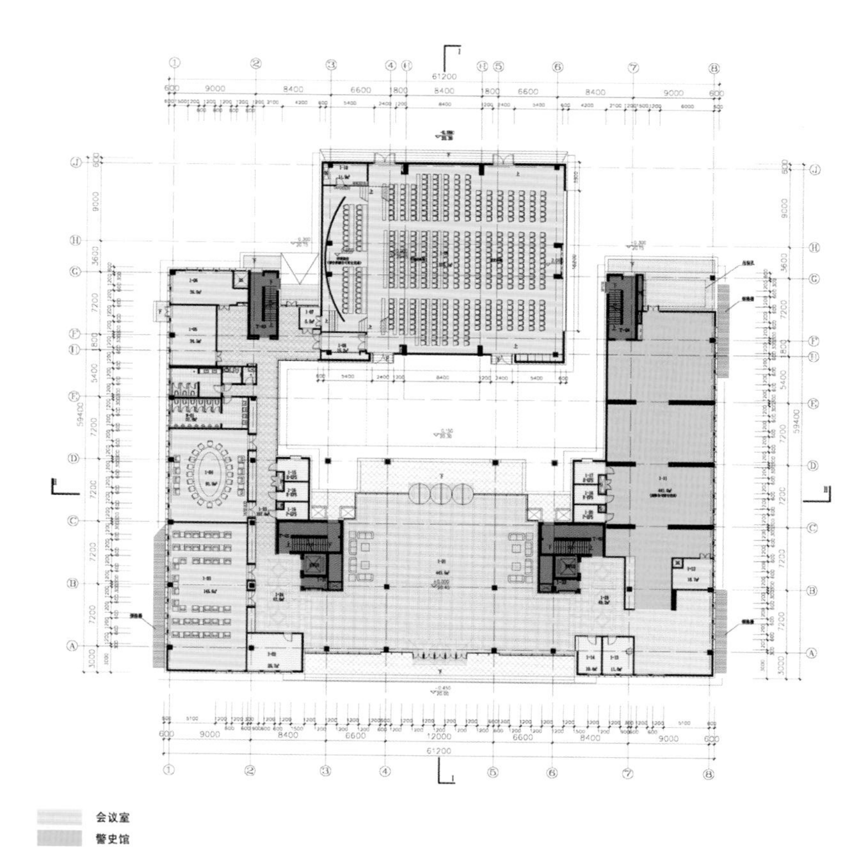

First Floor Plan 一层平面图

Second Floor Plan 二层平面图

West Elevation 西立面图

East Elevation 东立面图

North Elevation 北立面图

South Elevation 南立面图

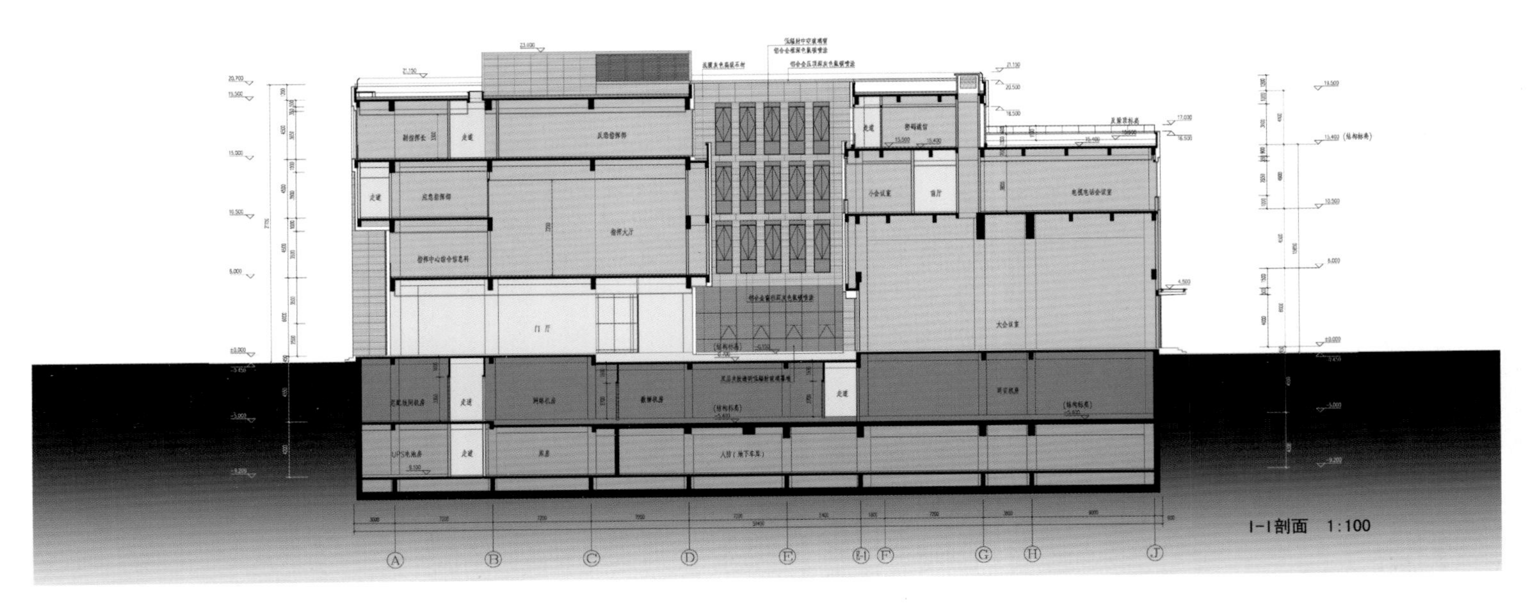

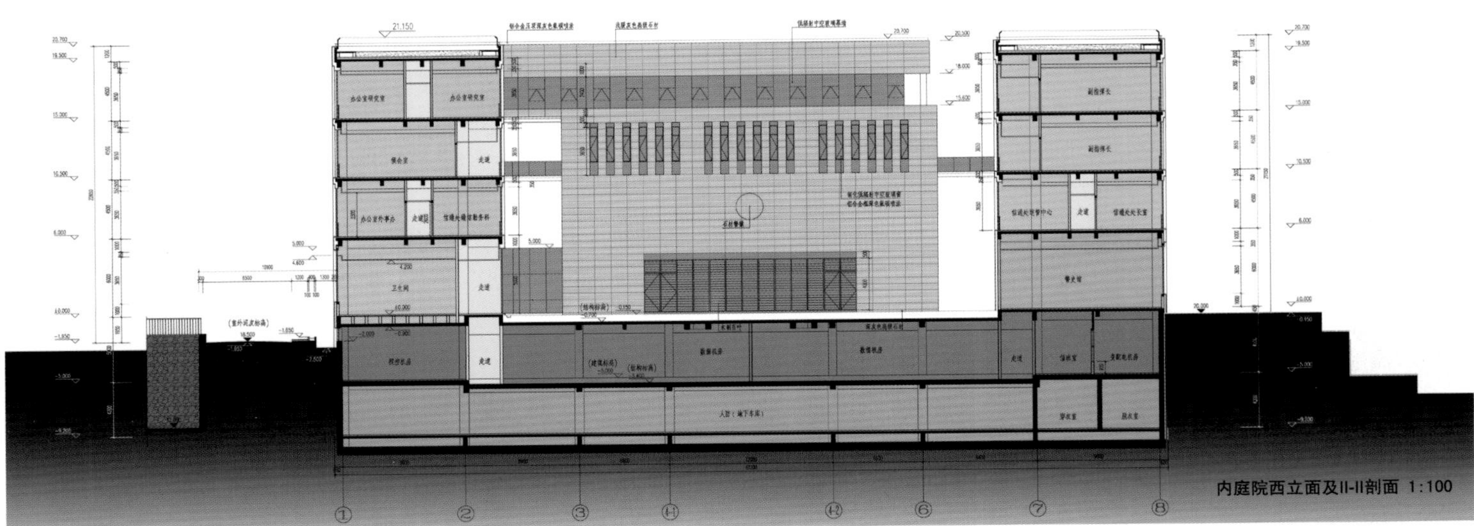

Sectional Drawing 剖面图

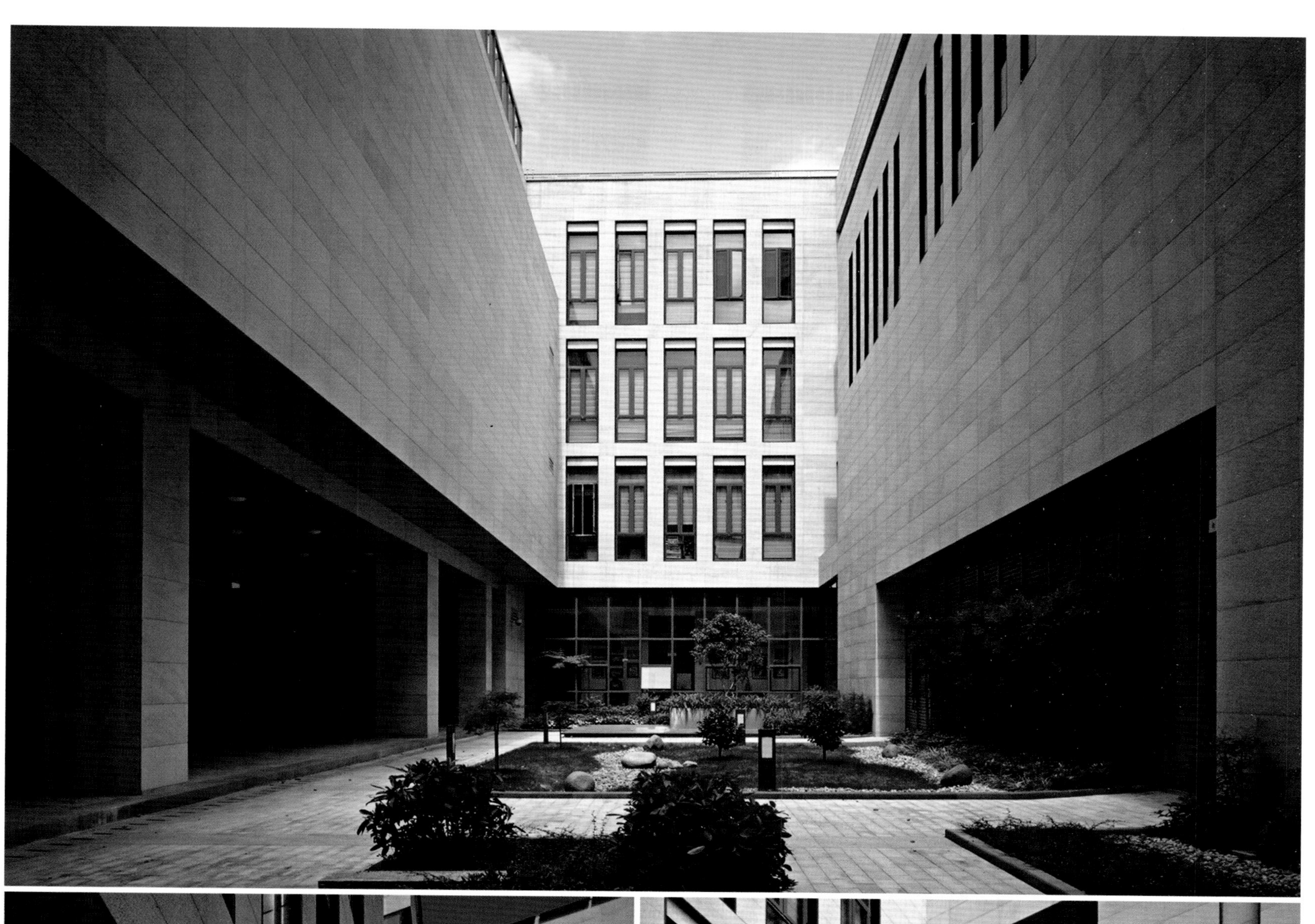

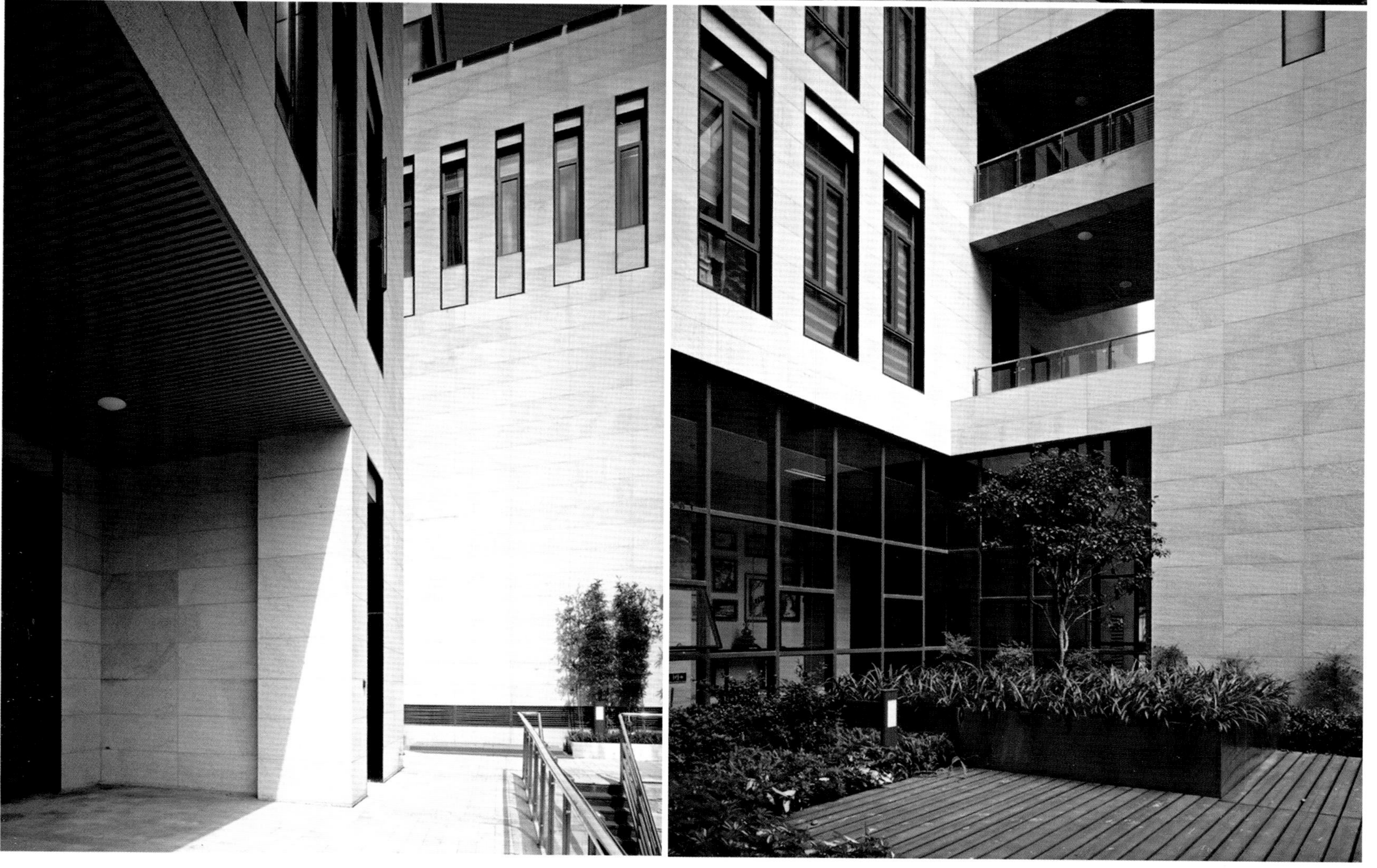

KEY WORDS
关键词

Concise Shape 造型简洁

Graceful Line 线条流畅

Rich Space 空间丰富

Fujian Ningde Nuclear Power Co., Ltd.
福建宁德核电有限公司综合办公楼

FEATURES
项目亮点

Front and back spaces are well arranged on the building facade which is dominated by modern structure and concise and lively form. And the smooth lines, rich space, floor-to-ceiling windows and round & free corner create a modern and outstanding image of an office & residential building.

建筑在立面上形成错落有致的前后空间，以现代构成为主，造型简洁、明快，以流畅的线条、丰富的空间、落地大玻璃窗、圆润自由的转角打造具有现代感、形象突出的办公居住建筑。

Location: Ningde, Fujian, China
Architectural Design: Fujian Architectural Design Institute
Total Land Area: 14,149.27 m^2
Total Floor Area: 43,073 m^2
Overground Floor Area: 39, 805 m^2
Underground Floor Area: 3, 268 m^2
Building Base Area: 4,626 m^2
Plot Ratio: 2.82
Green Coverage Ratio: 31.8%

项目地点：中国福建省宁德市
设计单位：福建省建筑设计研究院
总用地面积：14 149.27 m^2
总建筑面积：43 073 m^2
地上建筑面积：39 805 m^2
地下建筑面积：3 268 m^2
建筑基底面积：4 626 m^2
容积率：2.82
绿化率：31.8%

Overview

This project is a comprehensive office building that was constructed to meet the working and living need of Fujian Ningde Nuclear Power Co., Ltd. In order to match up the development of nuclear power industry in Ningde, and as an important nuclear power base in Ningde, it is an outcome tested by forward-looking, practical and economic scientific planning and argumentation.

项目概况

福建宁德核电有限公司综合办公楼是为了满足福建宁德核电有限公司在福鼎市内办公、生活需要而建的综合性大楼。为了配合宁德核电事业的发展规划，作为宁德核电的一个重要基地，该项目事先通过了前瞻性、实用性、经济性的科学规划和论证。

Planning

Located in Xingbating Plot by the west side of Tongjiang Stream in Northern Fuding City, the site of the project is a rectangle that is 180 m long in south-north direction and 80 m wide in east-west direction, which borders Tongjiang Stream to the east, boasts the best direction for landscape and enjoys convenient traffic for the adjacent Huangcheng East Road.

The project is comprised of a 16-storey office building, a 12-storey dormitory and a 2-storey auxiliary building with a total floor area of approximately 43,000 m^2. The 16-storey office building is in the south of the site, the 12-storey dormitory is in the north, and the 2-storey auxiliary building is playing a connector between the office building and the dormitory. In addition, a one-floor conjoined semi-basement is arranged between the three buildings according to the height difference.

核电大厦

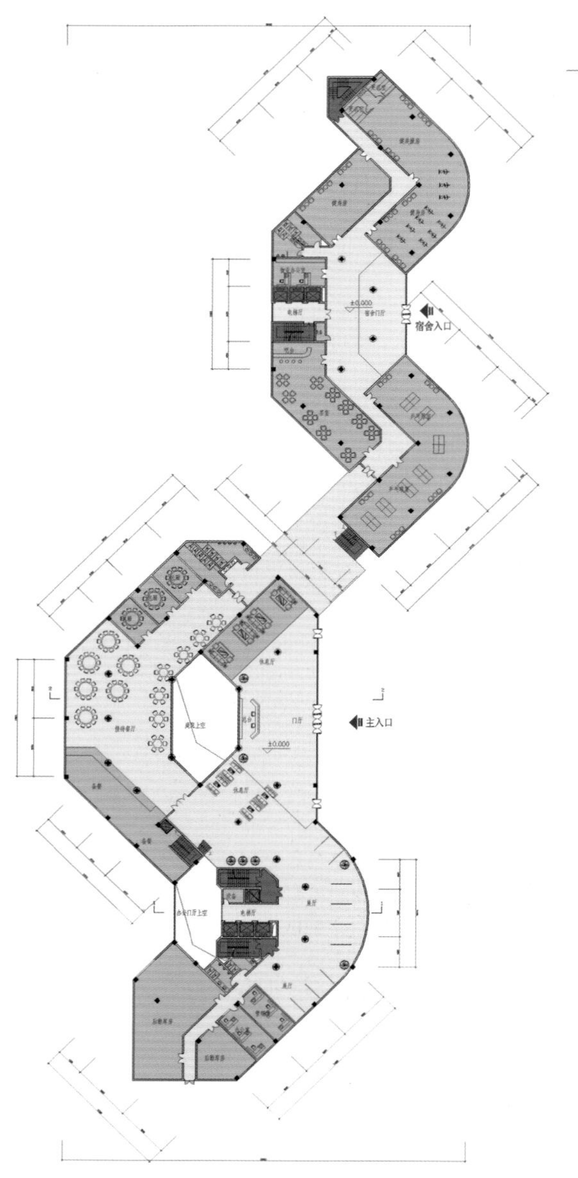

First Floor Plan 一层平面图

规划布局

本工程项目位于福鼎市北部桐江溪西侧新坝亭地块，地状呈长方形，南北长约为 180 m，东西宽约为 80 m。东侧为桐江溪，是本项目景观最好的方向，规划中的环城东路经过本地块。

本工程主要包括 1 幢 16 层办公楼、1 幢 12 层宿舍楼和 2 层裙房，总建筑面积约为 43 000 m²。16 层办公楼位于场地南部，12 层宿舍楼位于场地北部，2 层裙房位于办公楼和宿舍楼之间，为两幢楼的连接体。本工程结合场地高差情况，在办公楼、裙房和宿舍楼之间设 1 个 1 层联体半地下室。

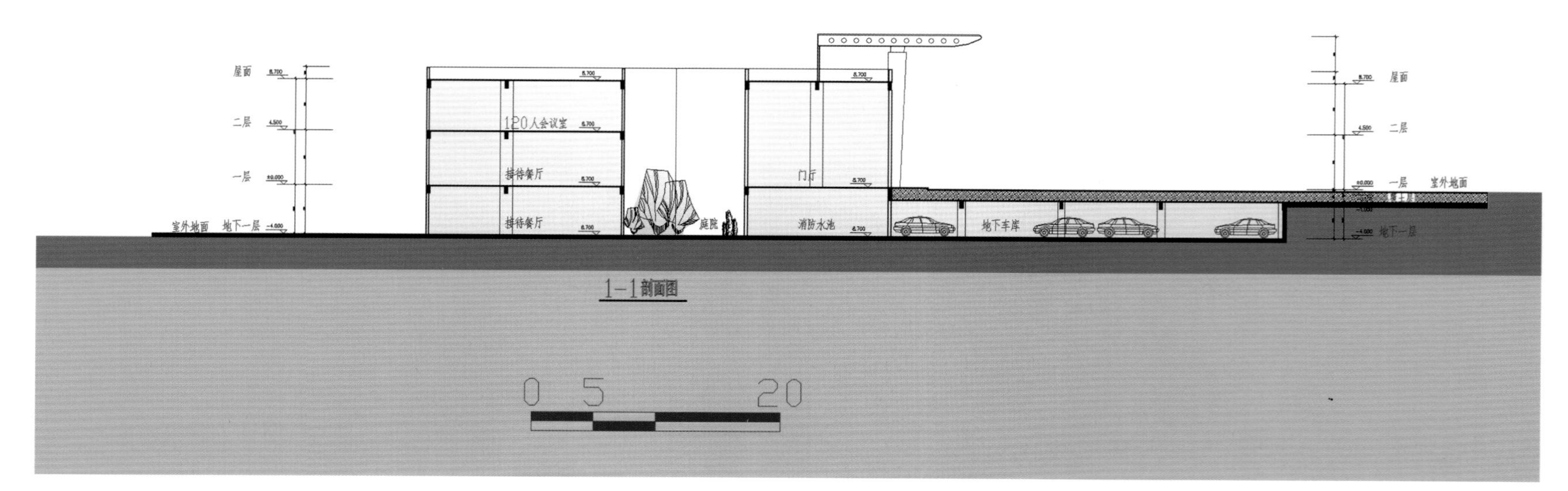

Sectional Drawing 剖面图

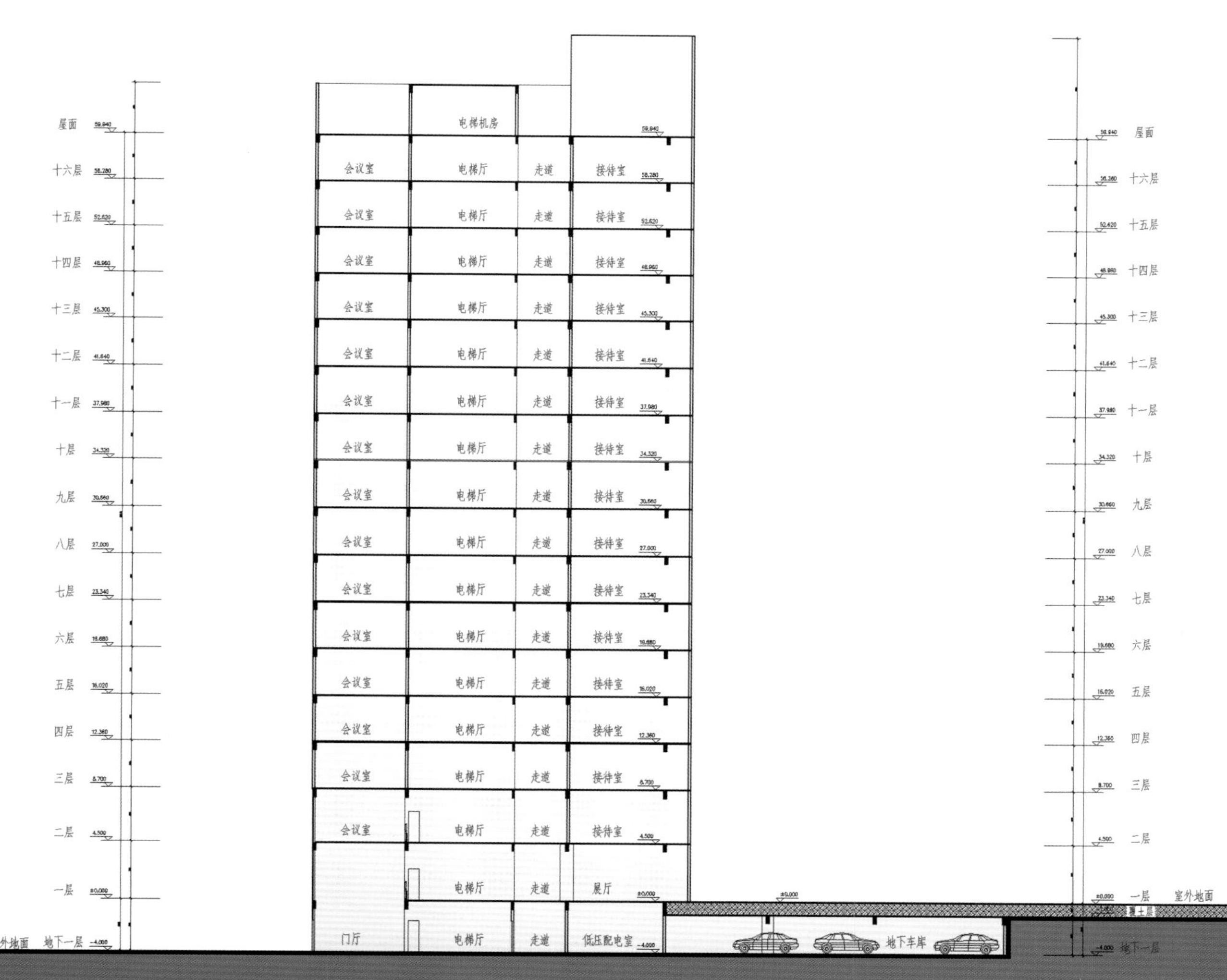

Sectional Drawing 剖面图

Facade Design

Front and back spaces are well arranged on the building facade which is dominated by modern structure and concise and lively form. And the smooth lines, rich space, floor-to-ceiling windows and round & free corner create a modern and outstanding image of an office & residential building.

立面设计

建筑在立面上形成错落有致的前后空间，以现代构成为主，造型简洁、明快，以流畅的线条、丰富的空间、落地大玻璃窗、圆润自由的转角打造具有现代感、形象突出的办公居住建筑。

KEY WORDS
关键词

Terrain-oriented 依山就势

Geometric Form 几何形体

Frame Structure 框架结构

Office Building of Housing & Urban-Rural Construction Department in Hunan Province
湖南省住房和城乡建设厅办公楼

FEATURES
项目亮点

The building is modeled on square geometry; after being extended, staggered and switched, not only dynamic and tensions are created, but also the rich space inside and characteristic appearance are highlighted, providing a shared place for human, building and nature and establishing a new spatial order and new friendly relationship.

建筑以方形几何体为原形，通过形体的伸展、交错、转接产生动感与张力，在构成内部丰富空间的同时也彰显出独特的体形特征，提供了人与人、人与建筑和自然交流共享的场所，建立了新的空间秩序和新的亲民关系。

Location: Changsha, Hunan, China
Architectural Design: Hunan Provincial Architectural Design Institute
Total Land Area: 33,037.87 m^2
Total Floor Area: 19,982 m^2
Plot Ratio: 0.49
Greening Coverage Ratio: 40.3%

项目地点：中国湖南省长沙市
设计单位：湖南省建筑设计院
总用地面积：33 037.87 m^2
总建筑面积：19 982 m^2
容积率：0.49
绿化率：40.3%

Overview

Located in Heping Village, Dongjing County, Yuhua District, Changsha, this project is an office building in reinforced concrete frame structure for Hunan Housing & Urban-Rural Construction Departmente. The entire office building is comprised of two parts: the main building and the auxiliary building. The main building has 7 floors and the auxiliary building has 4 floors and they are connected by a corridor.

项目概况

湖南省住房和城乡建设厅办公楼位于长沙市雨花区洞井镇和平村，属办公类建筑，结构形式为钢筋混凝土框架结构。整体建筑由主楼与附楼两部分组成。主楼为 7 层，附楼为 4 层，主楼与附楼通过连廊相连。

Architectural Design

Designers fully interpreted the terrain features and situated the building in line with local conditions. The building and the site got integrated, which minimized the amount of earthwork and protected the mountain and trees in the maximum level, fully embodying the concept of "harmonious co-existence of human, building and nature". The building is modeled on square geometry, after being extended, staggered and switched, not only dynamic and tensions are created, but also the rich space inside and characteristic appearance are highlighted, providing a shared place for human, building and nature and establishing a new spatial order and new friendly relationship.

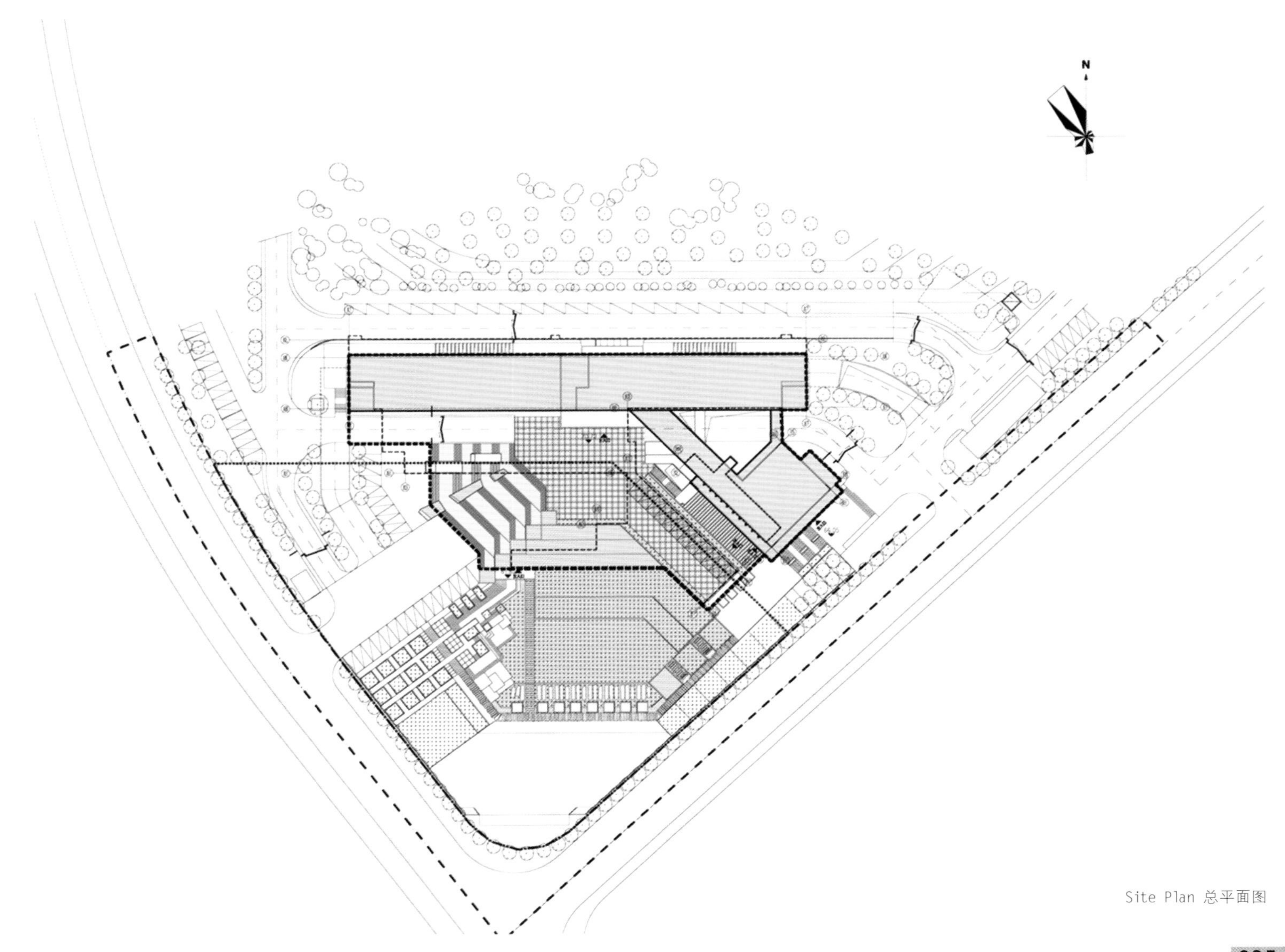

Site Plan 总平面图

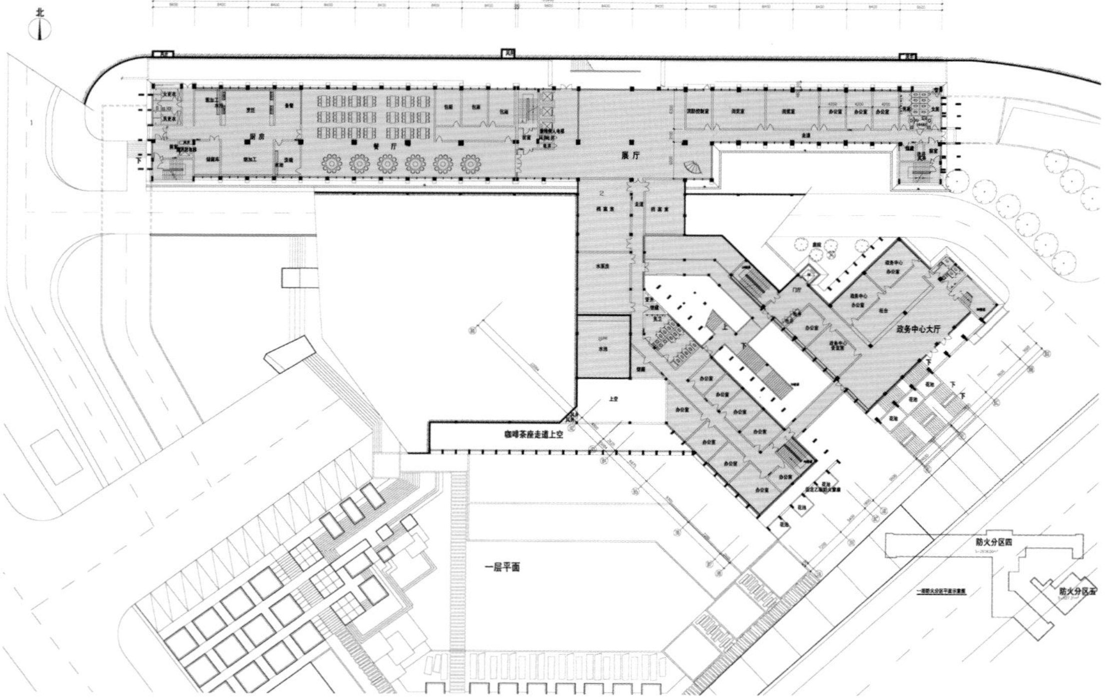

First Floor Plan 一层平面图

建筑设计

湖南省住房和城乡建设厅办公楼的设计充分解读了环境地形特征，因地制宜，依山就势而建。场地与建筑融为一体，最大限度地减小土方量，对山体树木进行了最大限度的保护，充分体现了“人与建筑及自然和谐共生”的理念。建筑以方形几何体为原形，通过形体的伸展、交错、转接产生动感与张力，在构成内部丰富空间的同时也彰显出独特的体形特征，提供了人与人、人与建筑和自然交流共享的场所，建立了新的空间秩序和新的亲民关系。

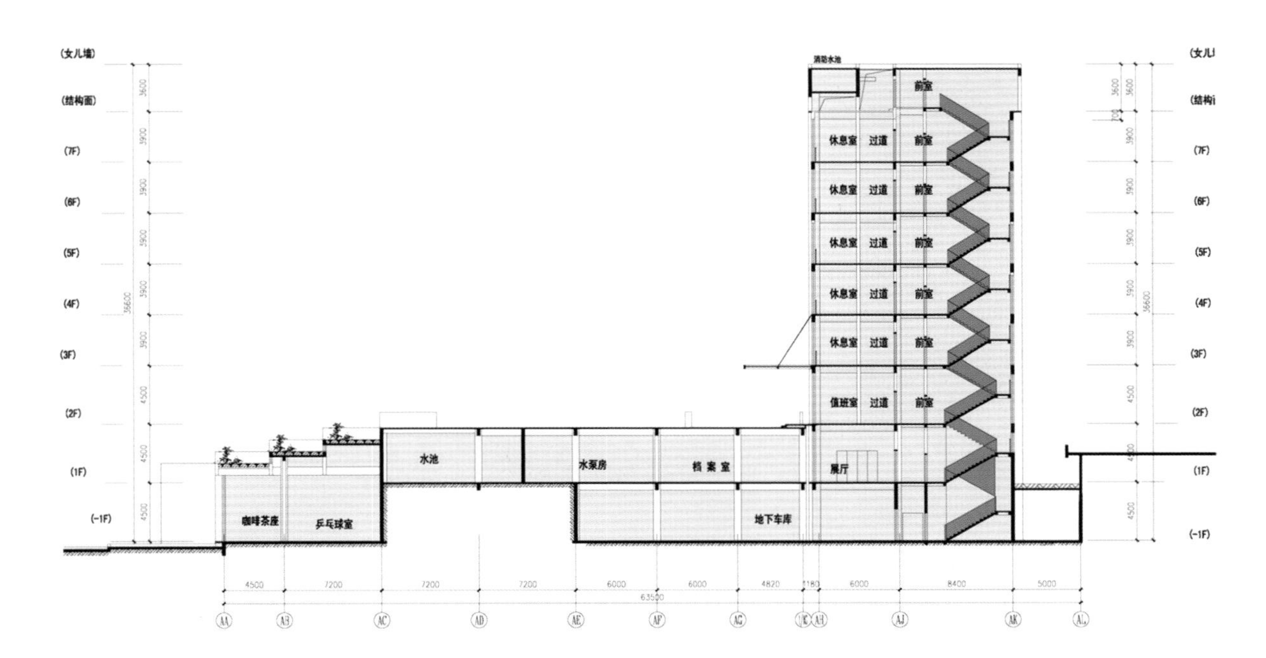

Sectional Drawing 剖面图

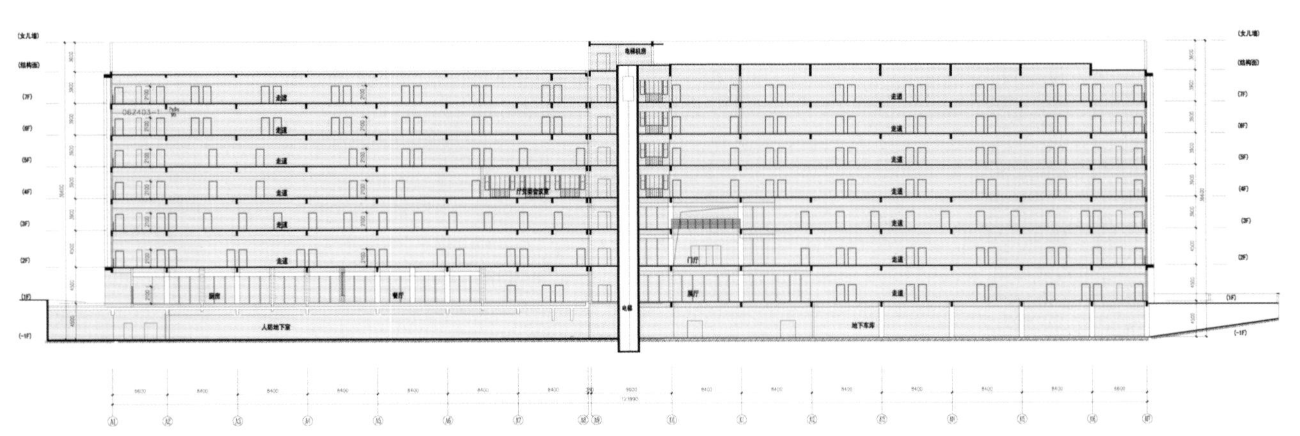

Sectional Drawing 剖面图

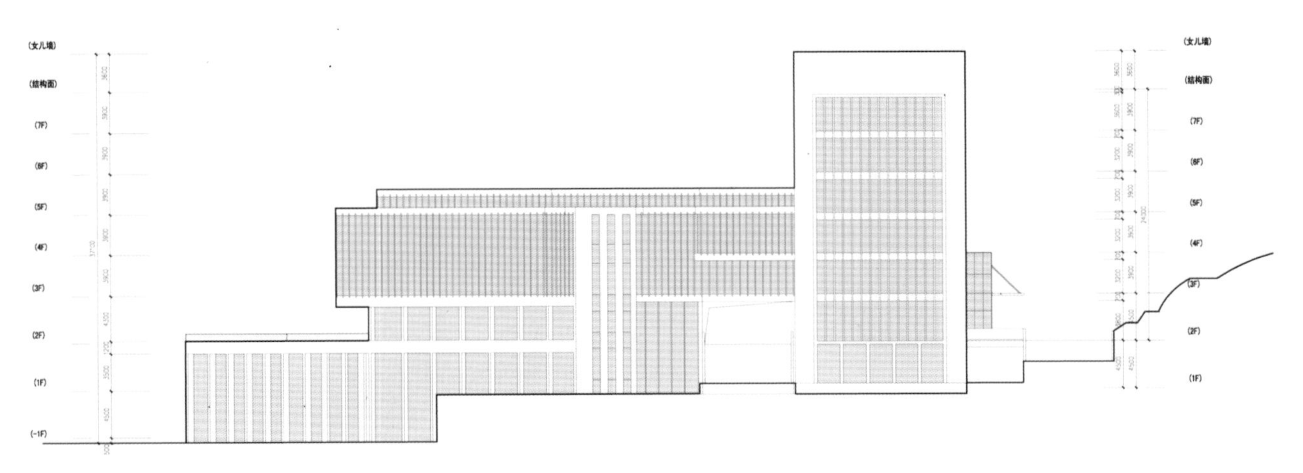

Elevation 立面图

湖南省

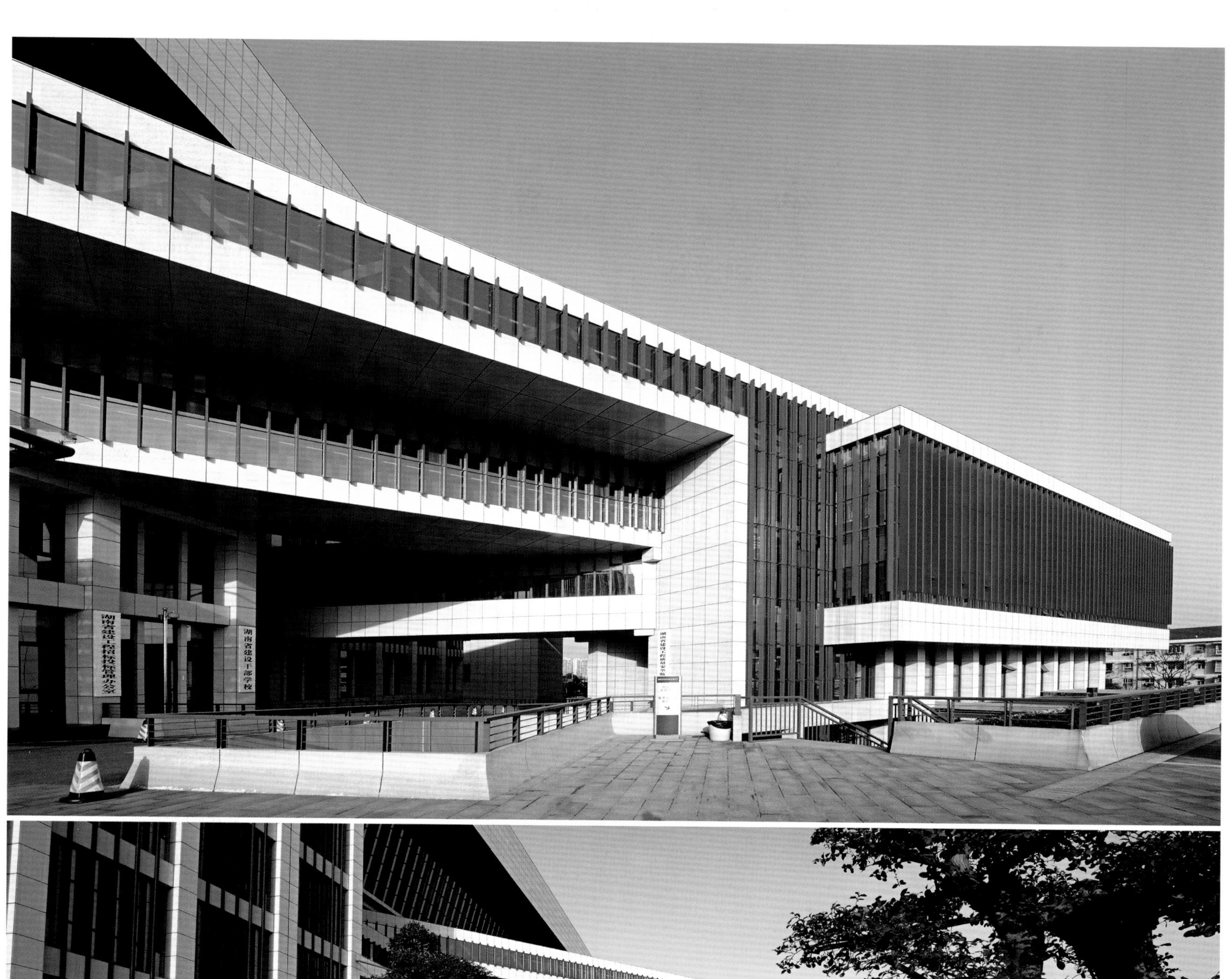
湖南省建设工程招标投标管理办公室
湖南省建设干部学校

KEY WORDS
关键词

Peace Layout 布局平和

Primitively Simple And Elegant 古朴典雅

Magnificent Momentum 气势宏大

The People's Procuratorate of Jilin
吉林省人民检察院

FEATURES
项目亮点

The coherent axis space, majestic big steps, sedate canopy and spacious hall all very accord with the functional requirements and building specific properties.

中轴线空间一气呵成，威严的大台阶、稳重的雨棚、宽敞的大厅，非常符合建筑的功能要求和特定的性质。

Location: Changchun, Jilin, China
Architectural Design: Architectural Design and Research Institute of HIT
Land Area: 106,900 m²
Total Construction Area: 49,468.10 m²
Building Height: 38.40 m
Plot Ratio: 0.4

项目地点：中国吉林省长春市
设计单位：哈尔滨工业大学建筑设计研究院
占地面积：106 900 m²
总建筑面积：49 468.10 m²
建筑高度：38.40 m
容积率：0.4

Architectural Design

The project design uses concise and elegant architectural vocabulary to show the complete image of the People's Procuratorate of Jilin, rich inscribed architectural details to express the exquisite style, and orderly concave-convex architectural lighting to convey its rich emotion. The architectural layout gives expression to the feelings of holding and possessing, which bears peace and stability. Looking from the south, the building seems primitively simple and elegant, well-ordered; looking from the north, it appears magnificent and freely stretching. The coherent axis space, majestic big steps, sedate canopy and spacious hall all very accord with the functional requirements and building specific properties. The west side completely shows the construction volume to respond to the landscape on the west side. Twelve pillars strengthen the sense of order and rich lighting of the main facade. The building also keeps harmonious and stretching proportion, well combining the demand of modern office buildings with classical scale.

项目概况

设计用凝练、典雅的建筑语汇展示吉林省人民检察院的完整形象；用刻画丰富的建筑细部诠释吉林省人民检察院的细腻作风；用凹凸有序的建筑光影表达吉林省人民检察院的丰富情感。建筑布局体现“坐拥”二字，气度平和，稳如泰山。从南侧看，建筑古朴典雅、秩序井然；从北侧看，建筑气势宏大、舒展开放。中轴线空间一气呵成，威严的大台阶、稳重的雨棚、宽敞的大厅，非常符合建筑的功能要求和特定的性质。西侧界面完整地呈现建筑体量，回应西侧景观要求。12 根列柱彰显出主立面的秩序感和丰富的光影。建筑比例和谐、舒展，把现代办公建筑的需求与古典的尺度相结合。

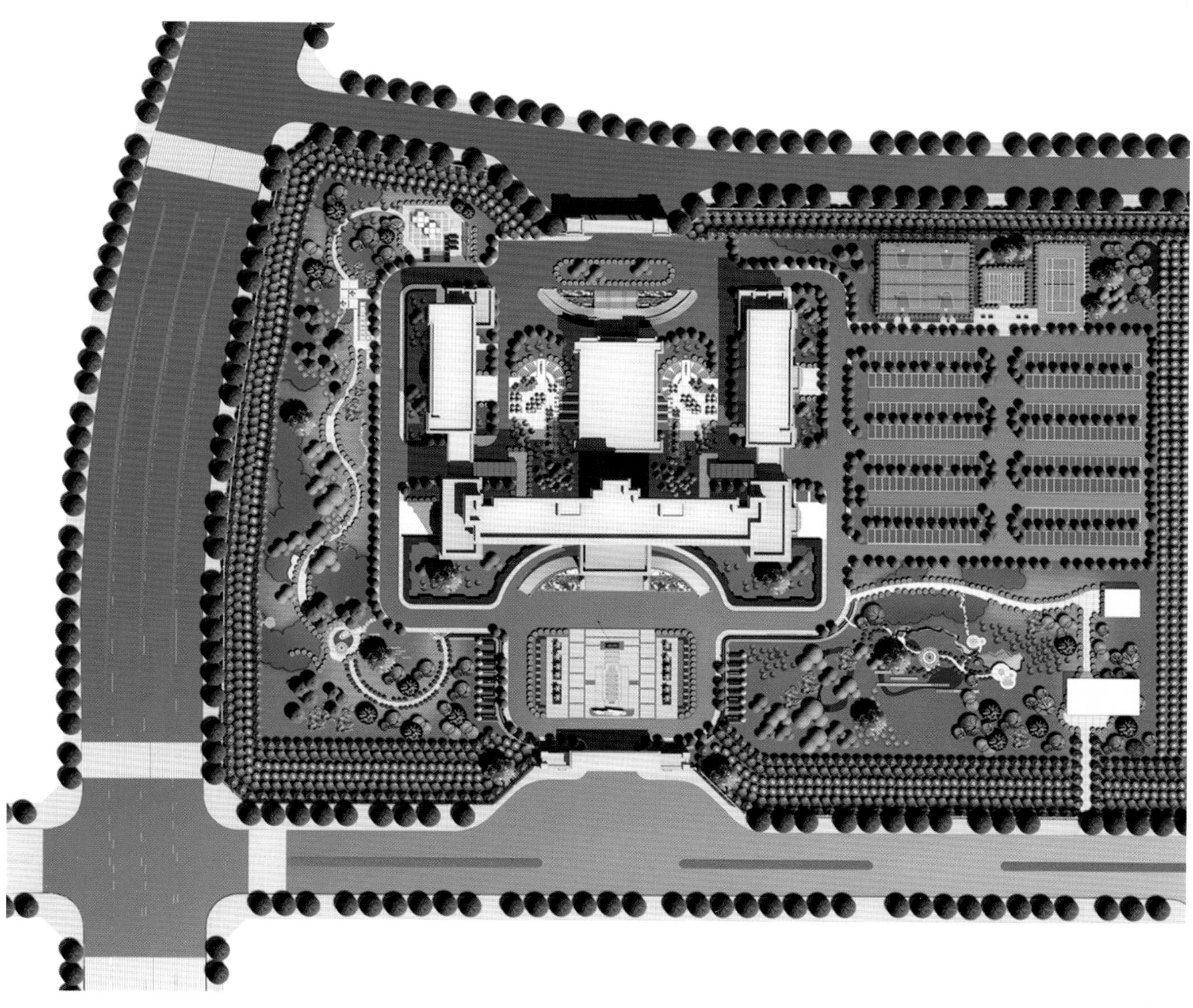

Site Plan 总平面图

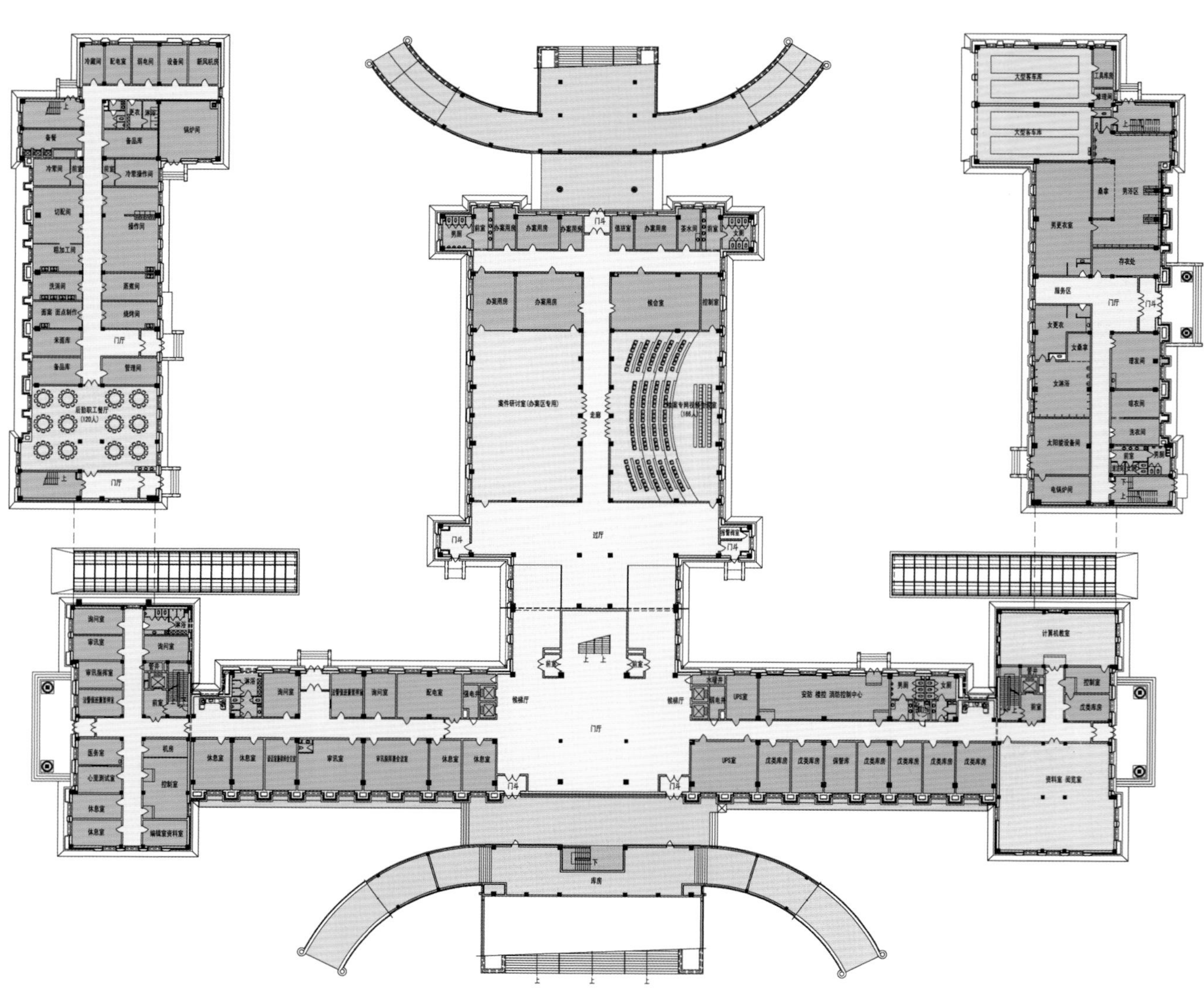

First Floor Plan　一层平面图

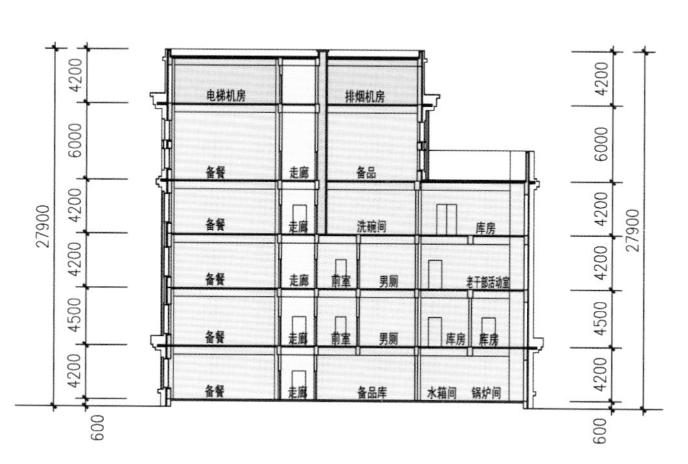

Sectional Drawing 剖面图

KEY WORDS
关键词

Imposing Volume 体量宏伟

Compact Layout 布局紧凑

Frame-shear Wall Structure 框架剪力墙结构

Service Center of Jinan Long'ao Assets Operation Co.,Ltd.
济南龙奥资产运营有限公司综合服务楼

FEATURES
项目亮点

Designed in traditional "courtyard" style, it features clear levels, imposing volume and compact layout. " 中 " shaped floor plan embodies the traditional culture of Shandong province and builds a strong connection between the building and the context.

工程采用中国传统建筑“院落”式布局，顶面看去呈“中”字形，外放内收，层次分明，体量宏伟，布局紧凑，“中”字形的平面形状赋予建筑厚重的文化内涵。

Location: Jinan, Shandong, China
Architectural Design: Shandong Tong Yuan Design Group Co.,Ltd.
Total Land Area: 267,000 m²
Total Floor Area: 365,340 m²
Overground Floor Area: 316,005 m²
Underground Floor Area: 49,335 m²
Building Height: 66.15 m
Building Density: 15%
Plot Ratio: 1.37

项目地点：中国山东省济南市
设计单位：山东同圆设计集团有限公司
总用地面积：267 000 m²
总建筑面积：365 340 m²
地上建筑面积：316 005 m²
地下建筑面积：49 335 m²
建筑高度：66.15 m
建筑密度：15%
容积率：1.37

Overview

Situated in Yanshan New District of Jinan City, the service center is near to the Olympic Sports Center in the north, and adjacent to the built Jingshi Road and Tourism Road in the south and north. The site is ideally located on the eastern extension axis of Jinan, playing an important role in the future development of the city. With the opportunity to host the 11th National Games, it has made Yanshan New District a landmark area for government affairs, cultural and sports activities.

项目概况

济南市龙奥资产管理运营有限公司综合服务楼坐落在济南燕山新区内，北邻奥体中心，与南北两侧已建成的经十路和旅游路两条城市东西大动脉相邻，地处济南主城区向东发展的主轴上，承接着济南城市发展总体战略中的“中疏”和“东拓”。由此，将带给济南一个崭新的标志性新城区，以承办第十一届全运会为契机，打造以政务、文化、体育为主的燕山新区。

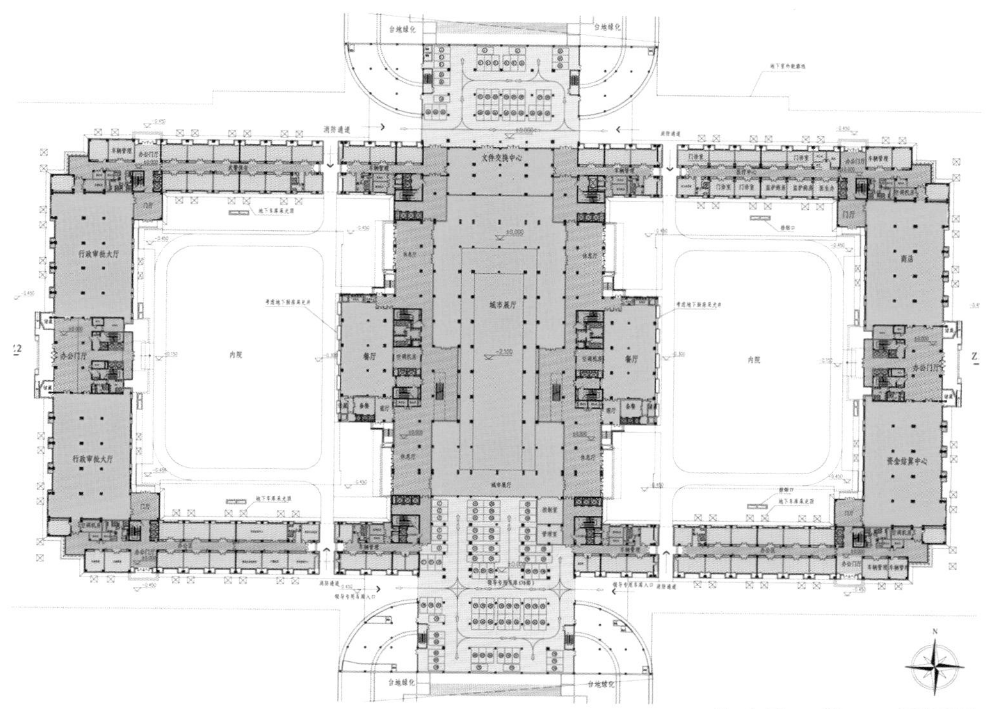

First Floor Plan 一层平面图

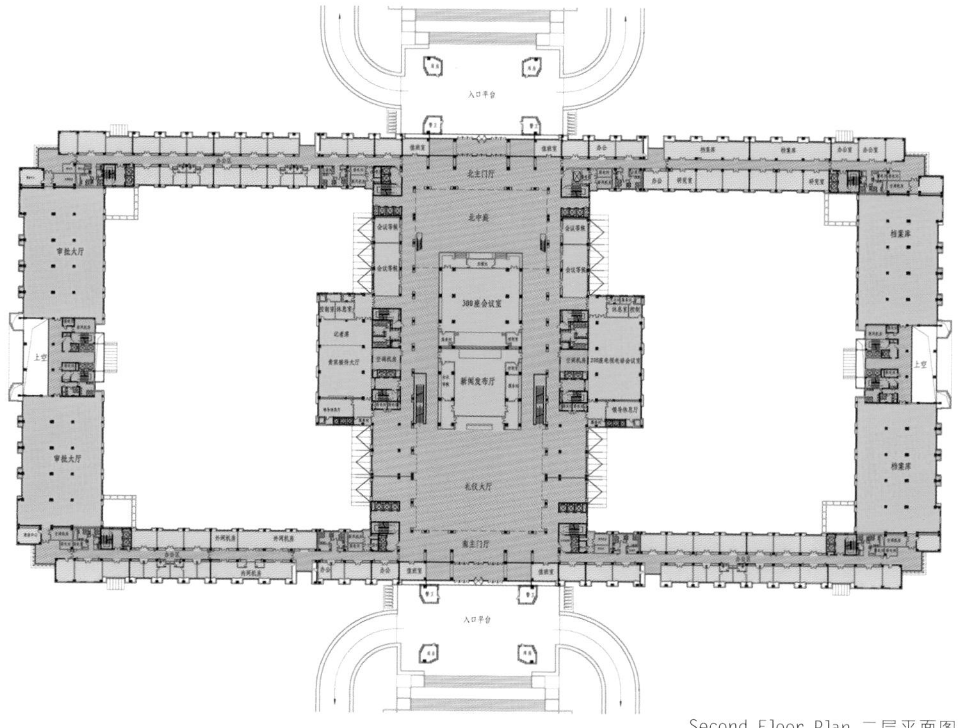

Second Floor Plan 二层平面图

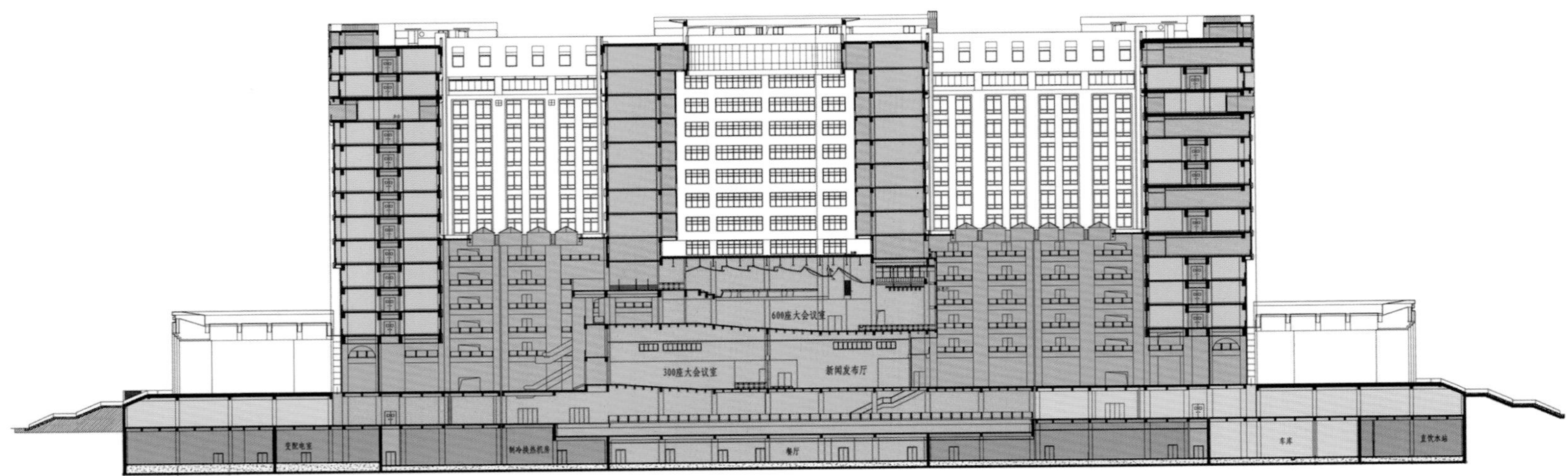

Sectional Drawing 剖面图

Architectural Structure

This office building is 288 m long and 144 m wide, having 15 floors over ground and one floor underground. Designed in traditional "courtyard" style, it features clear levels, imposing volume and compact layout. " 中 " shaped floor plan embodies the traditional culture of Shandong province and builds a strong connection between the building and the context. This project belongs to high-rise buildings and uses the frame-shear wall structure.

建筑结构

办公楼工程建筑总长度 288 m，总宽度 144 m，地上 15 层，地下 1 层。工程采用中国传统建筑“院落”式布局，外放内收，层次分明，体量宏伟，布局紧凑，“中”字形的平面形状赋予建筑厚重的文化内涵，展现了齐鲁文化博大精深、历久弥新的文化底蕴，实现了与泉城地域文脉文化的和谐统一，体现了对自然环境的尊重。此工程为高层一类建筑，框架剪力墙结构。

KEY WORDS
关键词

Glass Curtain Wall 玻璃幕墙

Mixed Functions 功能完善

Green and Ecological 绿色生态

Jianghai Business Tower
江海商务大厦

FEATURES
项目亮点

The facade uses simple and elegant glass curtain wall system, and the vertical components make the tower look tall and straight. Curtain wall design integrates with interior and lighting design to form complete and orderly facades.

立面采用了简洁的玻璃幕墙体系，主楼的竖向构件增加了向上的挺拔感。幕墙设计考虑了与内装及照明的一体化融合设计，使立面完整有序。

Location: Nantong, Jiangsu, China
Architectural Design: East China Architectural Design & Research Institute Co.,Ltd. (ECADI)
Land Area: 39,000 m^2
Total Floor Area: 66,000 m^2
Plot Ratio: 1.4

项目地点：中国江苏省南通市
设计单位：华东建筑设计研究院有限公司
用地面积：39 000 m^2
总建筑面积：66 000 m^2
容积率：1.4

Overview

The building stands on a square site with one floor underground and 21 floors on ground. The main structure is high to 96.5 m, and the total height of the building is 107 m.

项目概况

项目用地为正方形，大楼地下1层，地上21层，结构总高96.5 m，建筑总高度107 m。

Architectural Design

The facade uses simple and elegant glass curtain wall system, with the vertical components making the tower look tall and straight. Curtain wall design integrates with interior and lighting design to form complete and orderly facades. Large sliding windows are installed to provide fresh air and natural ventilation, which help to create a green and ecological working environment. The entrance of the tower is designed in modern glass structure, while the entry of the restaurant adopts grid steel framework to show the beauty of architecture in hi-tech era.

建筑设计

立面采用了简洁的玻璃幕墙体系，主楼的竖向构件增加了向上的挺拔感。幕墙设计考虑了与内装及照明的一体化融合设计，使立面完整有序。本项目立面结合使用功能，采用了大面积的平推窗。这样一方面保证了室内办公人员能呼吸到新鲜的空气，同时结合自然排烟及消防联动，创造了绿色生态办公的典范。主楼入口采用通高的玻璃体量，大气、简洁，彰显出鲜明的时代特征。餐饮区入口采用了网格式的钢构架，体现了高科技时代的建筑美学。

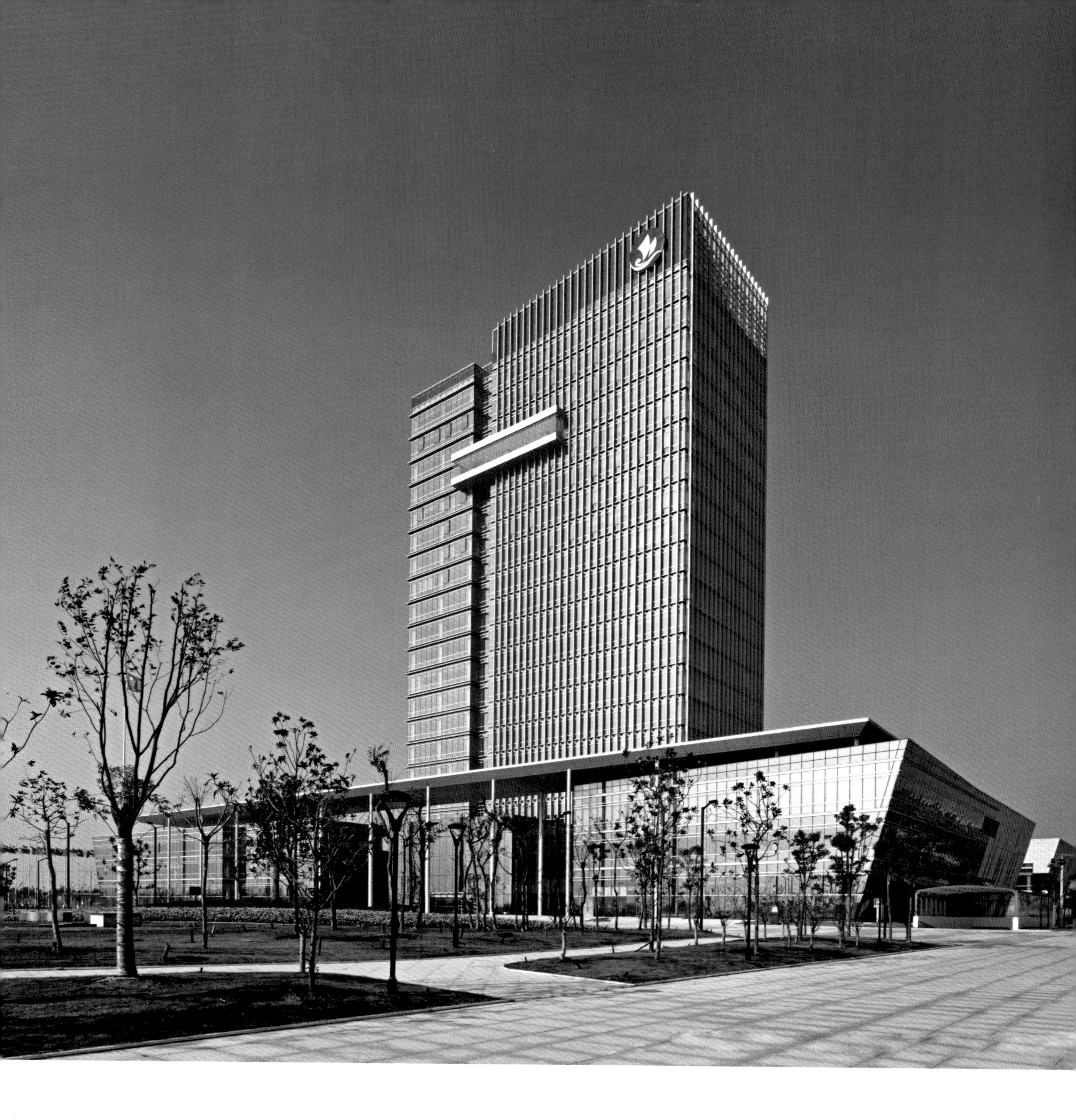

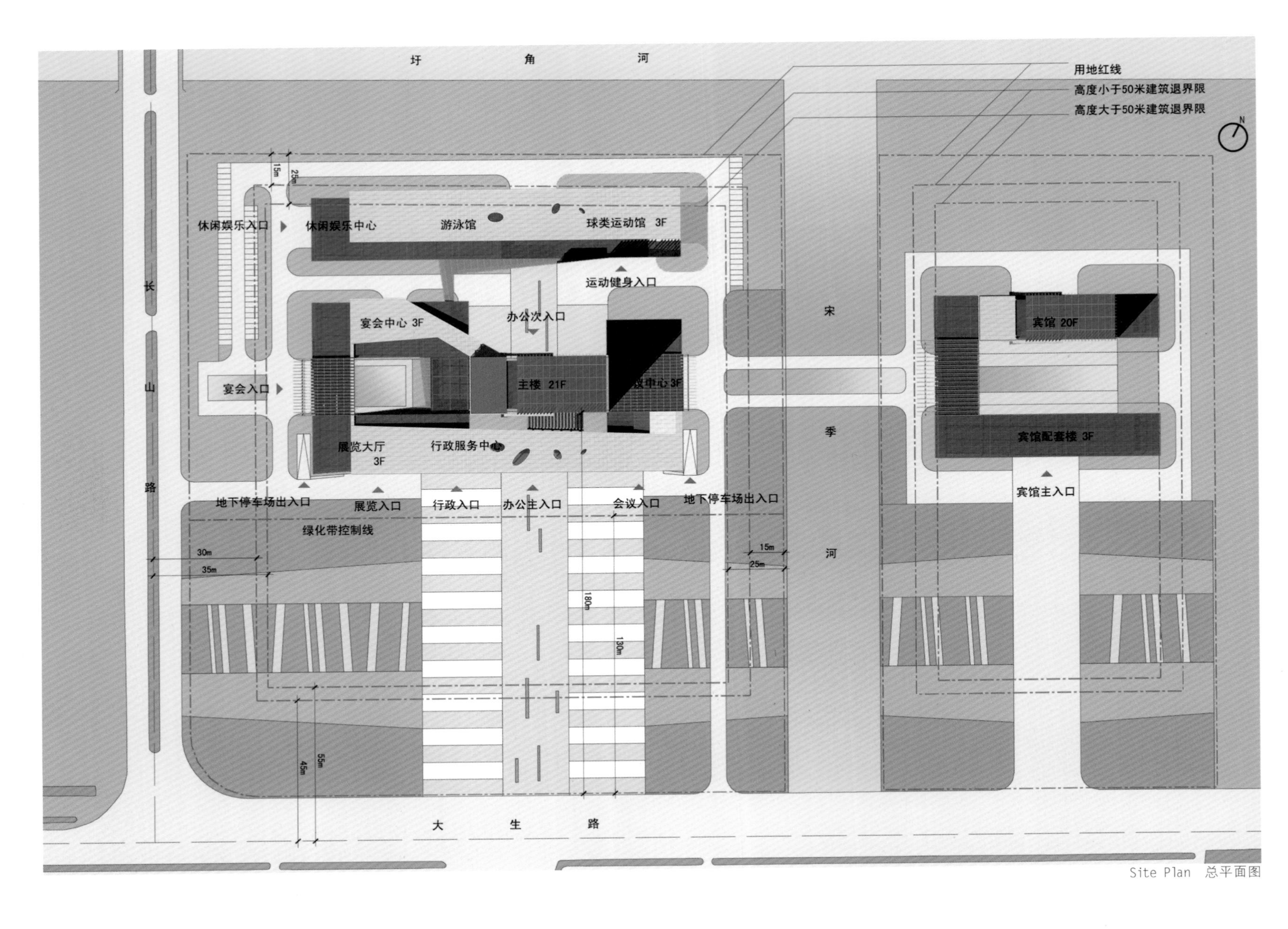

Site Plan 总平面图

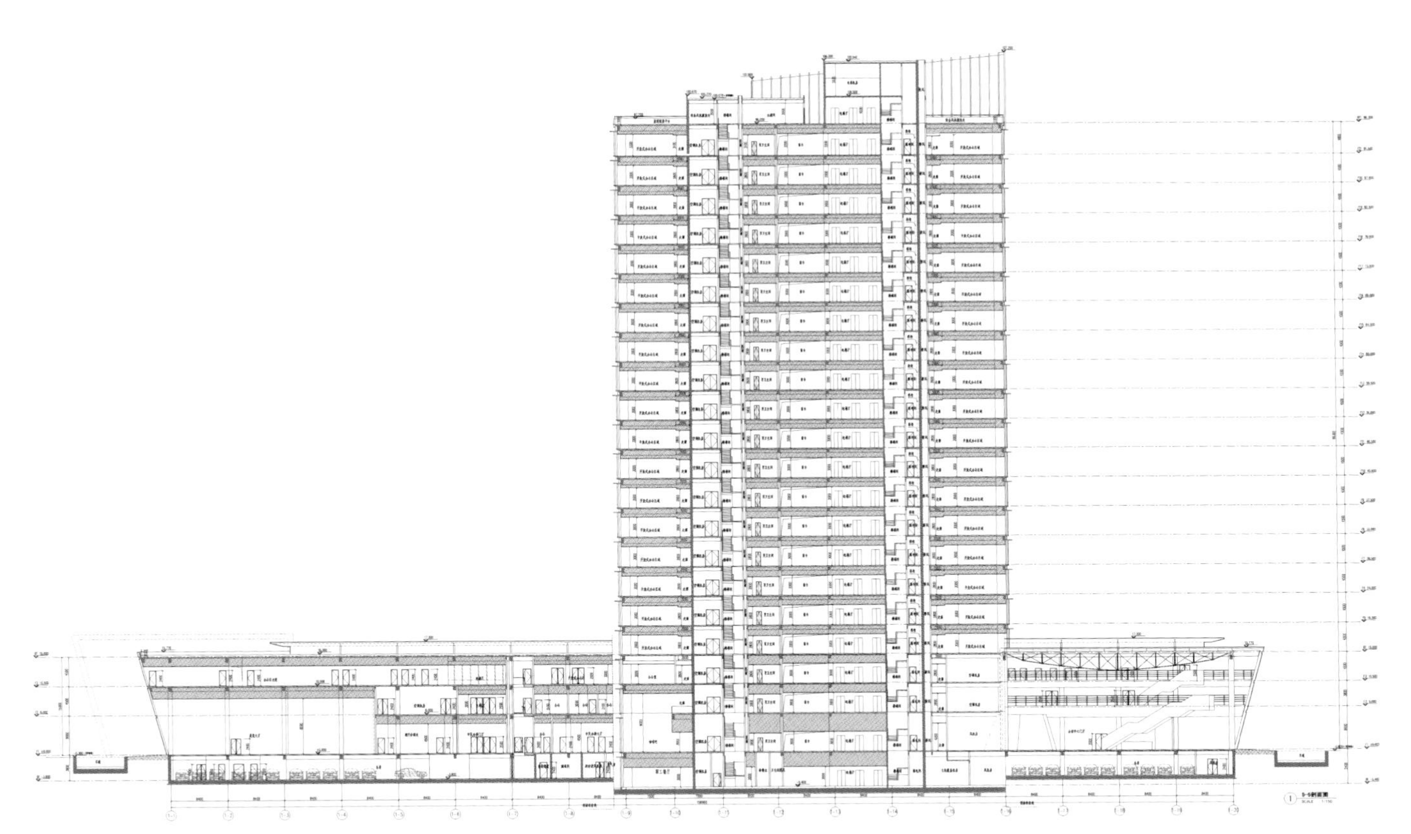

Sectional Drawing 剖面图

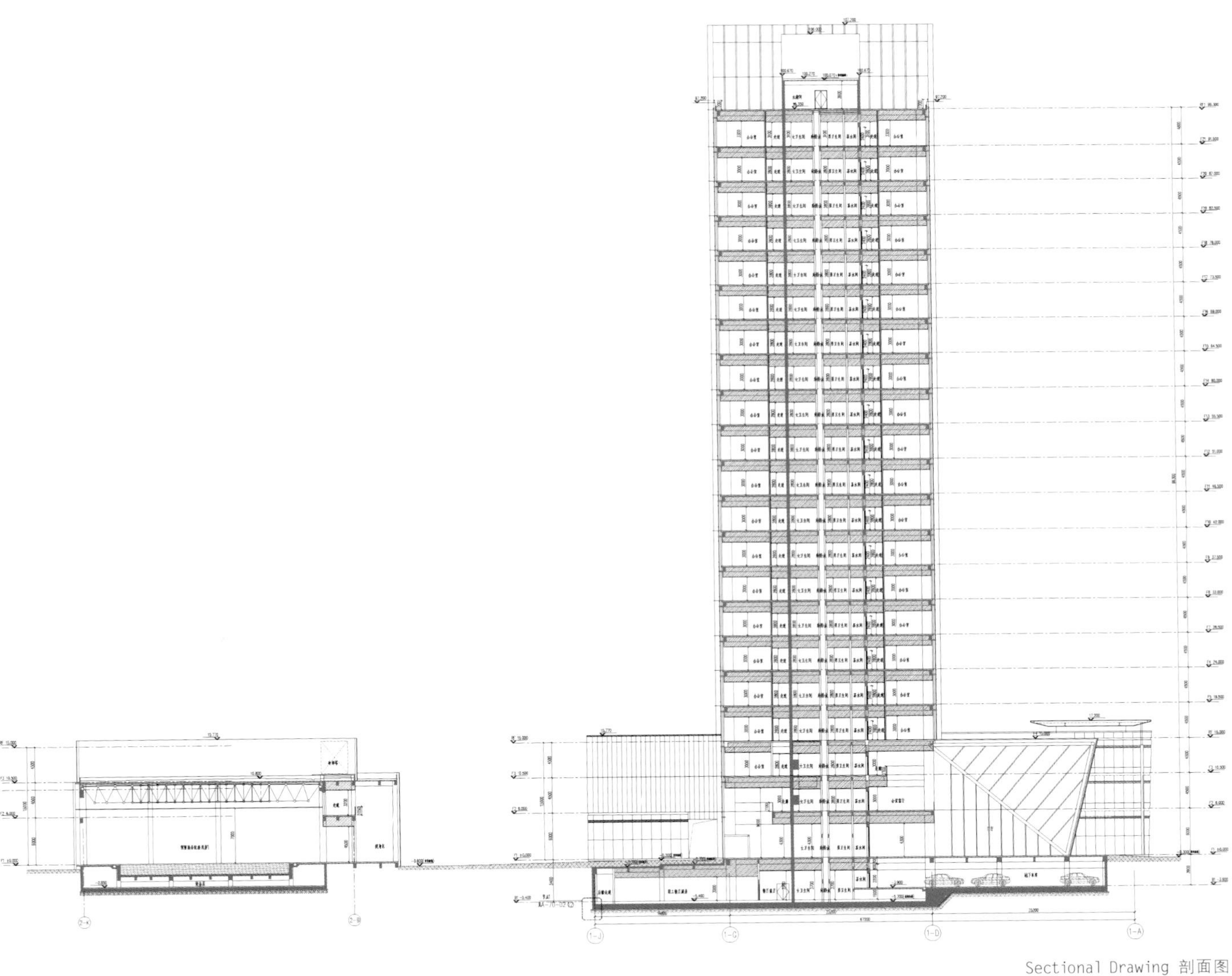

Sectional Drawing 剖面图

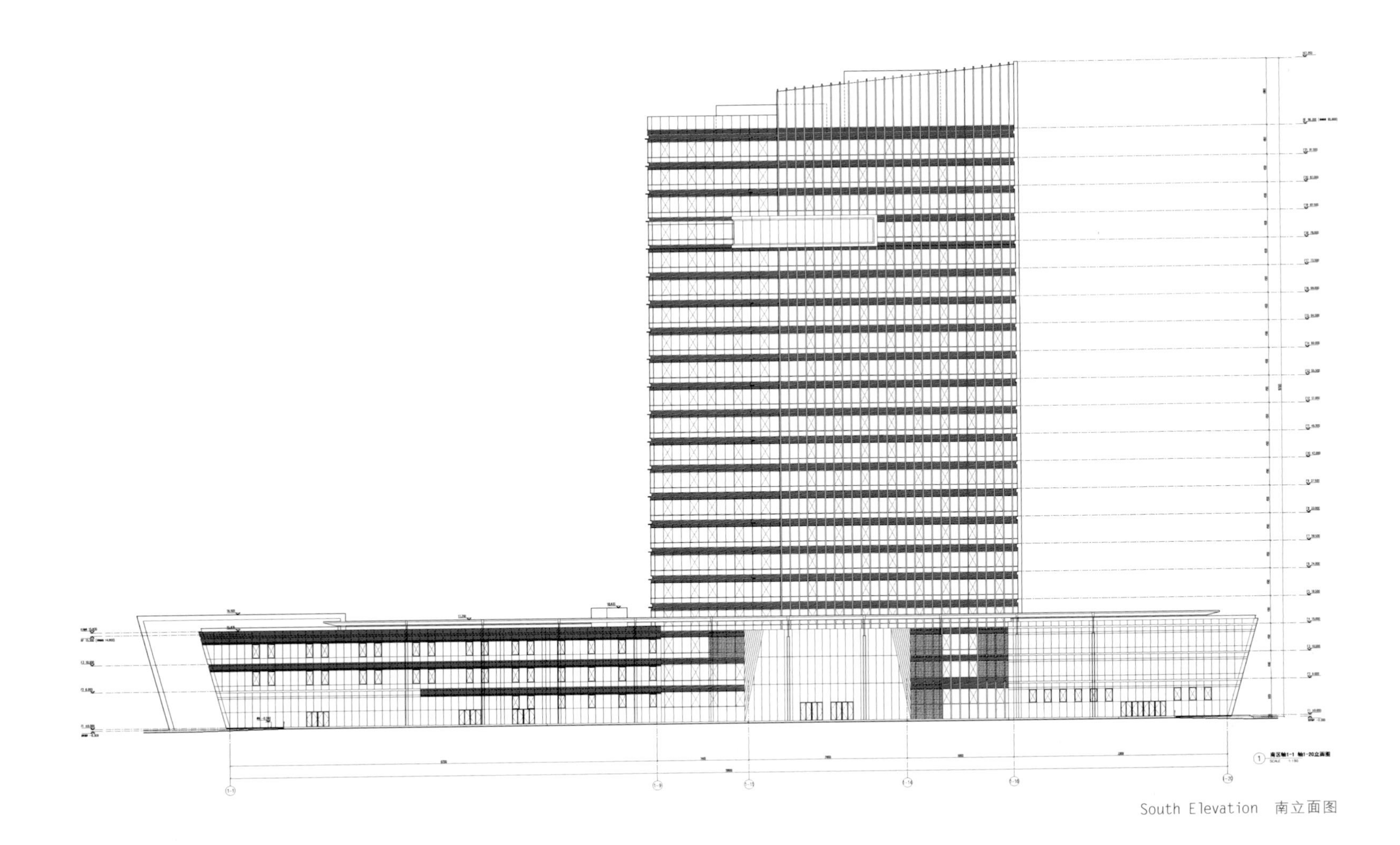

South Elevation 南立面图

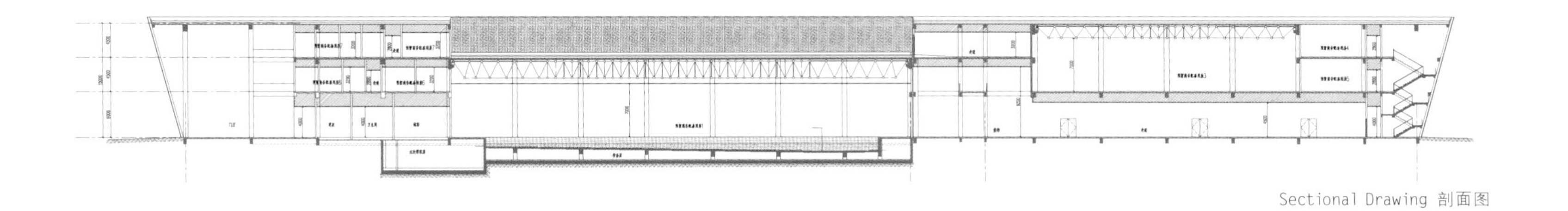

Sectional Drawing 剖面图

Sectional Drawing 剖面图

外景　JIANGHAI OFFICE BUILDING

餐饮入口　江海商务大厦　JIANGHAI OFFICE BUILDING

外景

KEY WORDS
关键词

Unique Shape 造型独特

Reasonable Organization 大气开合

Elegant Style 简洁明快

Standard Factory Building 1, 2 & 8 of Suzhou Science and Technology Town (SSTT)

科技城独立式标准厂房 1#、2#、8# 楼

FEATURES
项目亮点

Buildings in sci-tech park are usually designed in simple and elegant style. And to become a landmark of a park, the building shape and facade should be unique to stand out from the surroundings.

园区建筑风格讲究规整有序、简洁明快的建筑造型，不仅如此，由于城市界面对地块标志性作用的要求使得立面造型需要一定的唯一性，特殊的造型及变化丰富的立面元素与整体的均质化形成强烈的反差。

Location: Suzhou, Jiangsu, China
Architectural Design: Suzhou Institute of Architectural Design Co.,Ltd.
Total Land Area: 150,800 m^2
Total Floor Area: 127,438 m^2
Plot Ratio: 0.84
Green Coverage Ratio: 35%

项目地点：中国江苏省苏州市
设计单位：苏州设计研究院股份有限公司
总用地面积：150 800 m^2
总建筑面积：127 438 m^2
容积率：0.84
绿化率：35%

Overview

The project is an important part of the Chinese Academy of Sciences (CAS) Suzhou Biomedical Industrial Park, mainly undertaking the industrialization programs of Suzhou Institute of Biomedical Engineering and Technology, finishing the graduation works of the New Medical Center, and manufacturing other newly introduced medical apparatus and instruments as well as biological medicines.

Building 1 ~ 8 are developed first. No. 1 and 2 are five-story administrative buildings with a total floor area of 26,897 m^2; building 8 has only one floor and it is for the canteen and auxiliary rooms with a total floor area of 2,566 m^2.

项目概况

本项目是中国科学院苏州生物医学工程与生物医药产业化基地的重要组成部分，主要承接中科院生物医学工程技术研究所的产业化项目、苏州创业园新药中心毕业项目以及新引进的符合产业导向且对环境无重大影响的医疗器械和生物医药类项目。

1# ~ 8# 楼是本项目的首批开发工程，其中 1#、2# 楼为行政管理综合楼，建筑面积 26 897 m^2，地上五层；8# 楼为食堂及配套用房，建筑面积 2 566 m^2，地上一层。

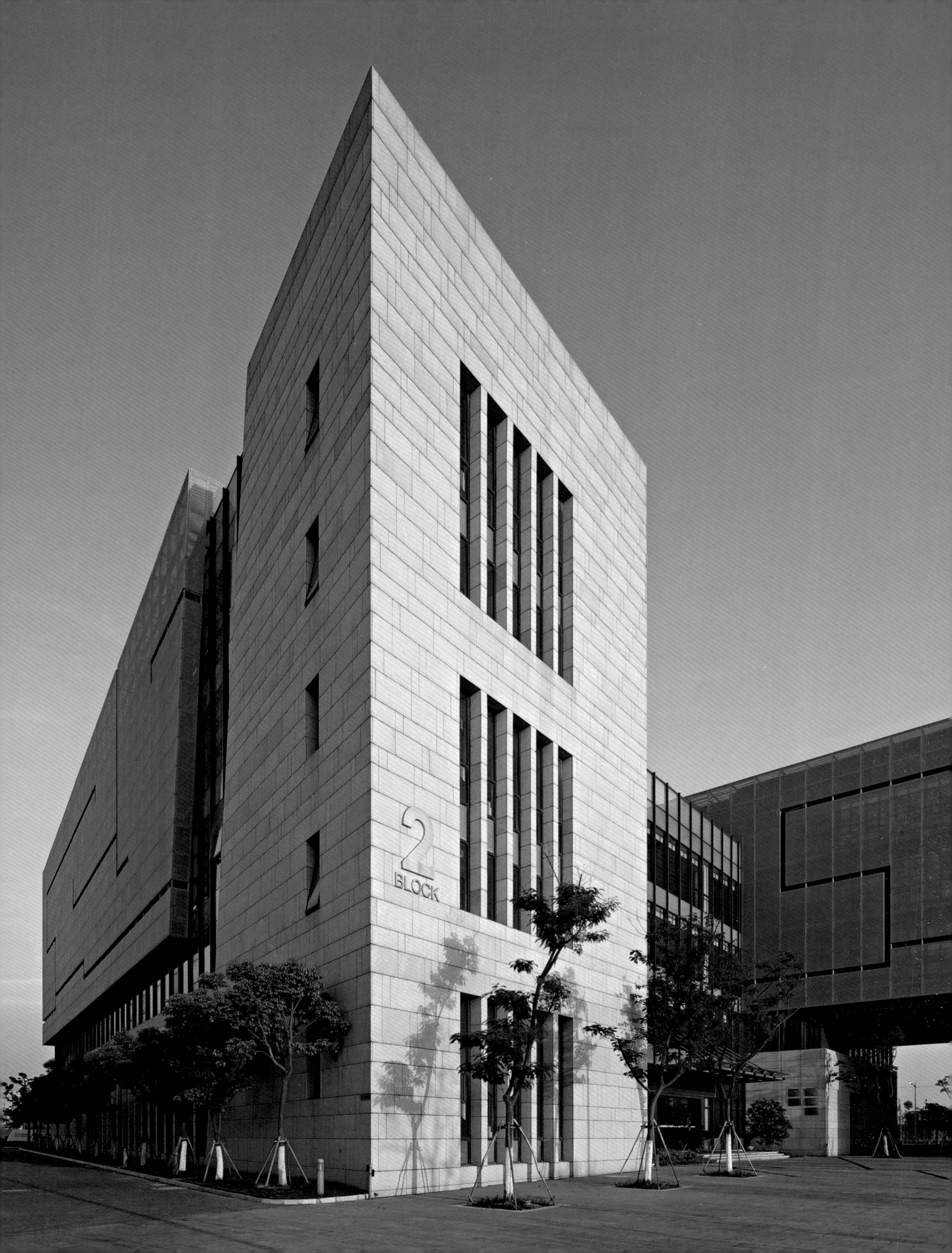
2
BLOCK

Site Plan　总平面图

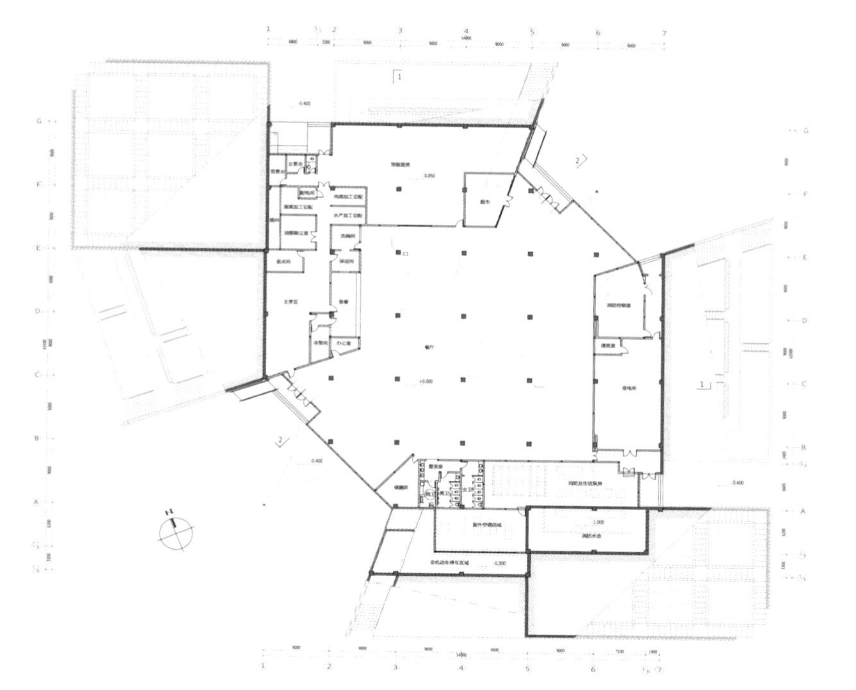

First Floor Plan for 8 #　8# 一层平面图

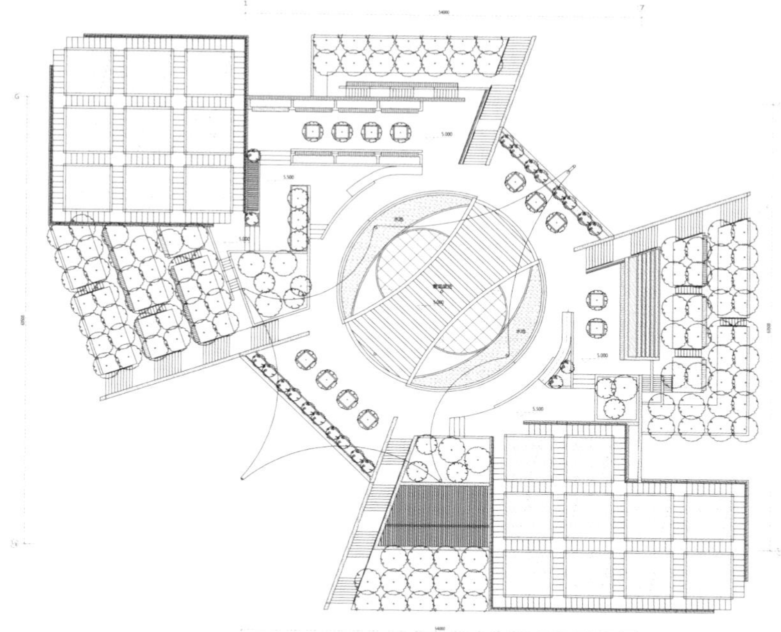

Roof Plan 屋顶层平面图

Architectural Design

Buildings in sci-tech park are usually designed in simple and elegant style. And to become a landmark of a park, the building shape and facade should be unique to stand out from the surroundings and provide different space experiences. With this in mind, the architects have created stretching and linear structures for the administrative buildings which result in flexible public spaces. And the contrast of convex and concave, shadow and light on the facade has made the buildings outstanding. At the same time, changeful and orderly horizontal lines shaped by the linear shapes have well interpreted the rational atmosphere of the R & D center.

建筑设计

园区建筑风格讲究规整有序、简洁明快的建筑造型，不仅如此，由于城市界面对地块标志性作用的要求使得立面造型需要一定的唯一性，特殊的造型及变化丰富的立面元素与整体的均质化形成强烈的反差，从而形成主次分明的整体风格、层次丰富的空间体验。在本项目中，唯一性表现在行政管理综合楼舒展的线性建筑形体得益于大气开合的变化而形成收放有度的公共空间中，也表现在符合其形式特质的立面上大面积虚实、明暗对比的曲折转承的线性表达中。与此同时，丰富的形体变化中，有条不紊的水平线条很好地诠释了逻辑稳定而又充满张力的理性研发气质。

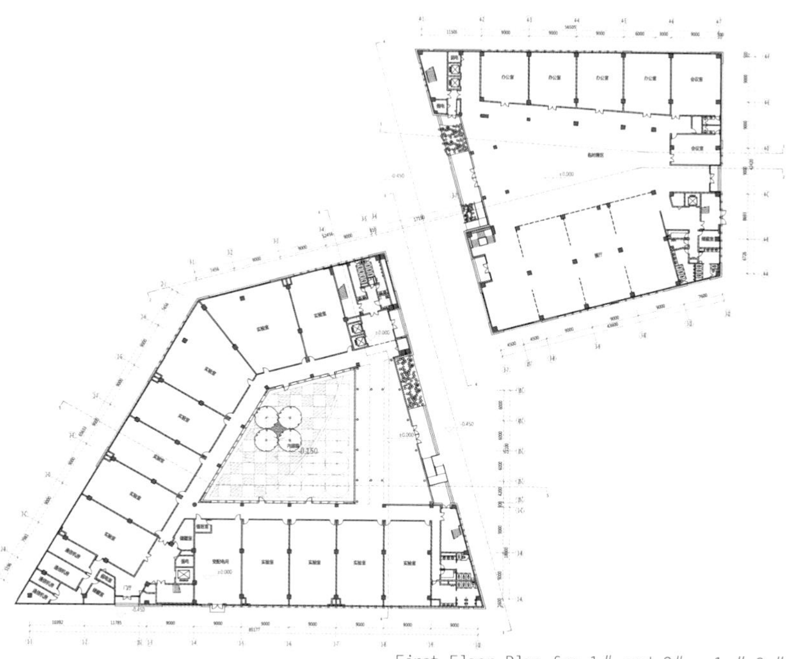

First Floor Plan for 1# and 2#　1 # 2 # 一层平面图

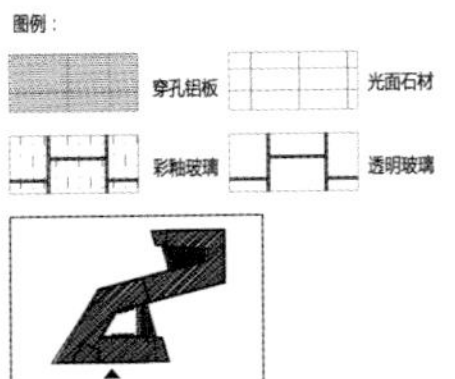

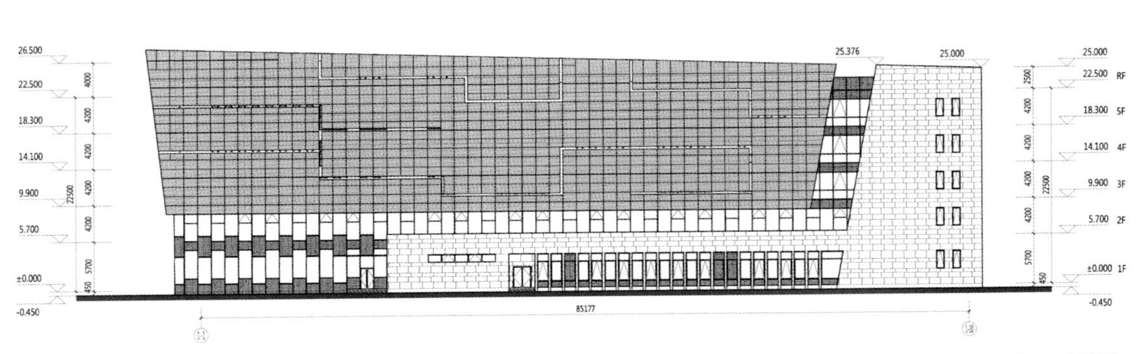

Elevation 立面图

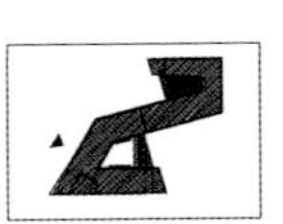

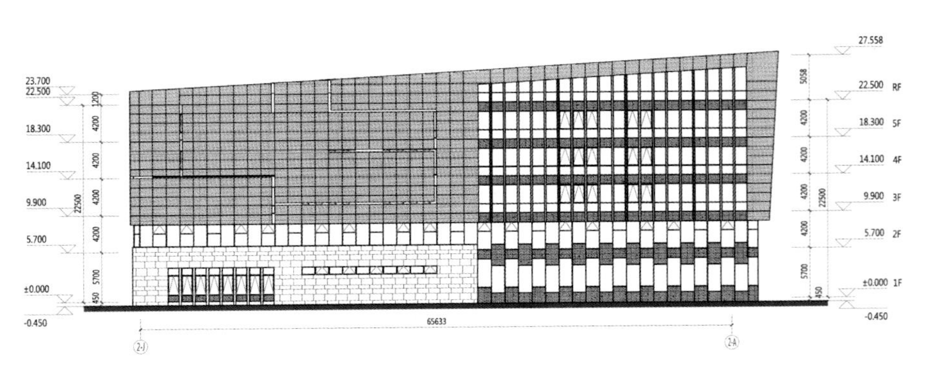

Elevation 立面图

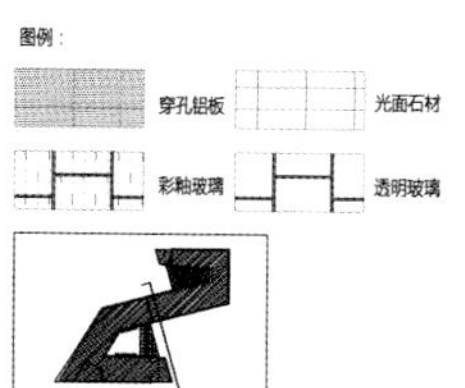

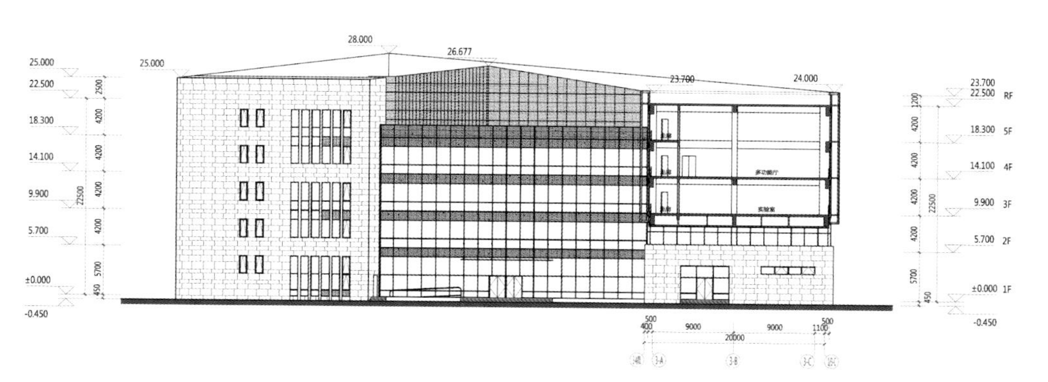

Sectional Drawing 剖面图

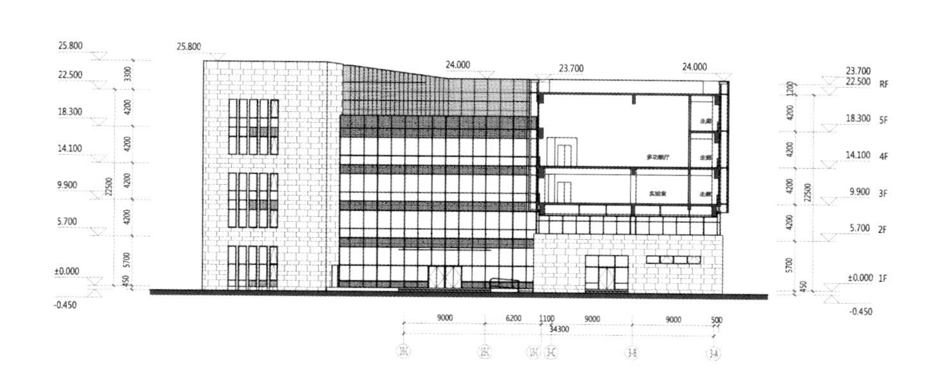

Sectional Drawing 剖面图

Medpark

3
BLOCK

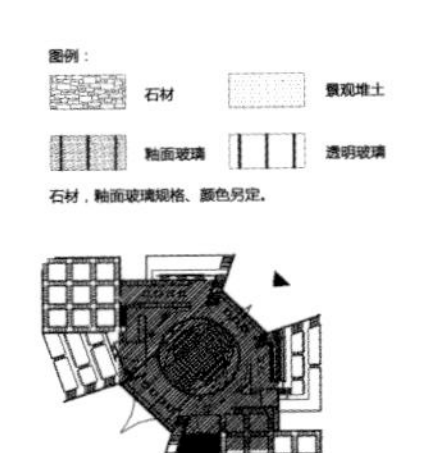

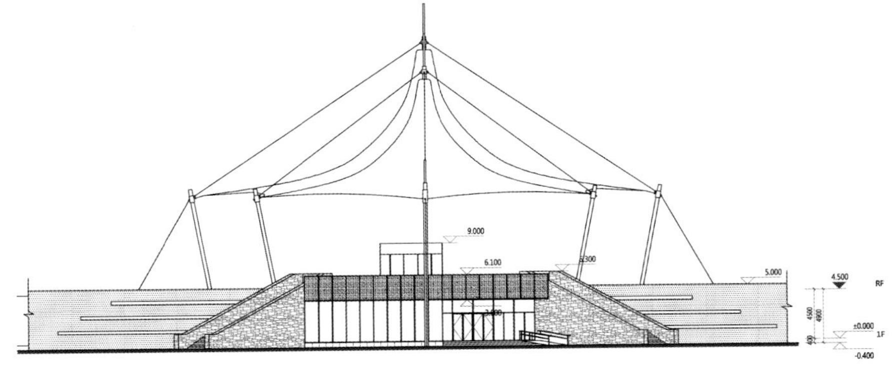

East Elevation 东立面图

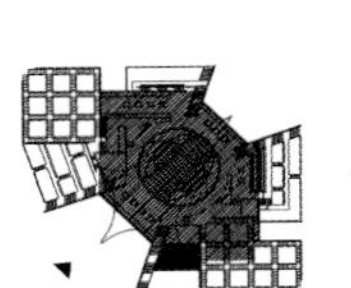

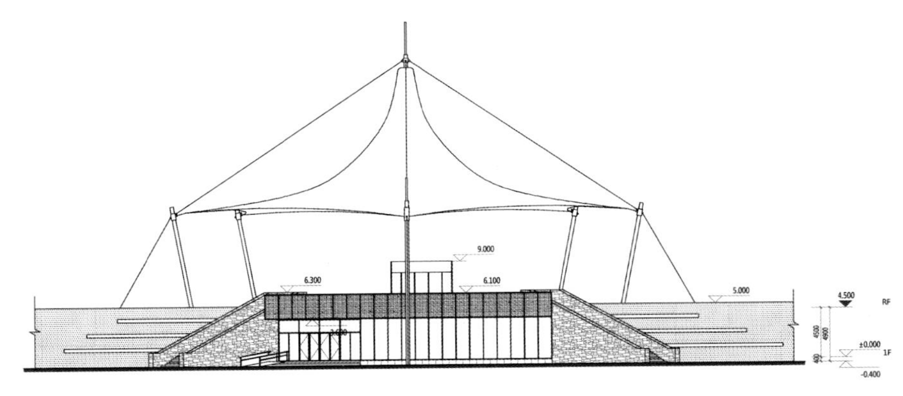

West Elevation 西立面图

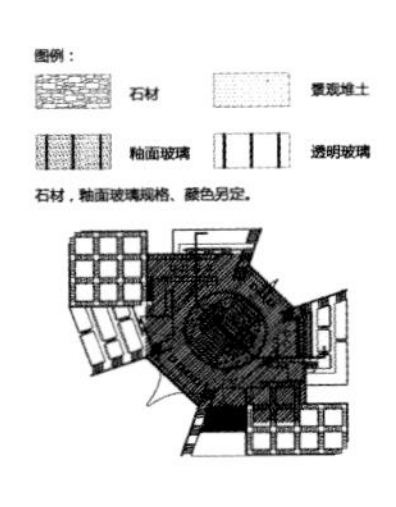

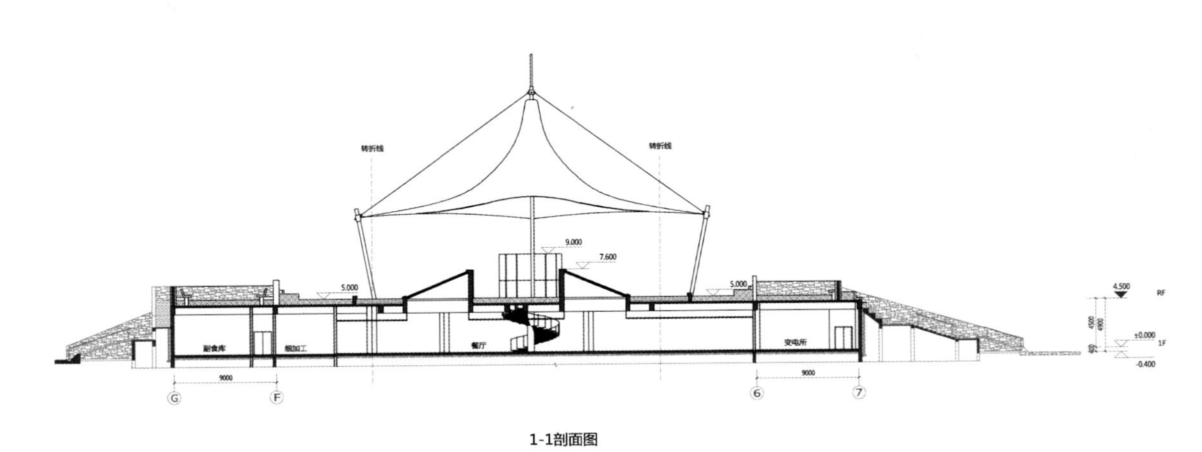

1-1剖面图

Sectional Drawing 剖面图

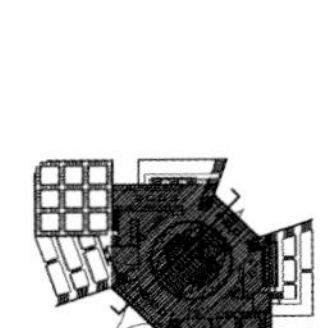

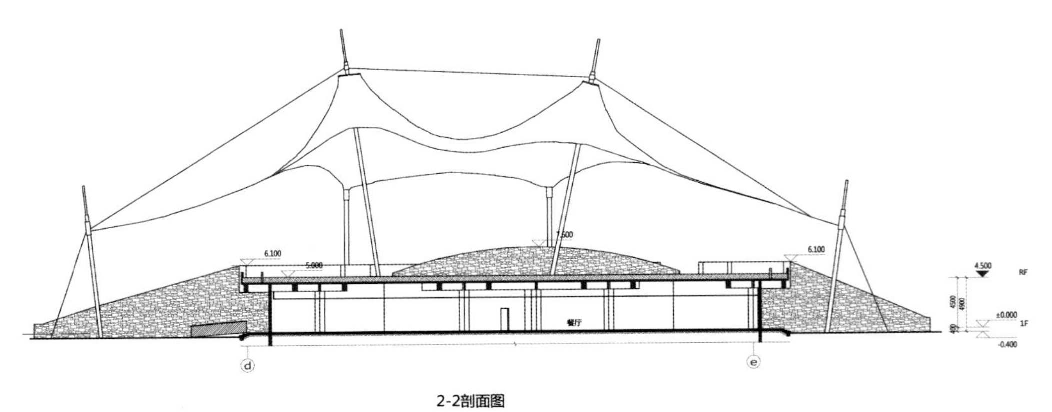

2-2剖面图

Sectional Drawing 剖面图

KEY WORDS
关键词

Combination of New and Old 新旧建筑融合

Unique Shape 造型独特

Elegant Appearance 外观优美

Nano Materials Research Building
纳米材料综合研究平台

FEATURES
项目亮点

The new complex is connected with the library building by the glass canopy and the east cambered wall. In this way, new and old buildings are combined together and create a public garden in between, which will benefit the communication between the staff and visitors.

新综合楼与图书馆通过玻璃顶棚与东侧弧形墙的连接而连为一体，新旧建筑之间成为庭院共享空间，为加强内部科研人员和外部来访人员的交流起到了重要的作用。

Location: Fuzhou, Fujian, China
Architectural Design: Fujian Architectural Design Institute
Land Area: 1,647.3 m^2
Total Floor Area: 15,690.6 m^2
Underground Floor Area: 2,136.5 m^2
Green Land Area: 2,619.7 m^2
Building Density: 30.0%
Plot Ratio: 2.45
Green Coveroge Ratio: 47.4%

项目地点：福建省福州市
设计单位：福建省建筑设计研究院
建筑占地面积：1647.3 m^2
总建筑面积：15 690.6 m^2
地下建筑面积：2 136.5 m^2
绿地面积：2619.7 m^2
建筑密度：30.0%
容积率：2.45
绿化率：47.4%

Architectural Design

The new complex is connected with the library building by the glass canopy and the east cambered wall. In this way, new and old buildings are combined together to create a public garden in between, which will benefit the communication between the staff and visitors.

The cambered square in the east serves as the sub entrance and the transition space between the living quarter and the research and library buildings. The space on the south of the library is designed to be a landscape garden which is an ideal place for relaxation after working and studying. In addition to the woods and lawns in the northeast of the site, it has formed an eco and green ring that surrounds the research complex and the library building.

建筑设计

新综合楼与图书馆通过玻璃顶棚与东侧弧形墙的连接而连为一体，新旧建筑之间成为庭院共享空间，为加强内部科研人员和外部来访人员的交流起到了重要的作用。

东侧的弧形广场成为次入口空间，呼应东侧的生活区，成为生活区进入综合楼和图书馆建筑的过渡空间，图书馆南侧的地块设计成后庭院景观空间，成为园区中轴线上的主要节点，也是园区科研人员工作学习生活之余的休闲之地，地块东北侧布局树林与大草坪以及后庭院，共同组成绿色环带，将综合楼和图书馆置于绿色生态景观之中，丰富了园区的办公环境。

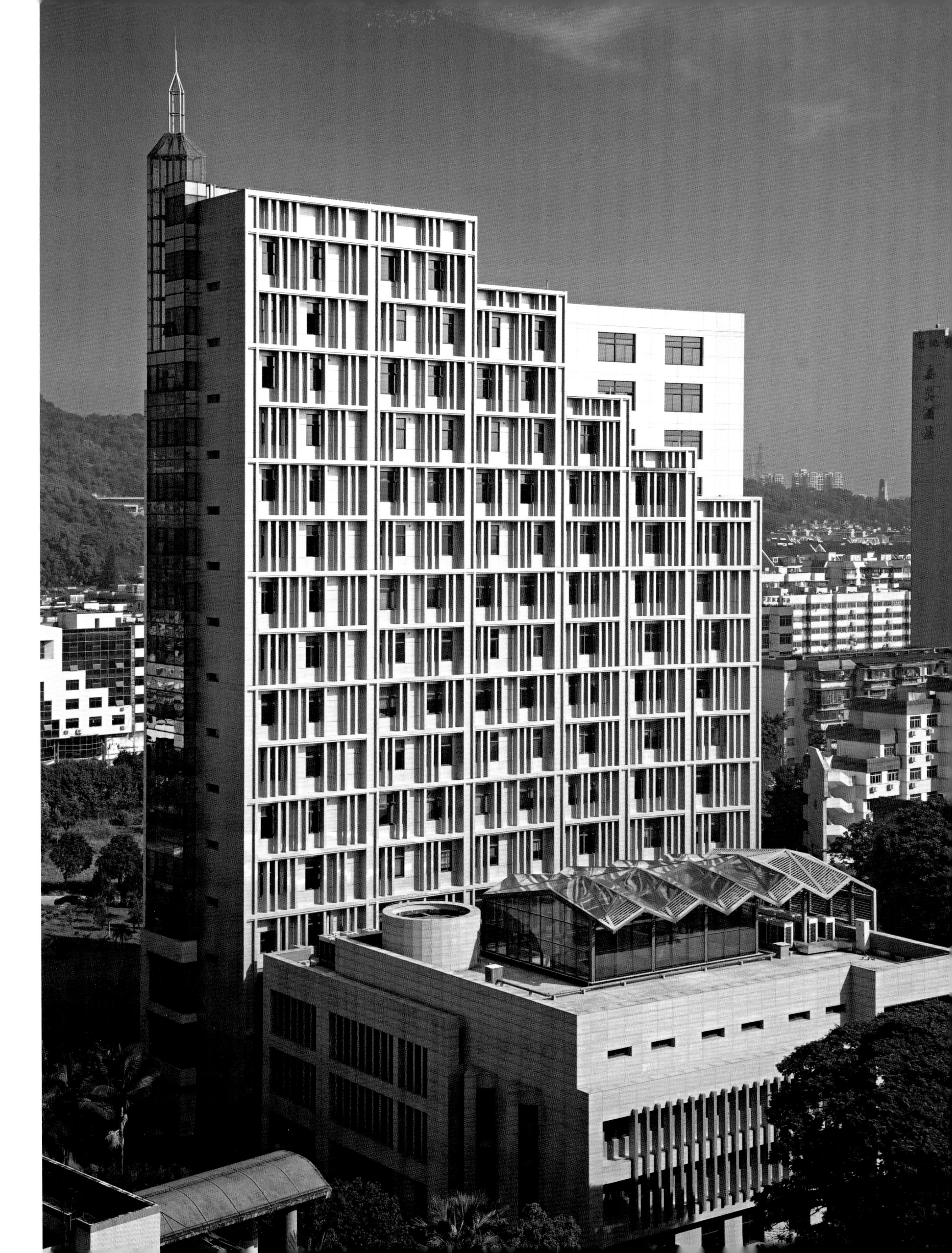

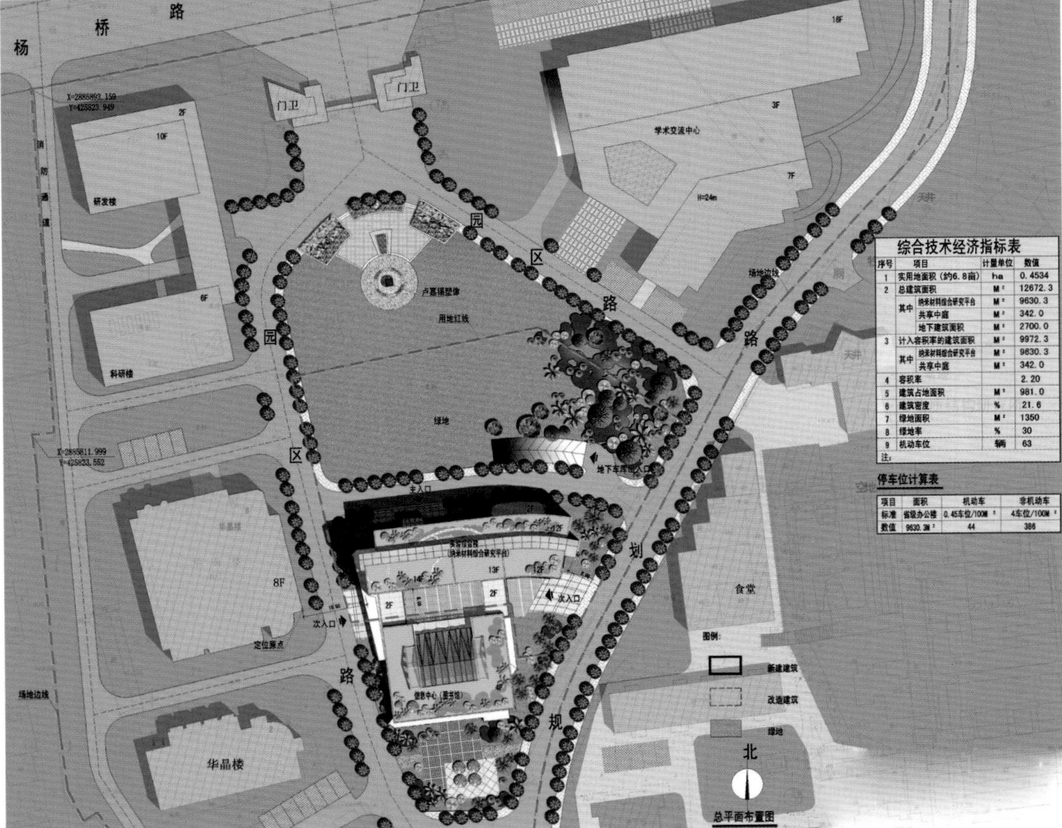

综合技术经济指标表

序号	项目		计量单位	数值
1	实用地面积（约6.8亩）		ha	0.4534
2	总建筑面积		M²	12672.3
	其中	纳米材料综合研究平台	M²	9630.3
		共享中庭	M²	342.0
		地下建筑面积	M²	2700.0
3	计入容积率的建筑面积		M²	9972.3
	其中	纳米材料综合研究平台	M²	9630.3
		共享中庭	M²	342.0
4	容积率			2.20
5	建筑占地面积		M²	981.0
6	建筑密度		%	21.6
7	绿地面积		M²	1350
8	绿地率		%	30
9	机动车位		辆	63
注：				

停车位计算表

项目	面积	机动车	非机动车
标准	省级办公楼	0.45车位/100M²	4车位/100M²
数值	9630.3M²	44	386

Site Plan 总平面图

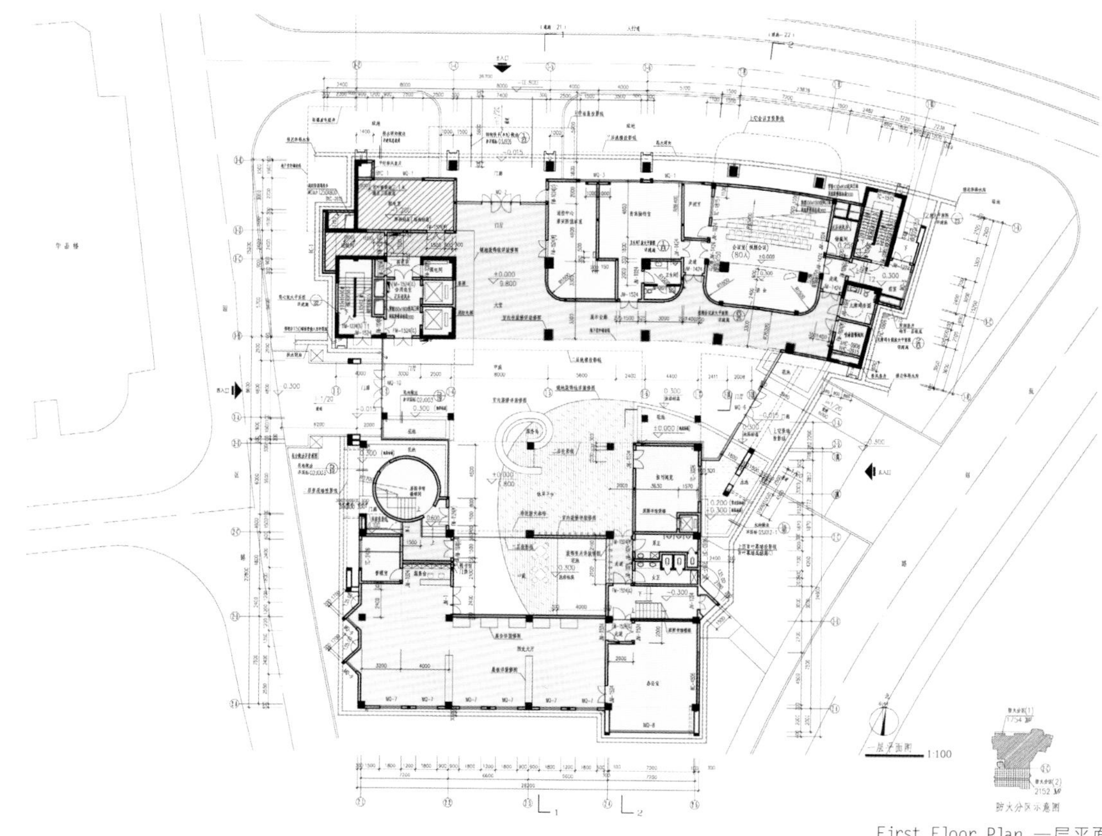

First Floor Plan 一层平面图

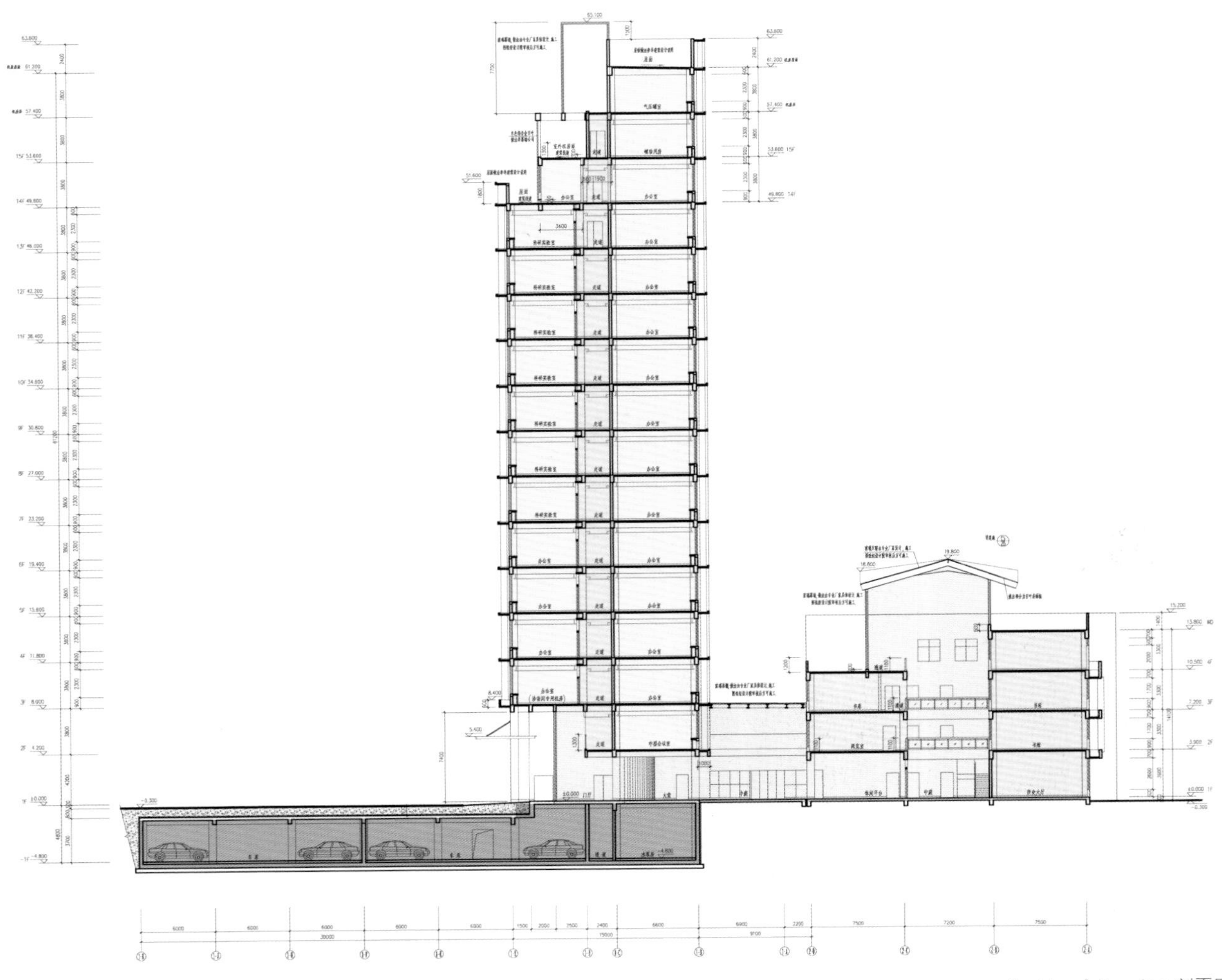

Sectional Drawing 剖面图

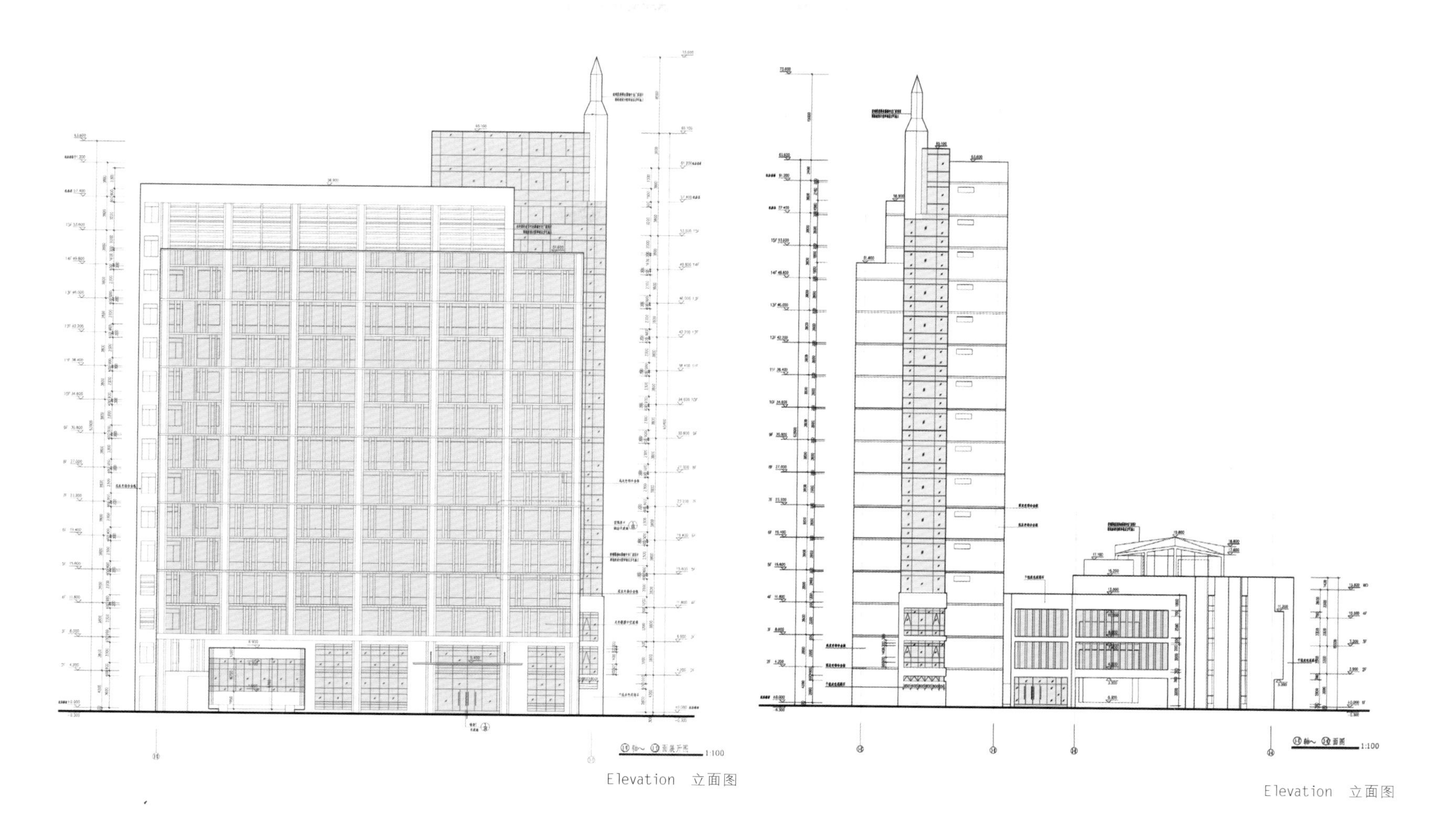

Elevation 立面图

Elevation 立面图

KEY WORDS
关键词

Green Open Floor 架空层绿化

Glass and Stone Facade 玻璃及石材立面

Space Style 空间意境

CHERVON Headquarters
泉峰集团总部办公楼

FEATURES
项目亮点

The wings of the buildings are designed with clear and logical facades. The north and south facades are composed of glass and solid walls, while the east and west facades are extended horizontally with rhythmic windows inlaid in stone walls.

建筑群各翼的立面处理逻辑非常清晰，南北向以大面玻璃和实墙穿插进行构图，而东、西立面则在石墙上以一系列富于韵律和节奏的竖向窗洞进行横向延展。

Location: Jiangning, Nanjing, Jiangsu, China
Architectural Design: Southeast University Architectural Design Institute Co.,Ltd.
Land Area: 26,460 m²
Total Floor Area: 37,380 m²
Building Height: 28.3 m
Plot Ratio: 1.02
Completion: 2009

项目地点：中国江苏省南京市江宁区
设计单位：东南大学建筑设计研究院有限公司
用地面积：26 460 m²
总建筑面积：37 380 m²
建筑高度：28.3 m
容积率：1.02
竣工时间：2009 年

Architectural Design

The building complex extends diagonally, with protruding eaves for the "flying" wings, feeling powerful and exquisite. The facades of the wings are designed with clear logic: the north and south facades are composed of glass and solid walls, while the east and west facades are extended horizontally with rhythmic windows inlaid in stone walls. The "S" shaped outdoor space enclosed by the buildings is divided into courtyards of different sizes and themes. These courtyard spaces are connected by corridors and bridges from south to north, quite and comfortable for people working here.

建筑设计

建筑体量呈斜向展开，各翼建筑物挑檐深远，轻盈欲飞，令人感觉既硬朗有力又玲珑剔透。建筑群各翼的立面处理逻辑非常清晰，南北向以大面玻璃和实墙穿插进行构图，而东、西立面则在石墙上以一系列富于韵律和节奏的竖向窗洞进行横向延展。贯穿南北的连廊、天桥将建筑群"S"形体量的户外空间划分成若干大小和主题不一的庭园空间，营造静谧而幽远的中式园林意境，创造舒适、宜人的办公空间。

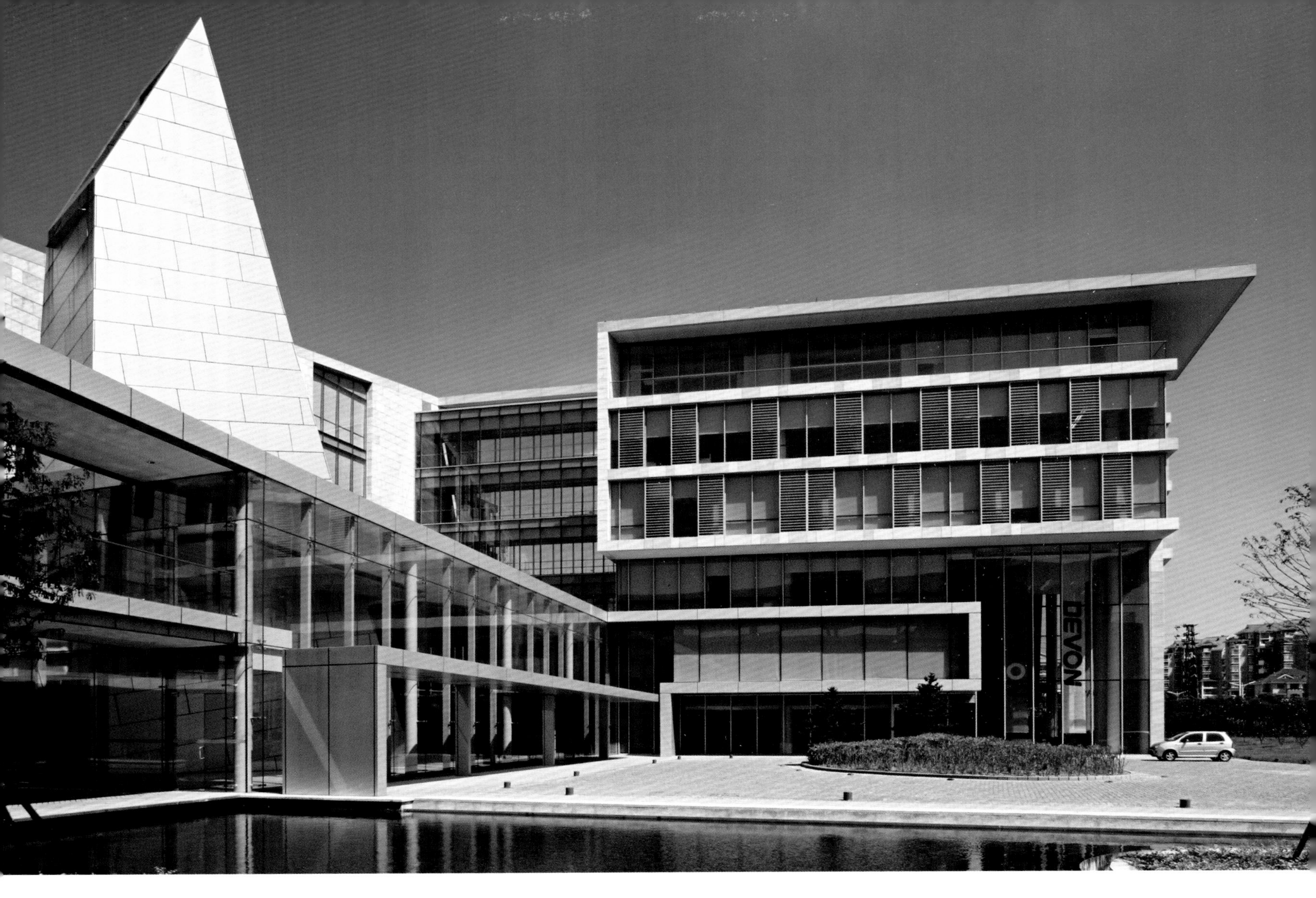

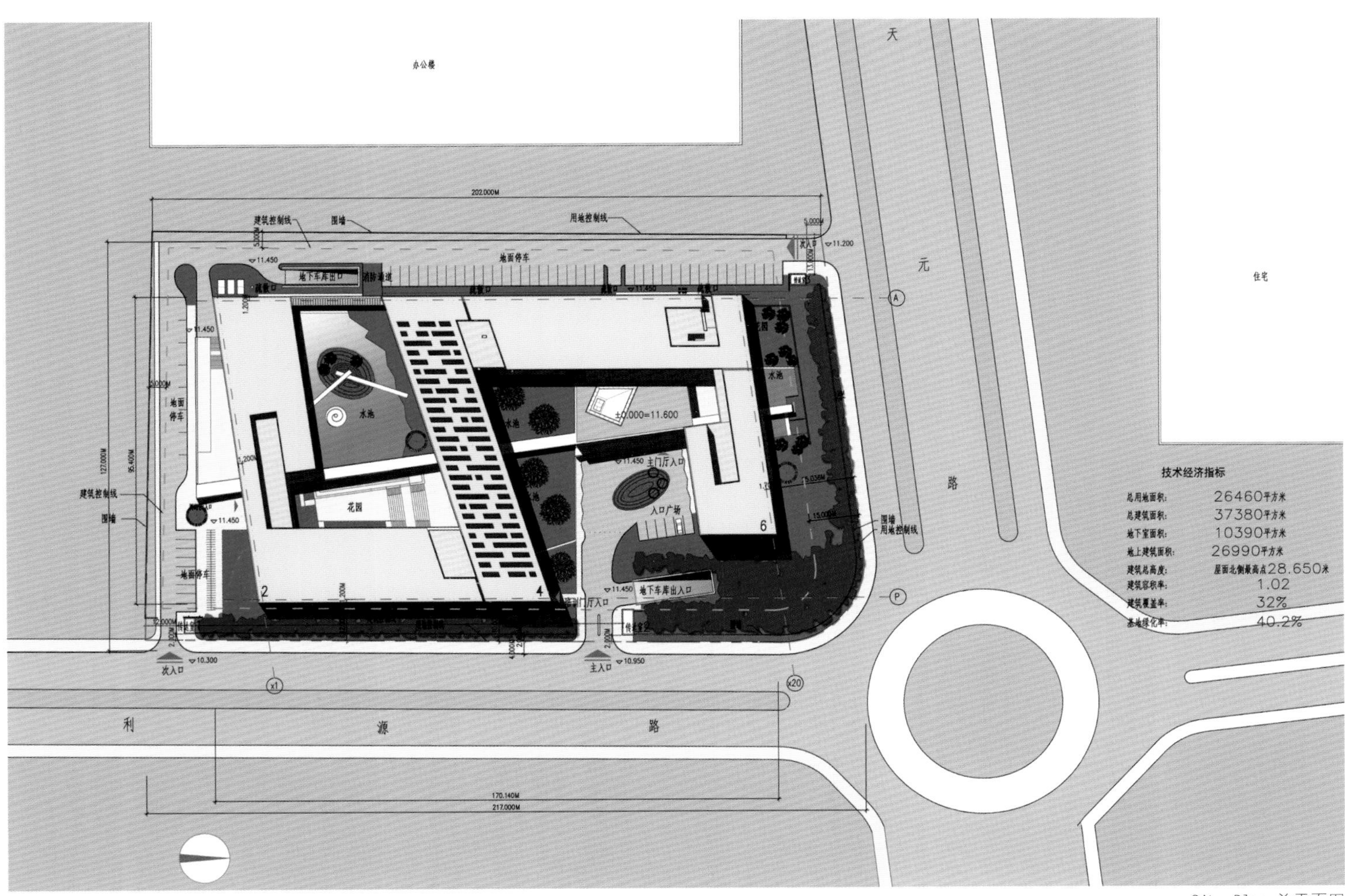

Site Plan 总平面图

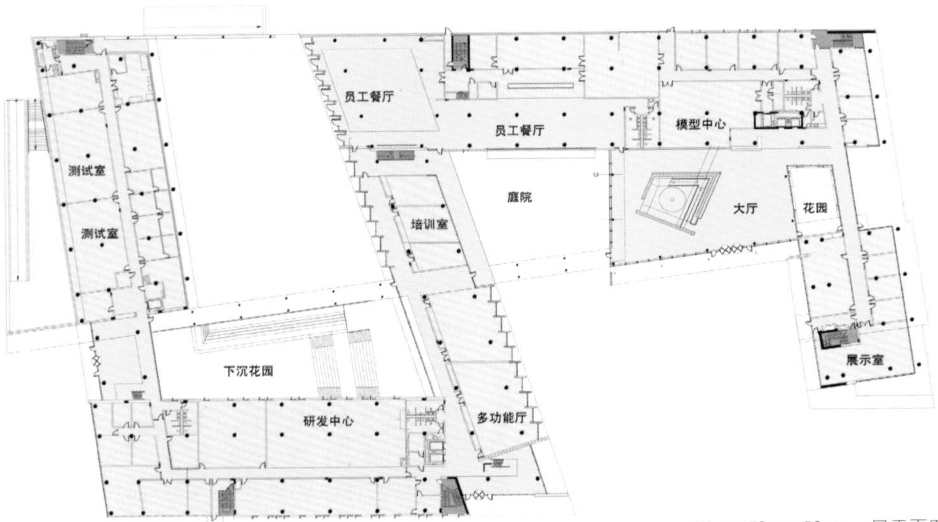

First Floor Plan 一层平面图

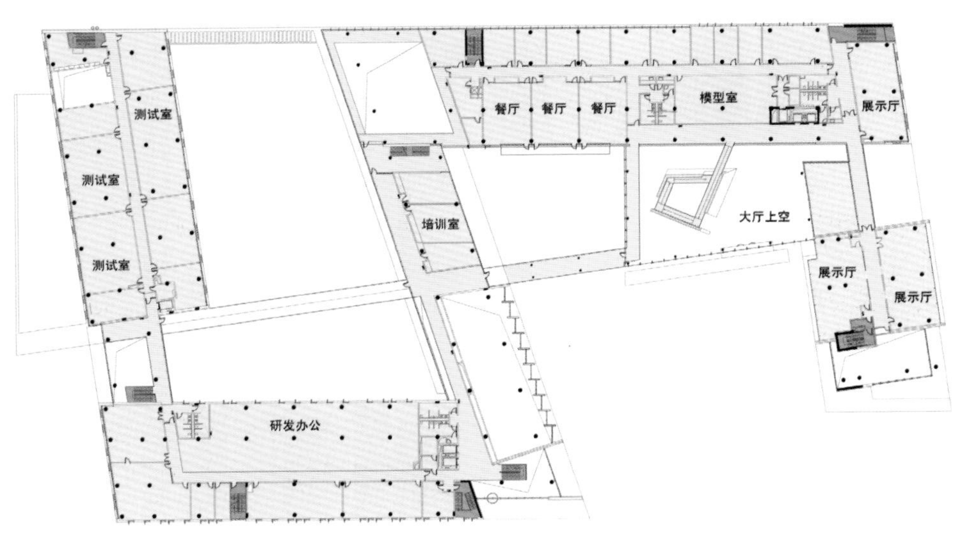

Second Floor Plan 二层平面图

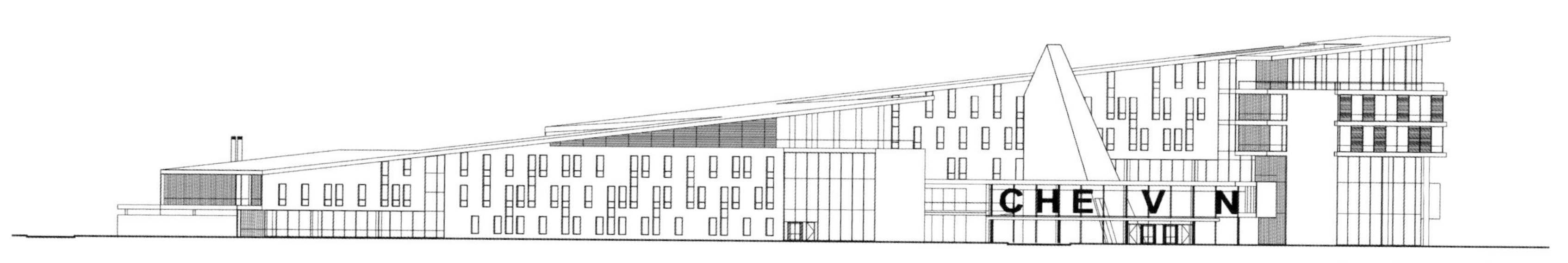

East Elevation 东立面图

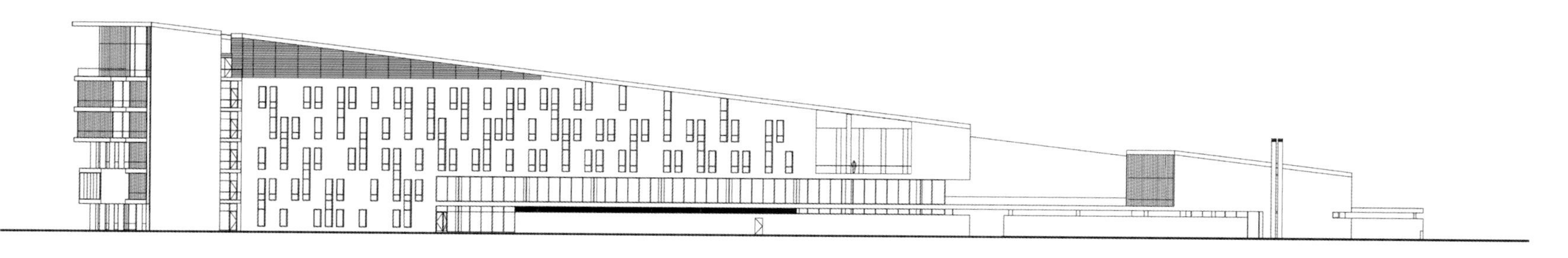

West Elevation 西立面图

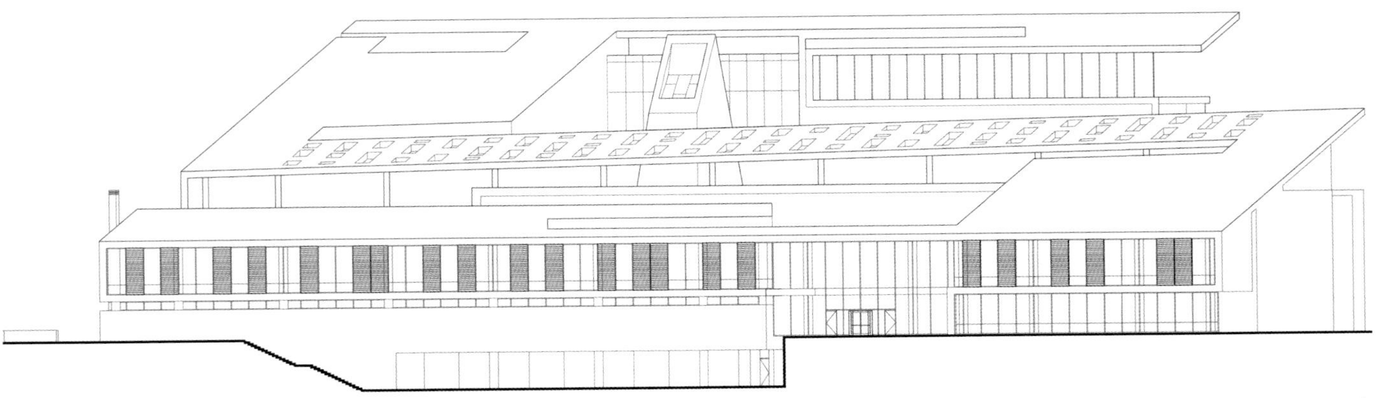

South Elevation 南立面图

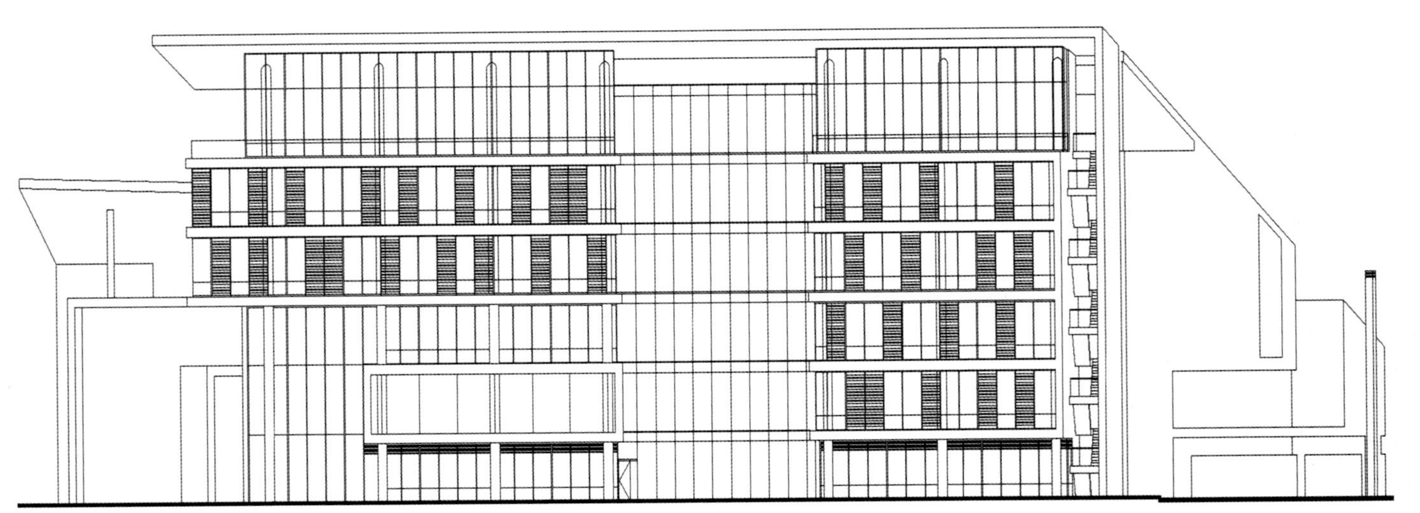

North Elevation 北立面图

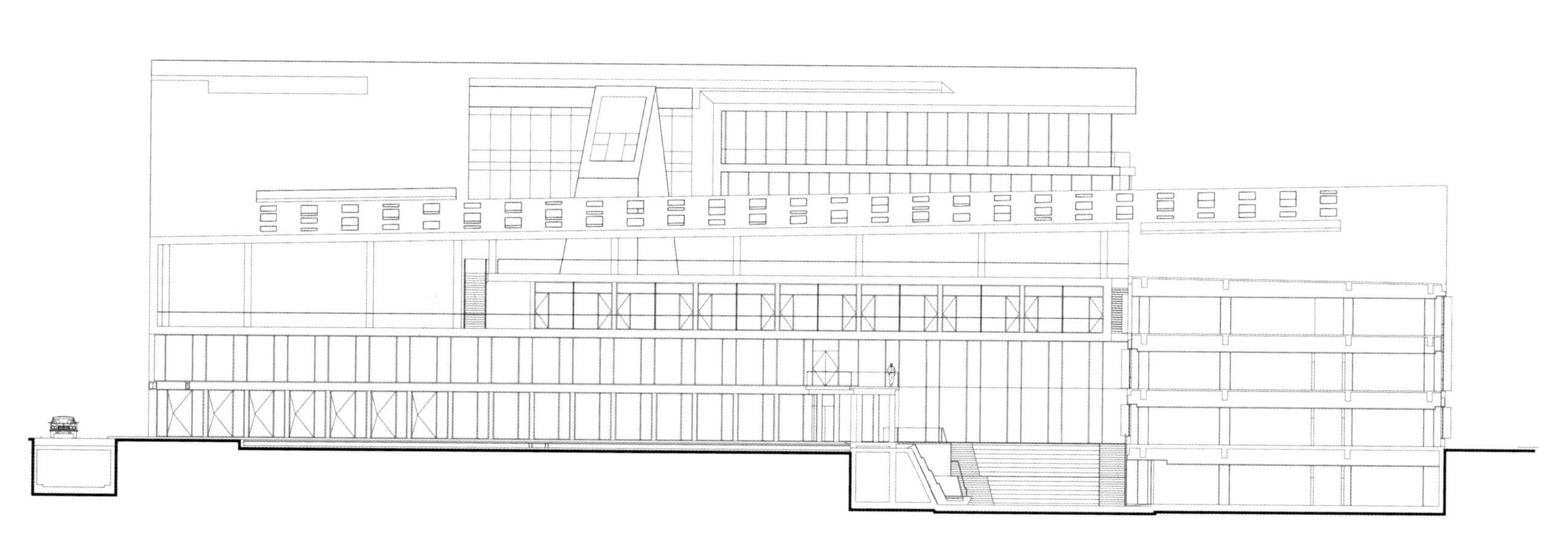

Sectional Drawing 剖面图

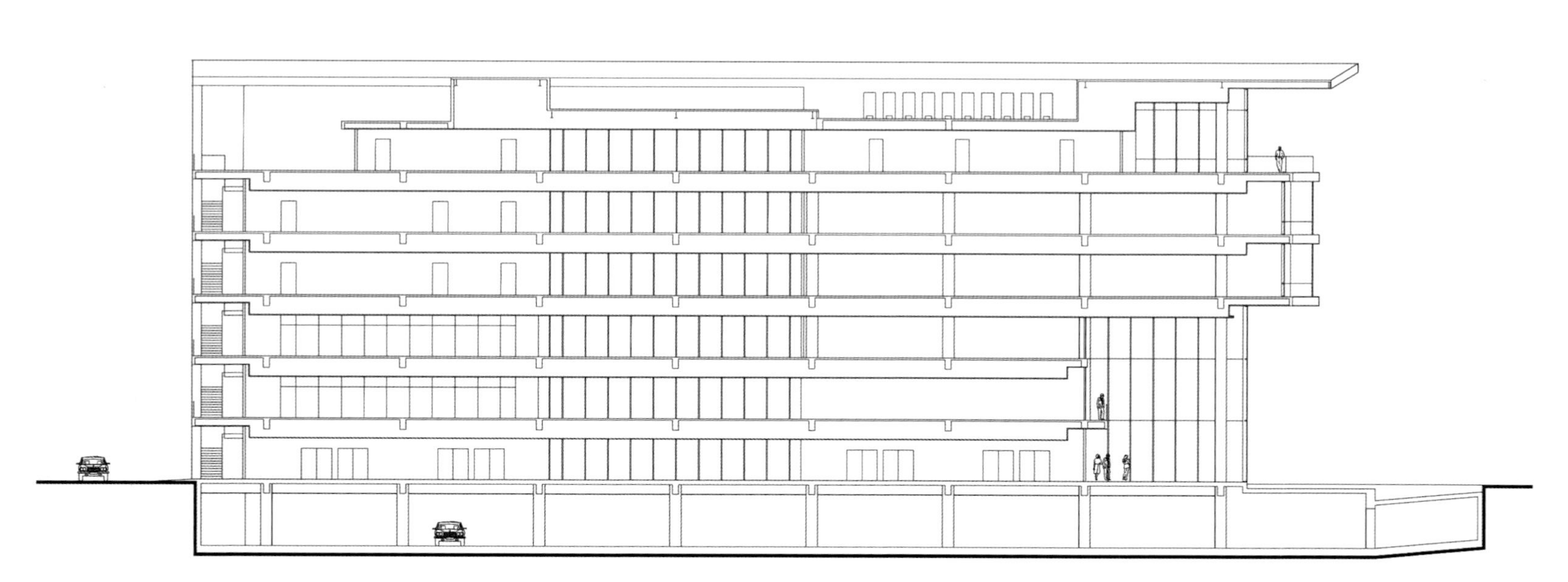

Sectional Drawing 剖面图

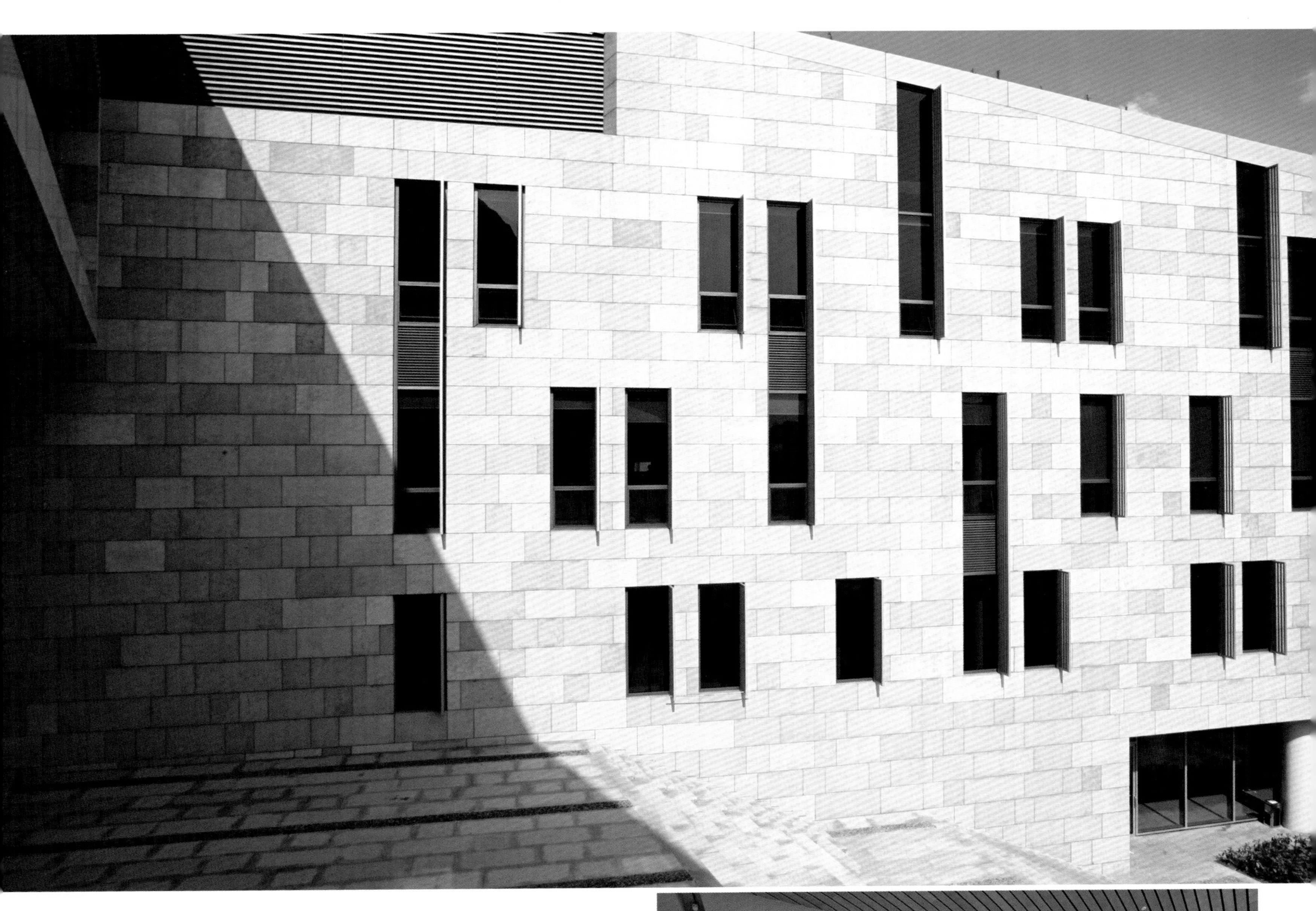

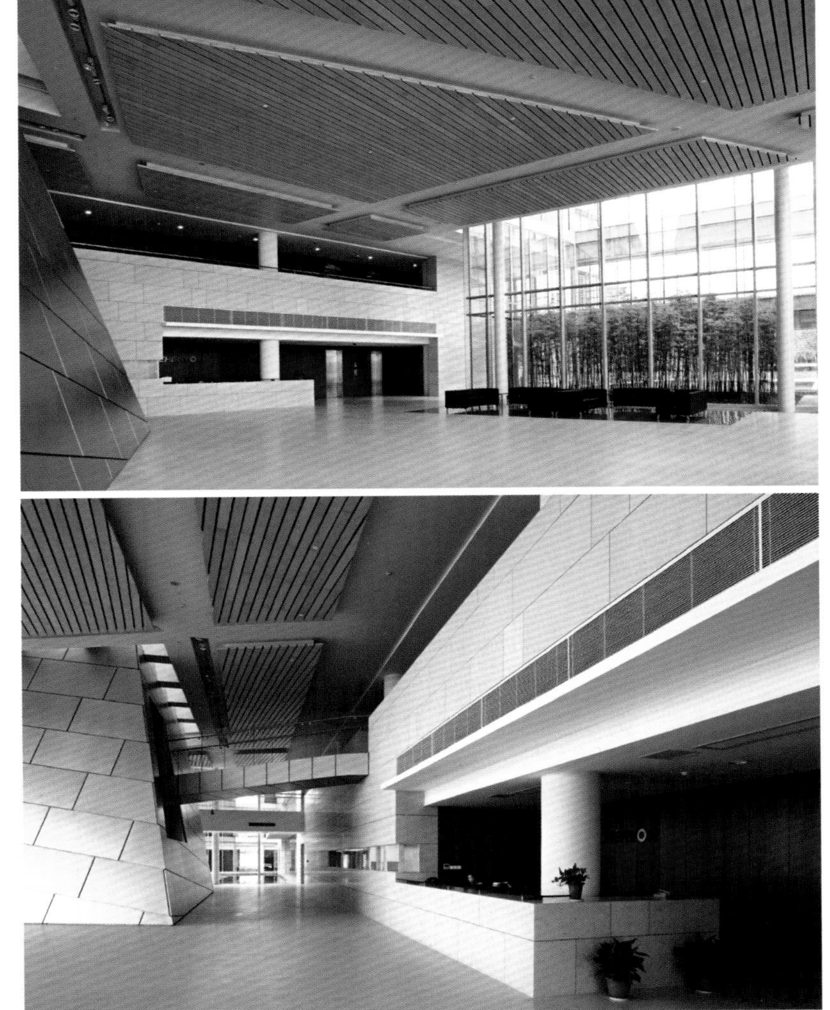

KEY WORDS 关键词

Modest Volume 造型稳重

Vertical Lines 竖向线条

Elegant Facade 简洁立面

Shandong Industry Association Building
山东省行业协会大楼

FEATURES 项目亮点

It decorates the north and south facade with crossing lines. In addition, the glass corners combine with the stone wall panels to present unique architectural impression.

南北立面上将方正的建筑体量适当打破，形成局部竖向线条与水平线条相结合的特征，转角处的玻璃体与贯穿上下的石墙板相穿插，丰富了建筑立面的造型语言。

Location: Jinan, Shandong, China
Architectural Design: Shandong Tong Yuan Design Group Co.,Ltd.
Land Area: 9,048.5 m²
Total Floor Area: 40,135 m²
Completion: 2010

项目地点：中国山东省济南市
设计单位：山东同圆设计集团有限公司
用地面积：9 048.5 m²
总建筑面积：40 135 m²
竣工时间：2010 年

Overview

Shandong Industry Association Building is located on Lishan Road, Lixia District of Jinan City. The tower has 14 floors and the podium has six, and the total floor area is 40,135 m².

项目概况

山东省行业协会大楼位于济南市历下区历山路，项目主体建筑地上十四层，裙房六层，总建筑面积为 40 135 m²。

Architectural Design

In consideration of the special location and its identity as an office building, the facade is designed to be simple and elegant, and the building volume well matches the internal functions. The south facade avoid large-area solid wall to allow great views of the Thousand Buddha Mountain. It also emphasizes the use of vertical lines and decorates the north and south facade with crossing lines. In addition, the glass corners combine with the vertical stone wall panels to give a unique architectural impression.

建筑设计

考虑到建筑所处的特殊地域空间位置和作为办公建筑的稳重、大方的外在气质要求，建筑立面处理采用明快、简洁的处理手法，建筑造型与内部使用功能相呼应。南向避免采用大面积实体墙面的处理手法，使得更多的办公空间获得良好的景观视野（南眺千佛山），强调竖向线条的运用，在南北立面上将方正的建筑体量适当打破，形成局部竖向线条与局部水平线条相结合的表现特征，转角处的玻璃体与贯穿上下的石墙板相穿插，丰富了建筑立面的造型语言。

BQD 青岛银行
CO-OP
百年人寿

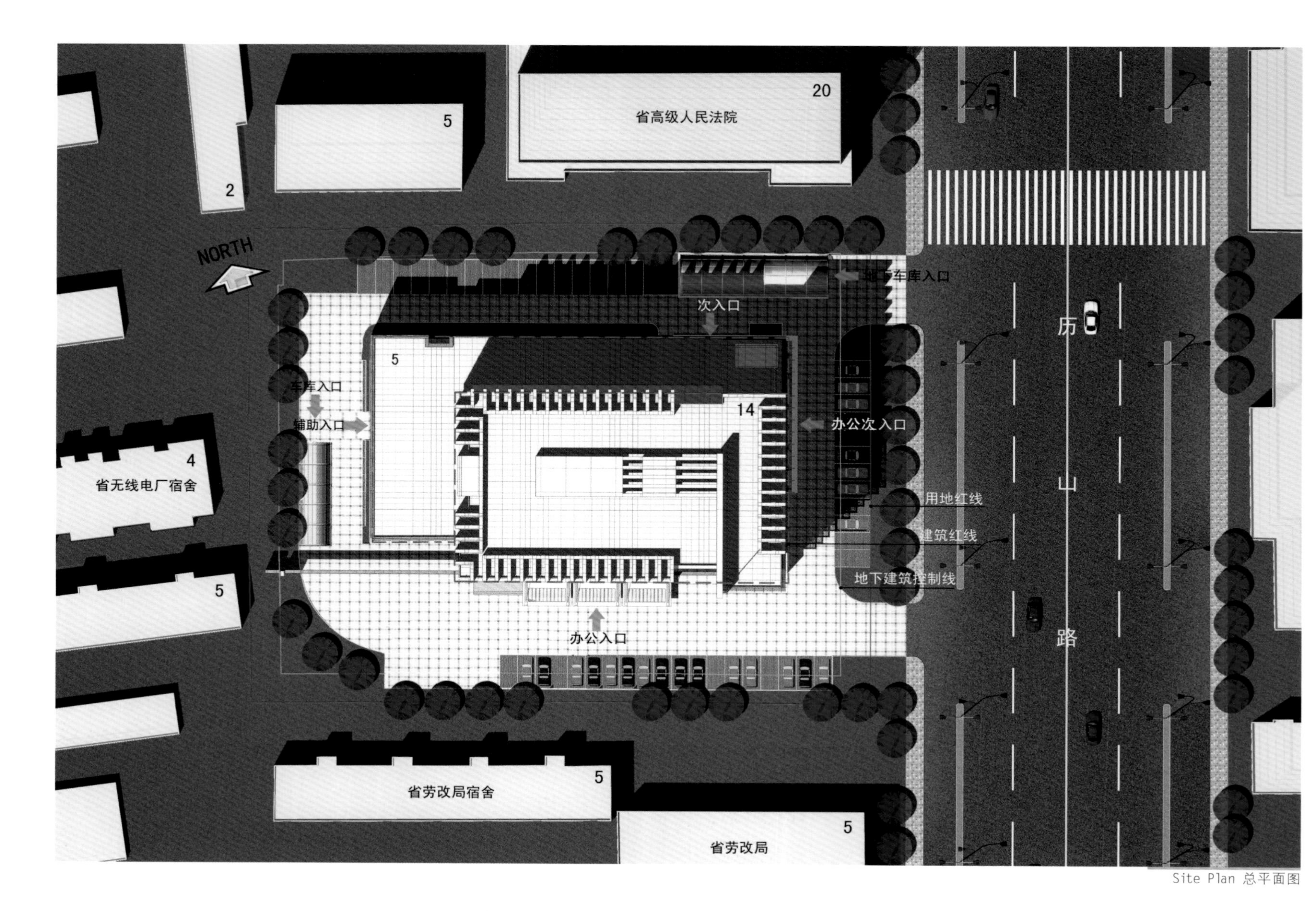

Site Plan 总平面图

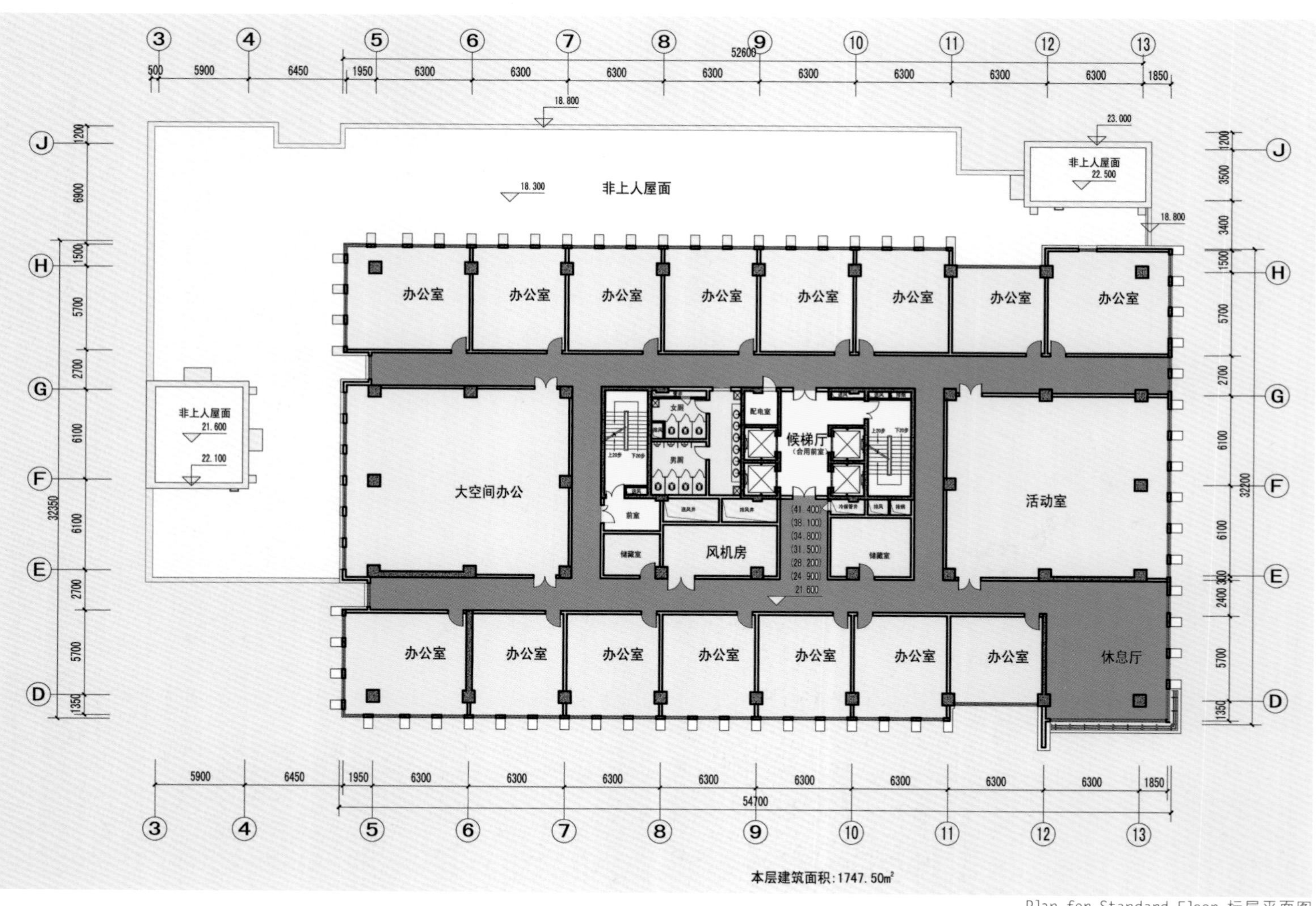

Plan for Standard Floor 标层平面图

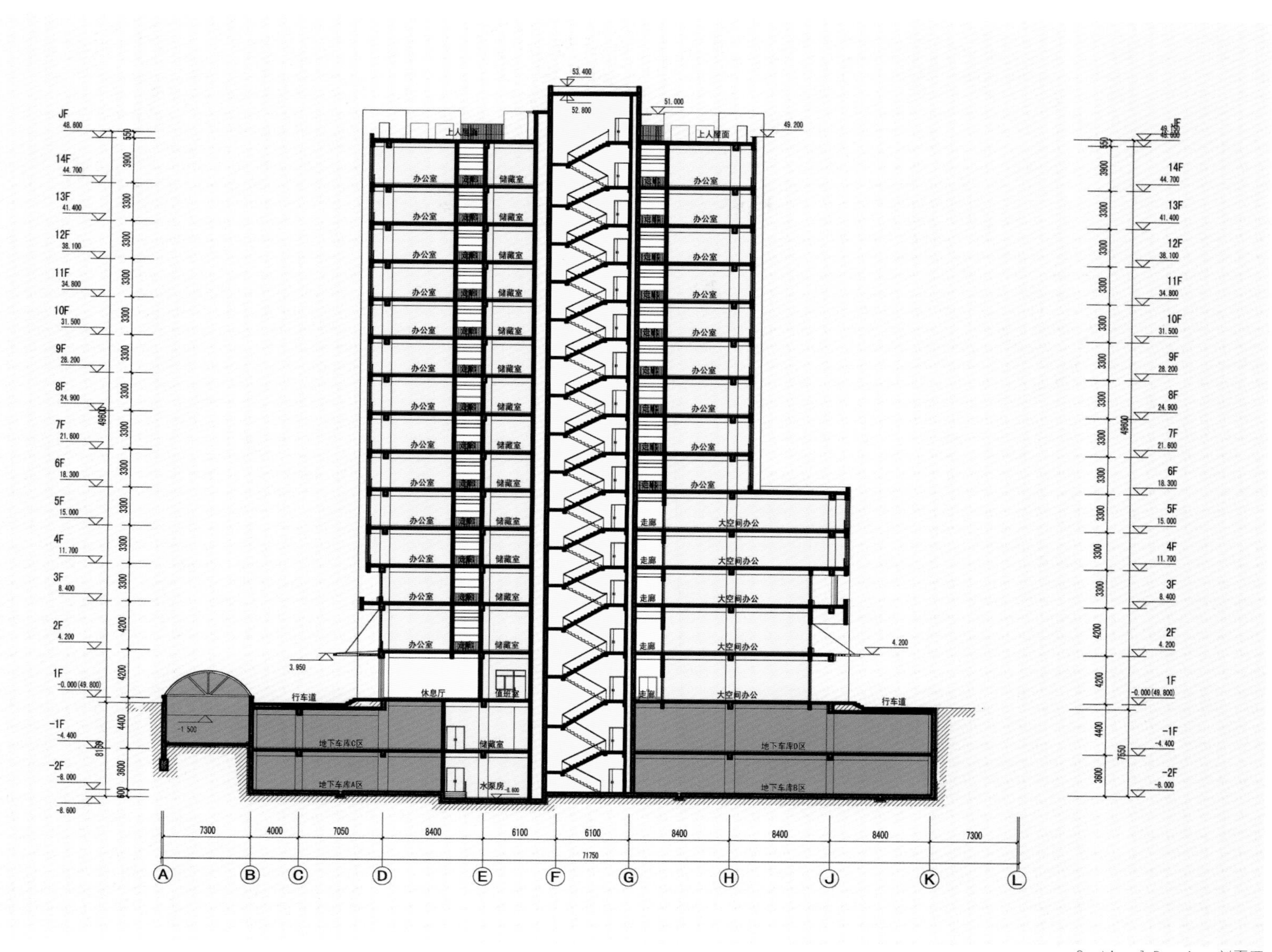

Sectional Drawing 剖面图

East Elevation 东立面图

KEY WORDS
关键词

Mixed Use 功能综合

Intelligent 智能化

Eco and Green 生态绿色

Office Complex of SKSHU Paint Co.,Ltd.
三棵树涂料股份有限公司办公研发楼

FEATURES
项目亮点

The complex looks like three huge leaves which echoes the enterprise' brand; operated and managed under advanced system, this mixed-use complex is highly intelligent, informationized and ecological.

建筑外形如三片巨大的树叶，呼应企业形象；整幢办公大楼采用国际先进的控制和管理系统，是高度智能化、信息化的多功能现代化综合生态办公楼。

Location: Putian, Fujian, China
Architectural Design: Fujian Architectural Design Institute
Total Land Area: 19,300 m²
Total Floor Area: 13,000 m²
Building Height: 47.7 m
Building Density: 10%
Plot Ratio: 0.66
Green Coverage Ratio: 48.2%
Completion: 2010

项目地点：中国福建省莆田市
设计单位：福建省建筑设计研究院
总用地面积：19 300 m²
总建筑面积：13 000 m²
建筑高度：47.7 m
建筑密度 10%
容积率：0.66
绿化率：48.2%
竣工时间：2010 年

Overview

The site is located on the east of an intersection of Putian, Fujian, where Liyuan North Road and Dongchuan Road cross. To the northeast across a river and a litchi forest, there are the staff dormitories and canteen; to the southeast, here is the front square and the green water features.

项目概况

建筑基地位于福建省莆田市新区荔园北路与东川路交汇处的东侧，其东北面隔着河流和荔枝林的地块为企业员工的宿舍和餐厅，东南面是宽敞的厂前广场和绿化水体景观。

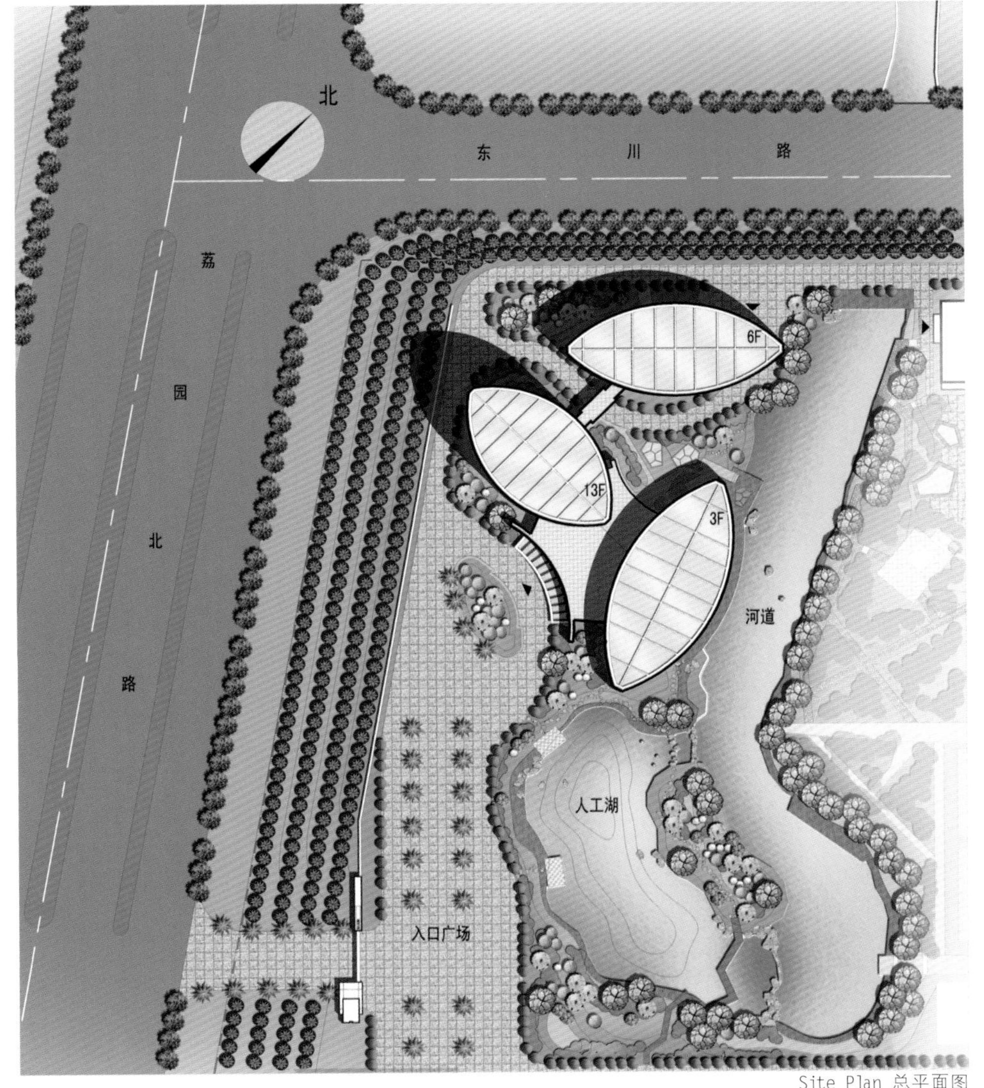

Site Plan 总平面图

Architectural Design

The intelligent office complex is composed of the central office building, the R&D center and the training center which extend like three huge leaves. The 13-storey office building is 47.7 m high to accommodate more than one thousand workers. The training center has four floors and it houses the first intelligent business school of this industry and thirteen more rooms for marketing wars, meetings and trainings. And the world-class R&D center, having six floors and featuring a total floor area of 5,000 m², is well equipped with a full set of advanced experimental facilities. It houses the post-doctoral research center. In addition, the complex also consists a 7-storey affiliated eco building. Operated and managed under international advanced system, this mixed-use complex is highly intelligent, informationized and ecological.

建筑设计

三棵树智能化办公楼群由办公中心大楼、研发中心大楼和培训中心大楼三部分组成，外形如三片巨大的树叶。办公中心大楼高 47.7 m，共 13 层，能容纳上千人同时办公；培训中心楼 4 层，设有业界首家智能化商学院，另有营销作战室以及各类大小会议室、培训室共 13 间；研发中心楼 6 层，总面积近 5000 平方米，是国际顶级研发中心，配备全套先进试验检测设备，设有国家级博士后流动站；另有 7 层附属生态式综合楼。整幢办公大楼采用国际先进的控制和管理系统，是高度智能化、信息化的多功能现代化综合生态办公楼。

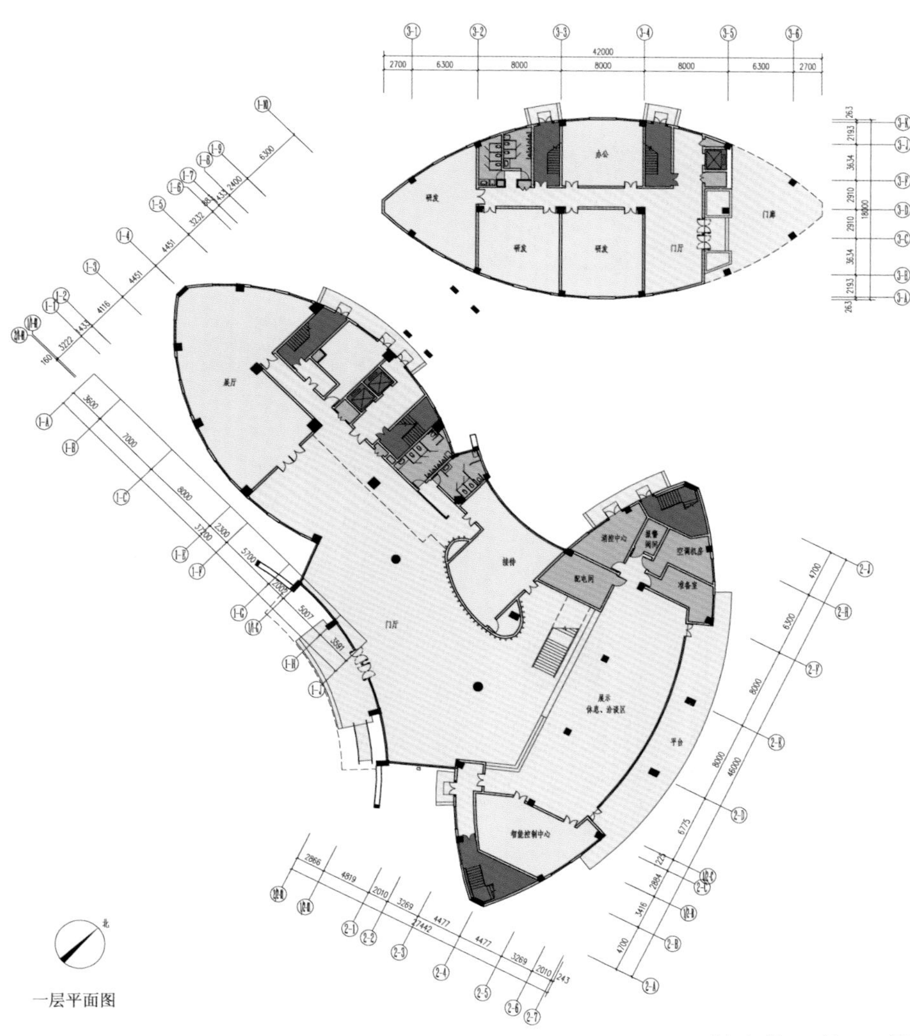

First Floor Plan 一层平面图

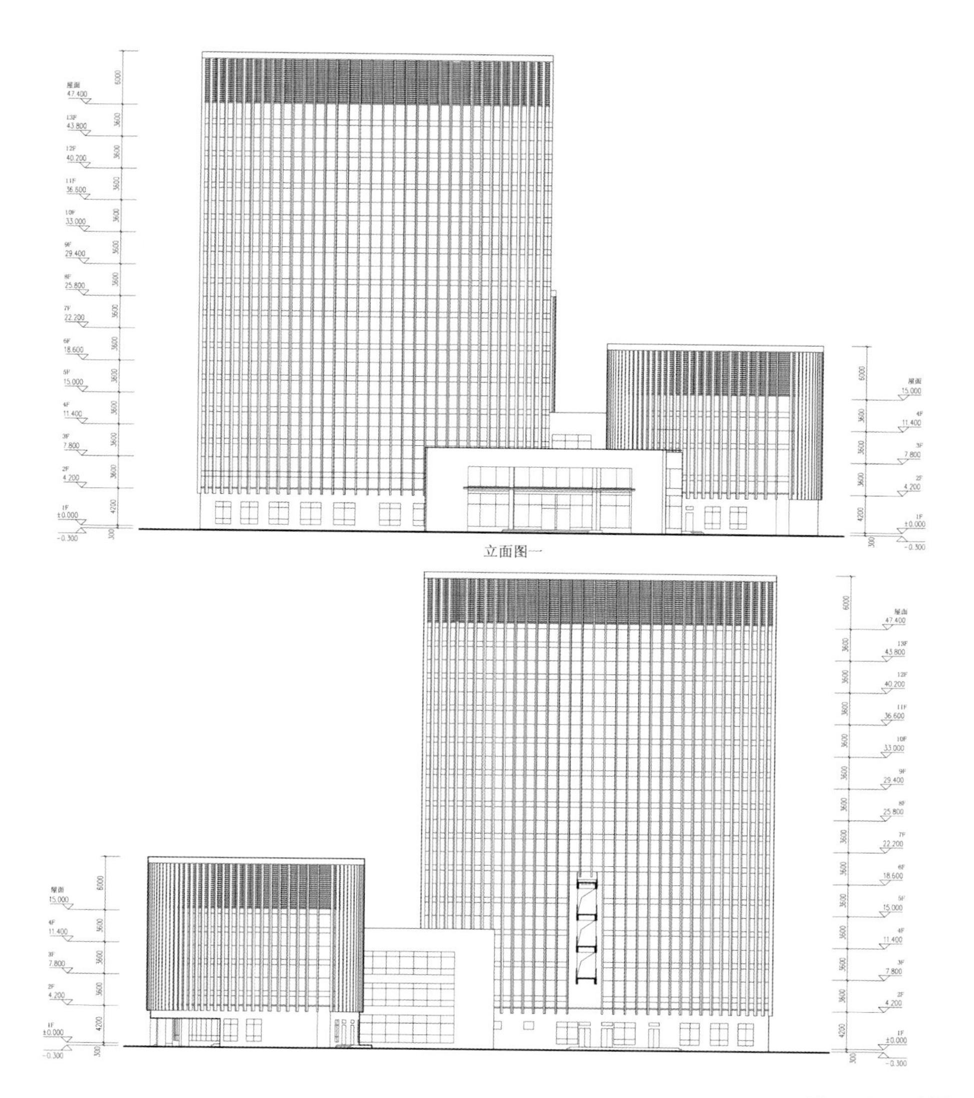

Elevation 立面图

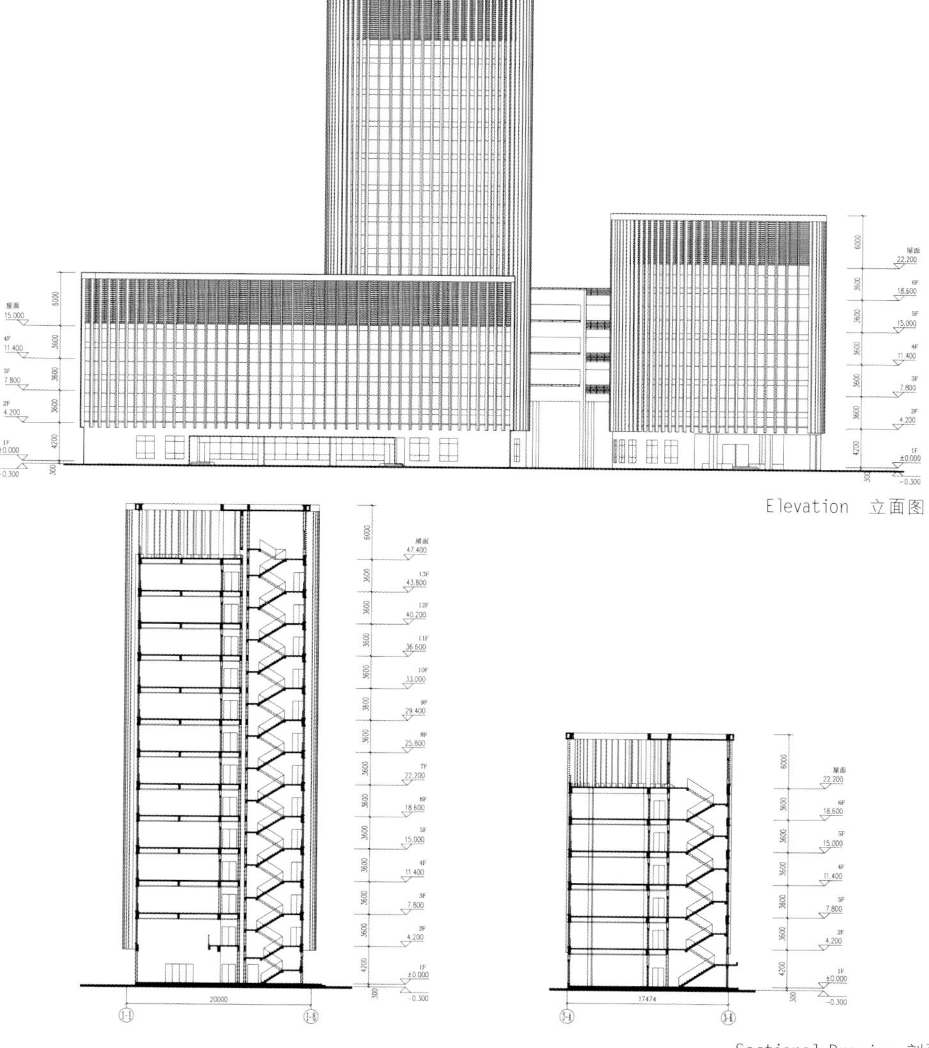

Elevation 立面图

Sectional Drawing 剖面图

KEY WORDS
关键词

Barrier-free Design 无障碍设计

Mixed Use 功能综合

Detail 细部

Tanggu Service Center for the Disabled, Tianjin
天津市塘沽区残疾人综合服务中心

FEATURES
项目亮点

Design of the details shows great humanistic care and well interprets the idea of "barrier free". Many advanced technologies are used to build a leading and professional facility in China.

细部设计充分体现人文关怀，处处体现“无障碍”这一理念，并采用了多种先进技术，居国内领先水平。

Location: Tianjin, China
Architectural Design: Tianjin University Research Institute of Architectural Design
Total Land Area: 6,002.4 m²
Total Floor Area: 16,962 m²
Plot Ratio: 2.26
Completion: 2009

项目地点：中国天津市
设计单位：天津大学建筑设计研究院
总用地面积：6 002.4 m²
总建筑面积：16 962 m²
容积率：2.26
竣工时间：2009 年

Overview

This 31.8 m high building has eight floors over ground and one floor underground. As the biggest service center for the disabled among the prefecture-level cities' in China, it features the most comprehensive functions to meet high standard. Because of its location in the center of Tanggu District, the land is limited for construction. Thus, the architects take advantage of the surrounding environment and arrange different functions around the business hall in a short distance, which will provide great convenience for the disabled.

项目概况

本项目地上八层，地下一层，建筑高度 31.80 m。作为目前全国地市级城市中规模最大、功能最全的残疾人综合服务中心，综合性要求比较高，而项目选址又位于塘沽区中心地段，用地紧张。设计充分利用现有环境，以办事大厅为核心，将多种功能空间以最短的交通路线结合起来，方便残疾人使用。

Floor Plan

The floor plan aims to create a "sharing space" for both the disabled and the able-bodied. It well considers the physical needs and psychological feelings of the disabled, and tries to make them feel equal with rational designs and measures.

平面设计

平面设计遵循“残健共享”的原则，设计过程中关注残疾人状况的点点滴滴，掌握各类残疾人的基本行为特征和心理感受，对各类残疾人和健全人不同的环境需求加以协调，采取合理的设计措施，力求做到使残疾人和健全人平等地感受建筑，平等地参加相关的活动。

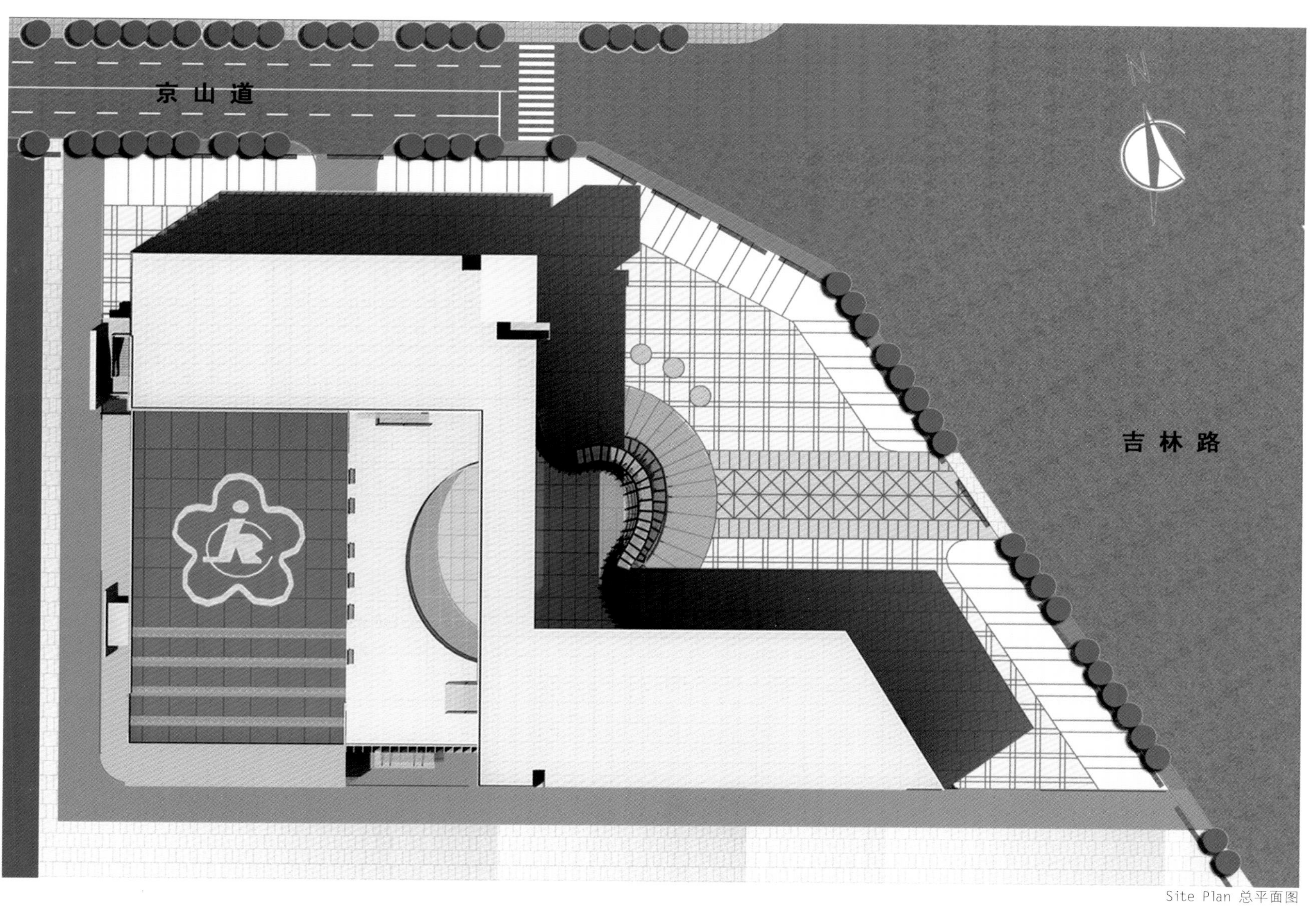

Site Plan 总平面图

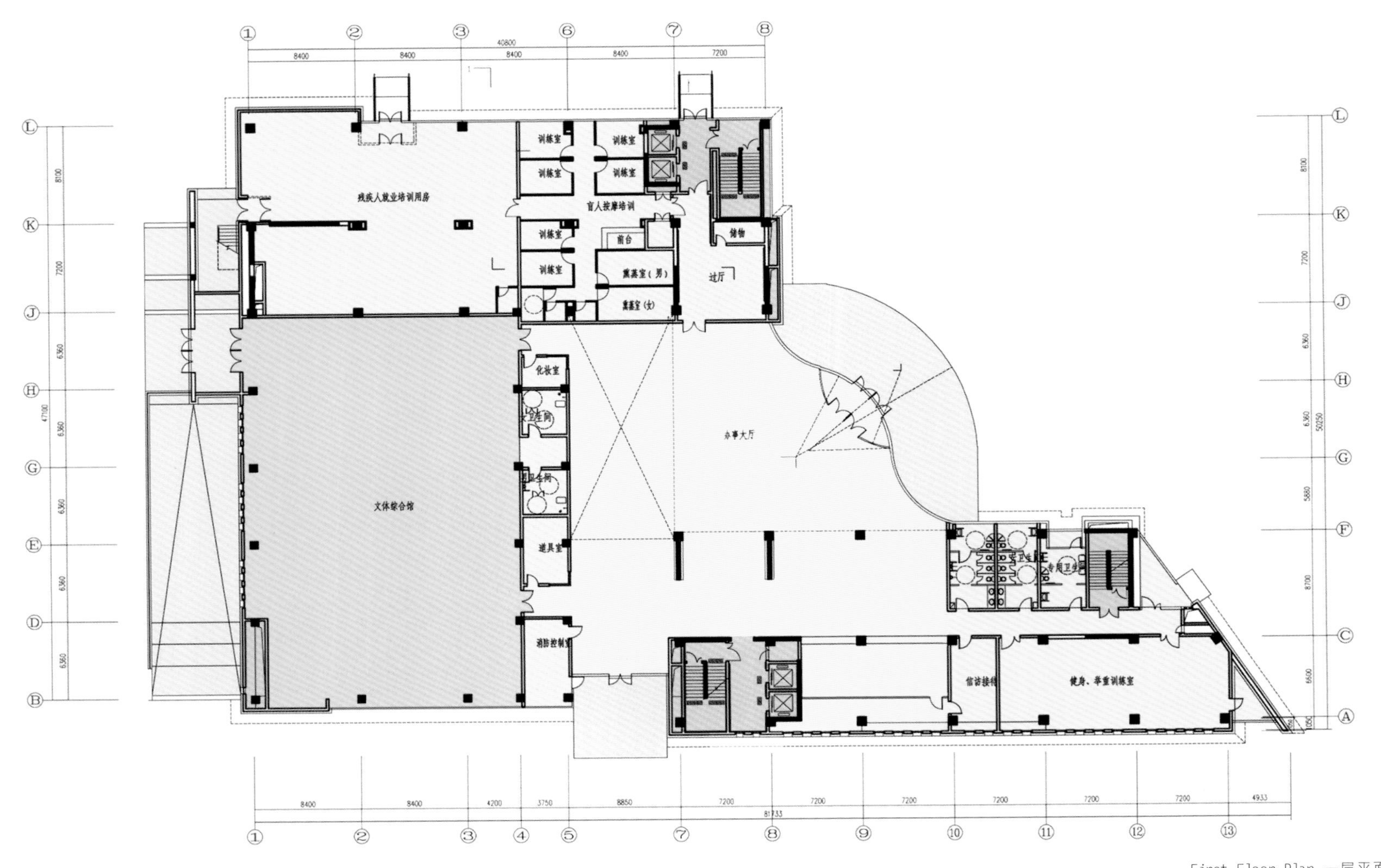

First Floor Plan 一层平面图

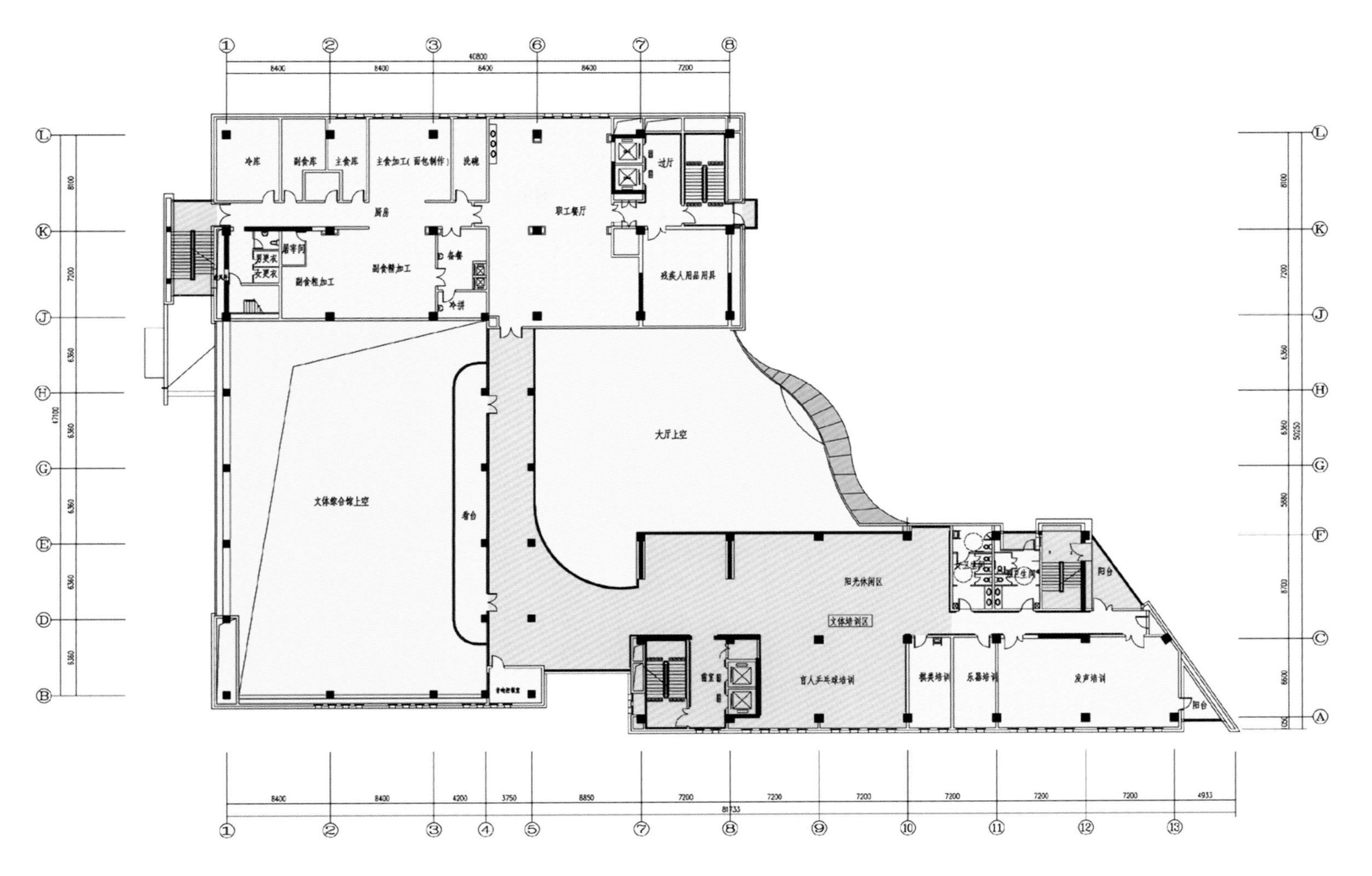

Second Floor Plan 二层平面图

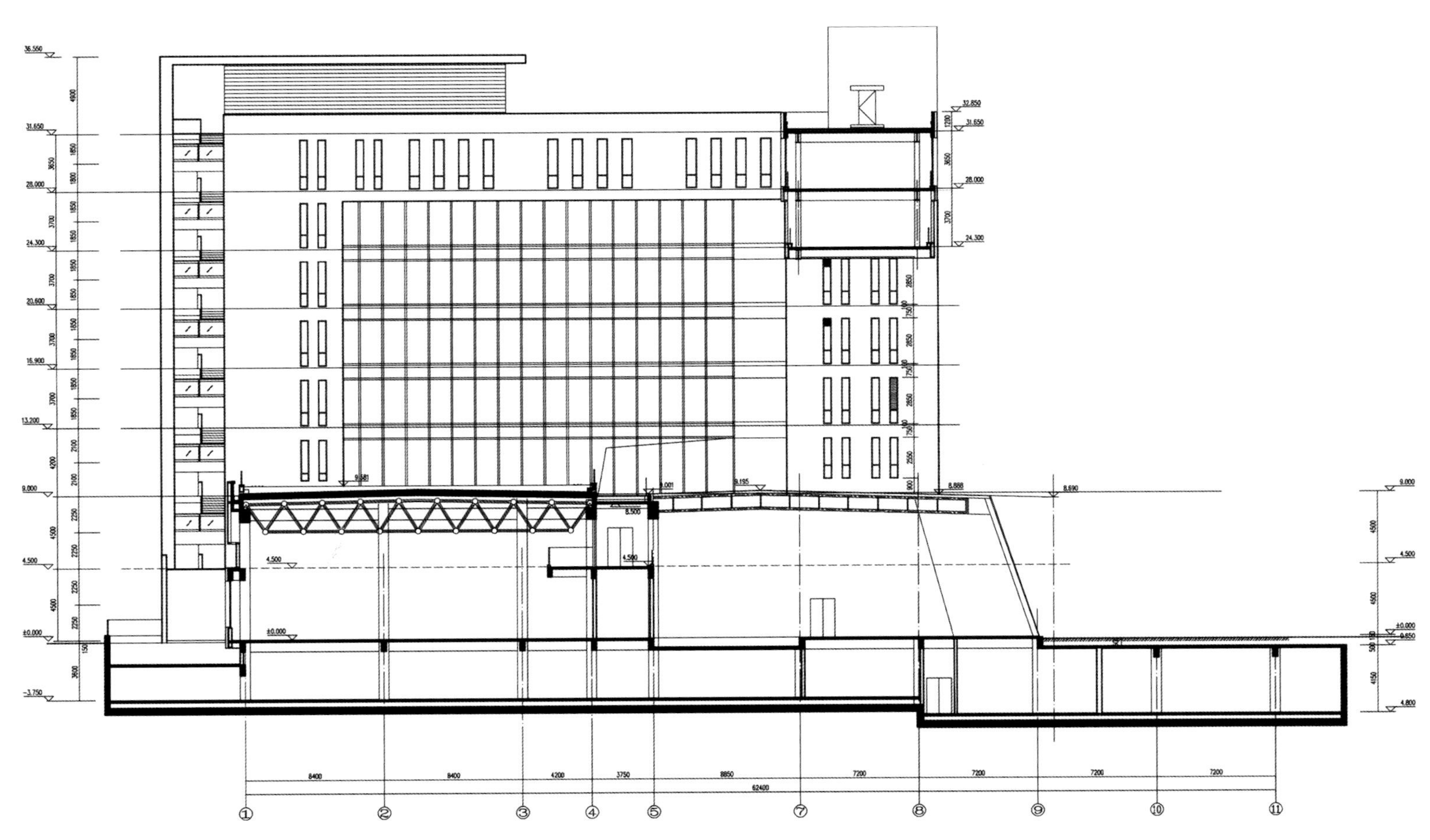

Sectional Drawing 剖面图

Facade Design

The facade is designed in elegant color to shape a new style for the architecture of this category. It also echoes the prosperous development of Tianjin Binhai New Area.

立面设计

立面设计采用简洁的色彩组合，以强烈的时代感塑造了残疾人建筑的新风格，并与天津滨海新区蒸蒸日上、蓬勃发展的现状相合拍。

Detail Design

Design of the details shows great humanistic care and well interprets the idea of "barrier free". Many advanced technologies are used to build a leading and professional facility in China.

细部设计

细部设计充分体现人文关怀，处处体现“无障碍”这一理念，并采用了多种先进技术，居国内领先水平。

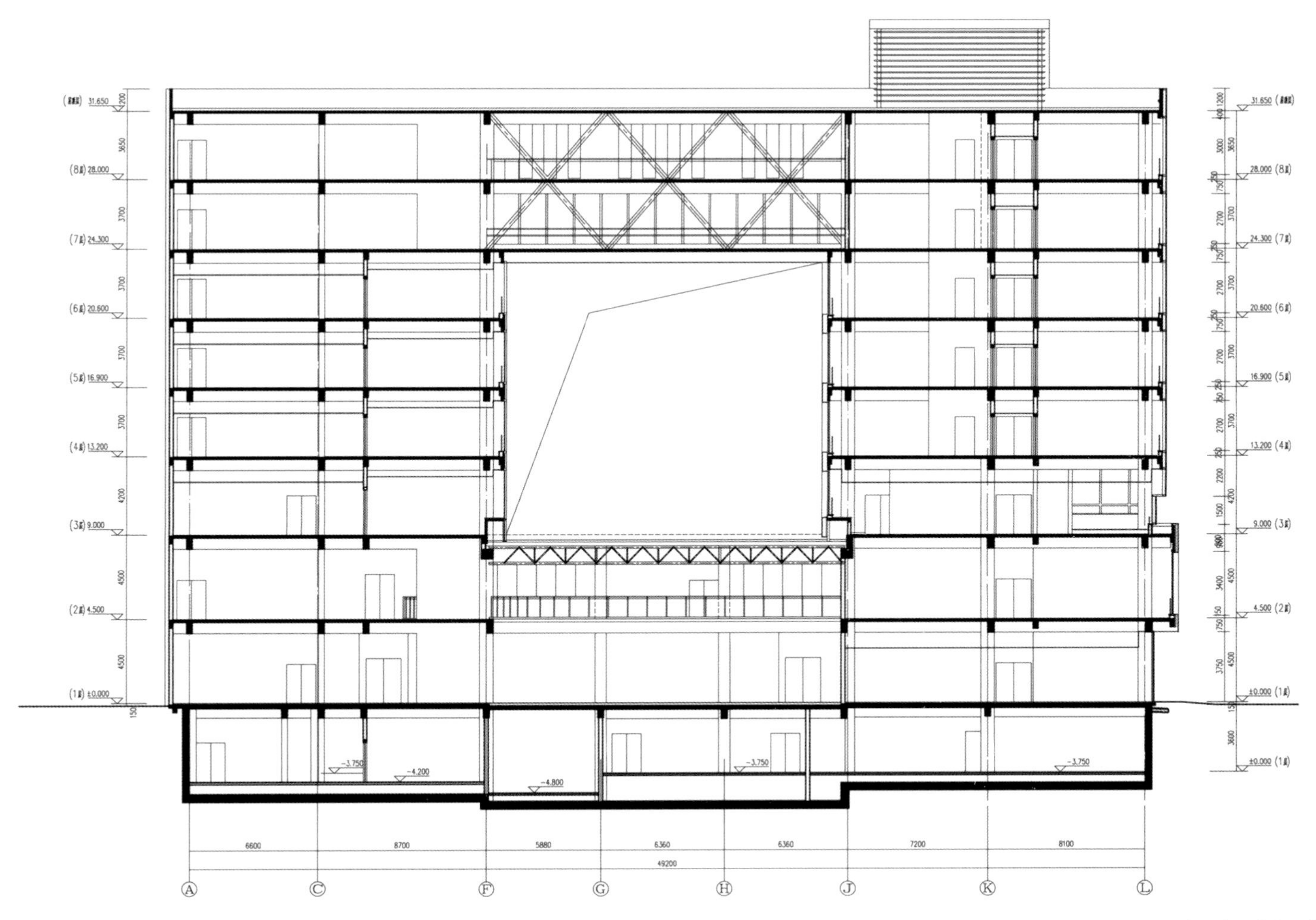

Sectional Drawing 剖面图

KEY WORDS
关键词

Broadcasting Design Element 广播设计元素

Media Characteristic 传媒特征

Cultural Connotation 文化内涵

Broadcast and Dubbing Building of Xinjiang People's Broadcasting Station

新疆人民广播电台播控译制楼

FEATURES
项目亮点

The project extracts the architectural design elements from the beating audio frequency and the appearance of receiving objects (ratio as example) and presents them in the design scheme; its novel design and strong sense of times highlight the broadcasting media culture characteristics.

项目从跳动的广播音频和收音机等接收物体的外形提炼出建筑设计的各种元素，体现在设计方案中，构思新颖，富于时代感，突出了广播传媒文化特征。

Location: Urumqi, Uygur Autonomous Region, Xinjiang, China
Architectural Design: Xinjiang Institute of Architectural Design and Research
Completion: 2008
Total Land Area: 4,367 m^2
Total Floor Area: 19,539.56 m^2 (17,352.56 m^2 above the ground, 2,161.26 m^2 under the ground)
Height: 75.6 m
Building Plot Ratio: 3.97
Green Coverage Ratio: 35%

项目地点：中国新疆维吾尔自治区乌鲁木齐市
设计单位：新疆建筑设计研究院
竣工时间：2008 年
总用地面积：4 367 m^2
总建筑面积：19 539.56 m^2
（其中地上面积为 17 352.56 m^2，地下面积为 2 161.26 m^2）
建筑高度：75.6 m
容积率：3.97
绿化率：35%

Overview

The project extracts the architectural design elements from the beating audio frequency and the appearance of receiving objects (ratio as example) to present them in the design scheme and increase its personality, which is precisely the design conception of this project. Its novel design and rich sense of times highlight the broadcasting media culture characteristics. Besides, it fully excavates the cultural conncotation of the broadcasting as a public venue media carrier, and translates them into architectural language performing in the design, so as to reflect the development and advance of the radio era.

项目概况

项目从跳动的广播音频和收音机等接收物体的外形吸收设计元素，提炼出建筑设计的各种元素，体现在设计方案中，增加项目的个性，这就是本方案设计的立意，构思新颖，造型整合，富于时代感，突出广播传媒文化特征，充分挖掘广播作为大众会场传媒载体的文化内涵，并将其转化为建筑语言表现在方案设计中，体现广播发展的时代性及先进性。

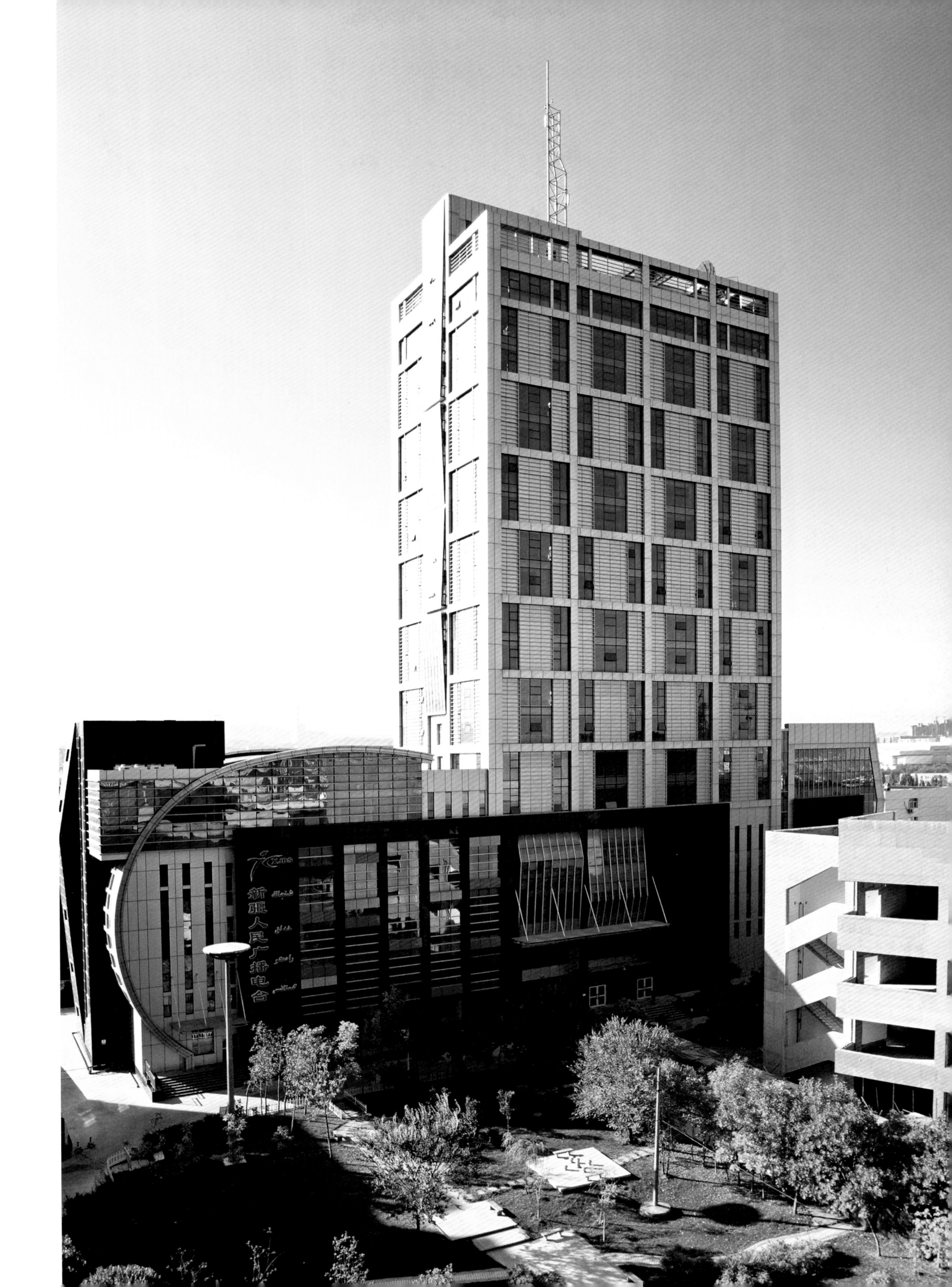
新疆人民广播电台

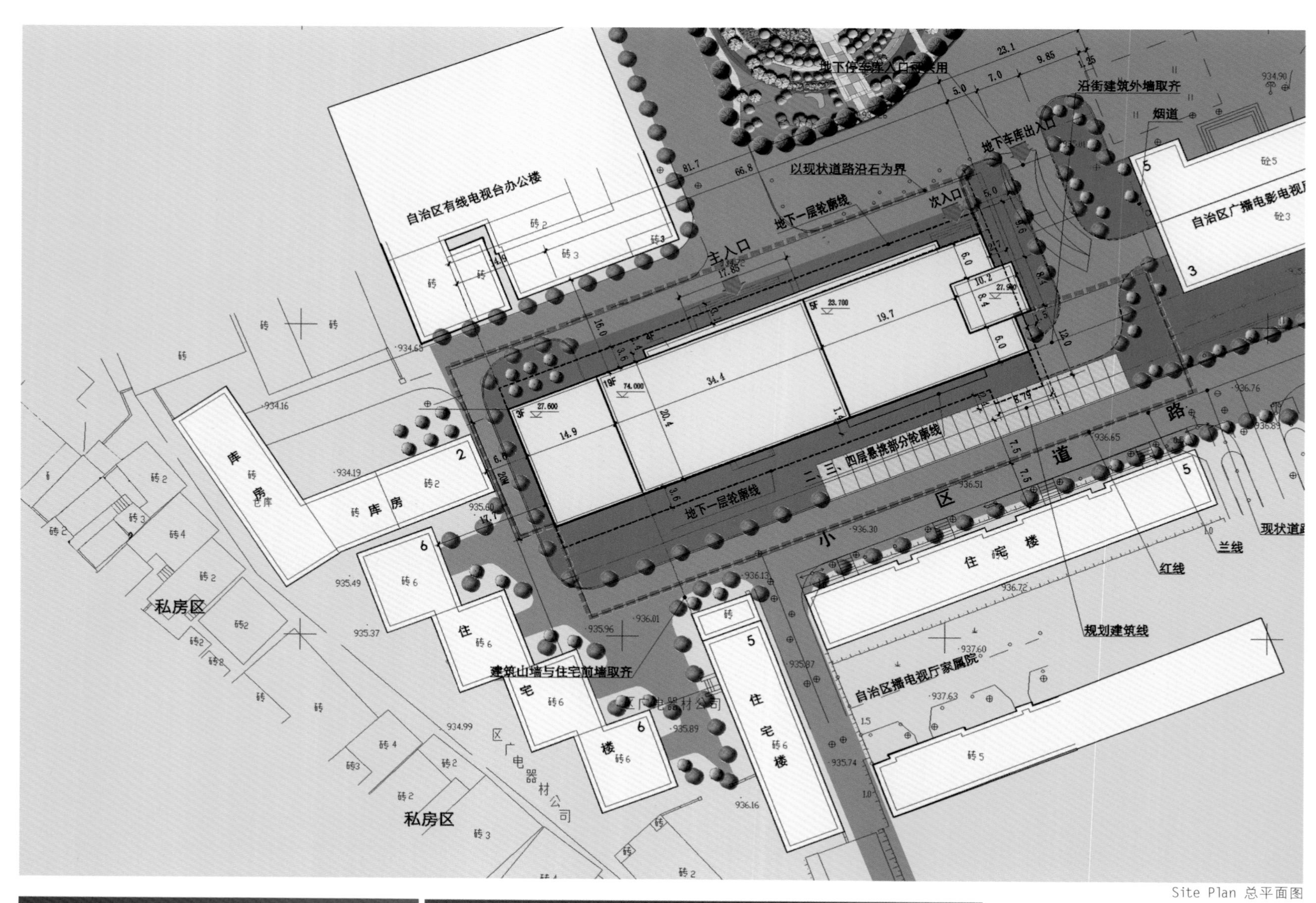

Site Plan 总平面图

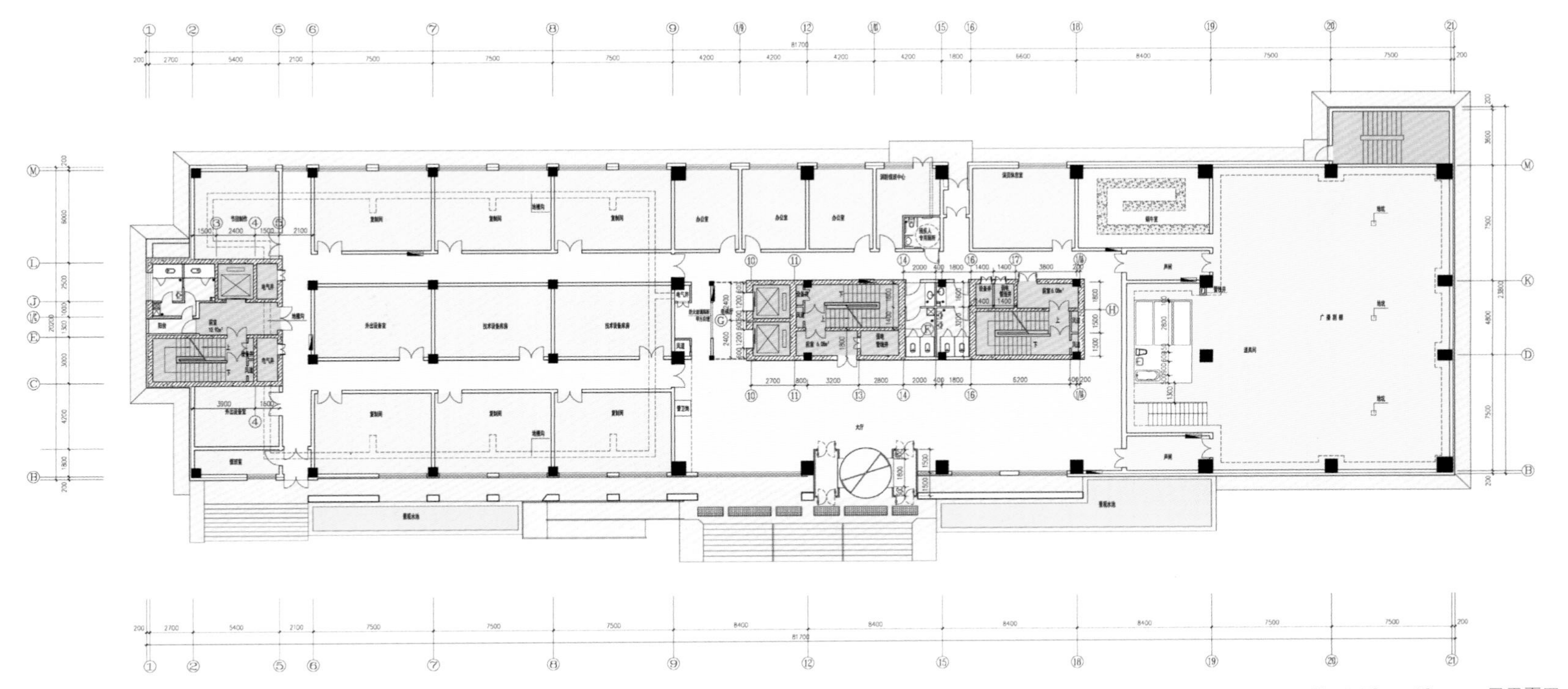

First Floor Plan 一层平面图

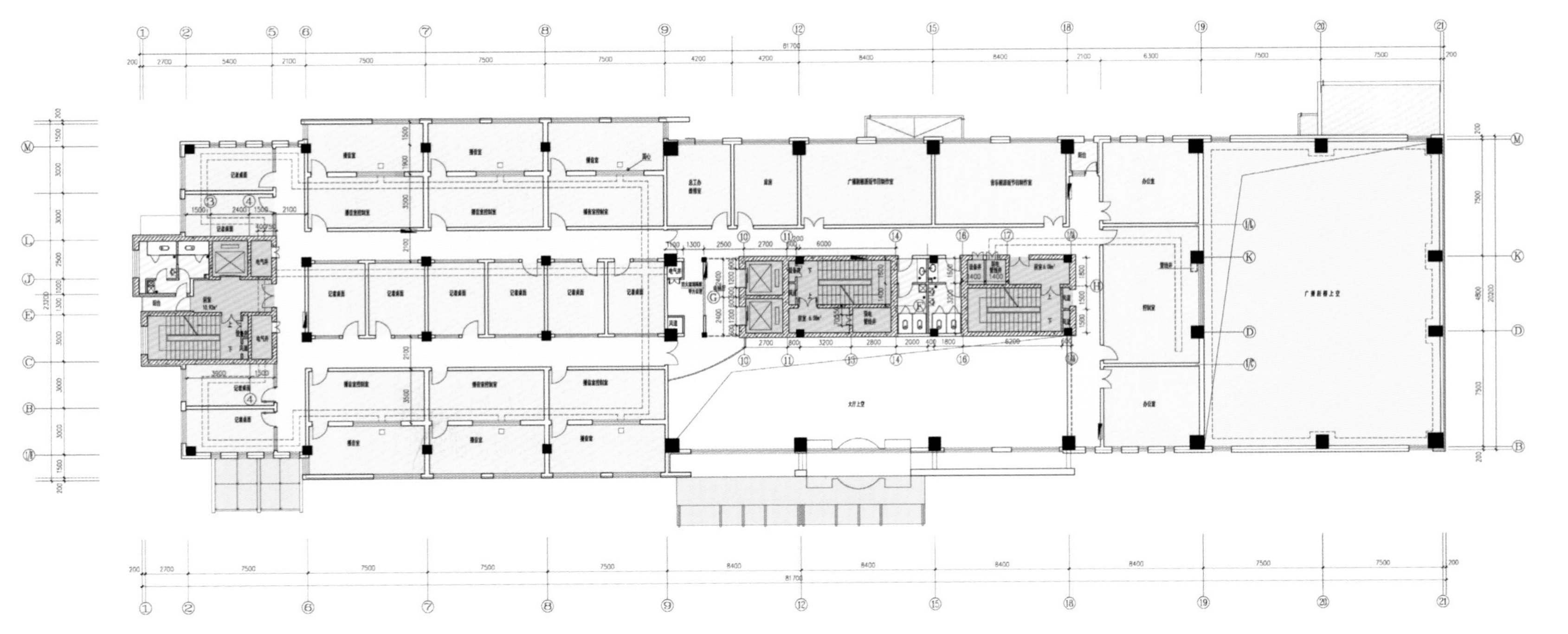

Second Floor Plan 二层平面图

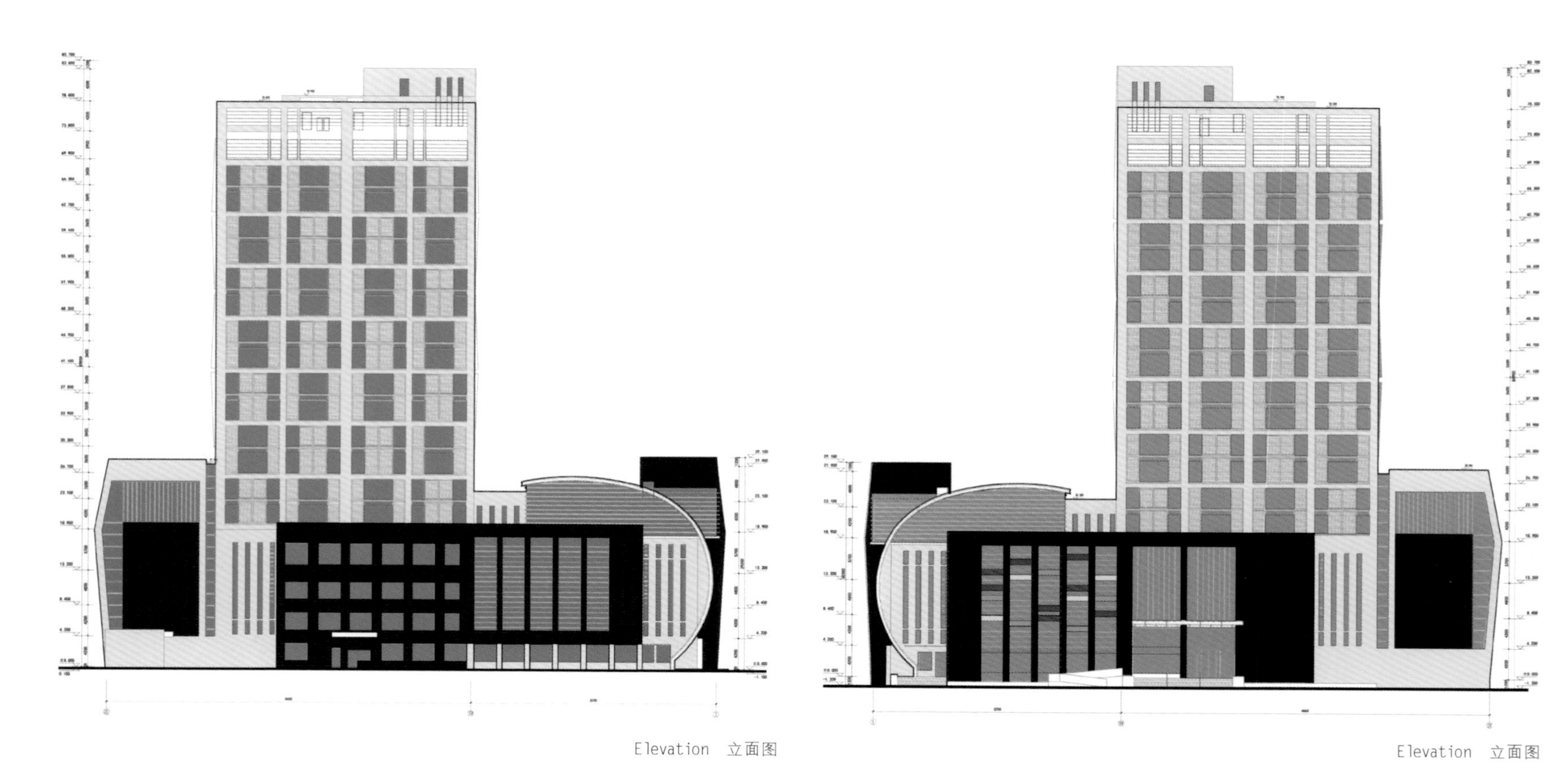

Elevation 立面图

Elevation 立面图

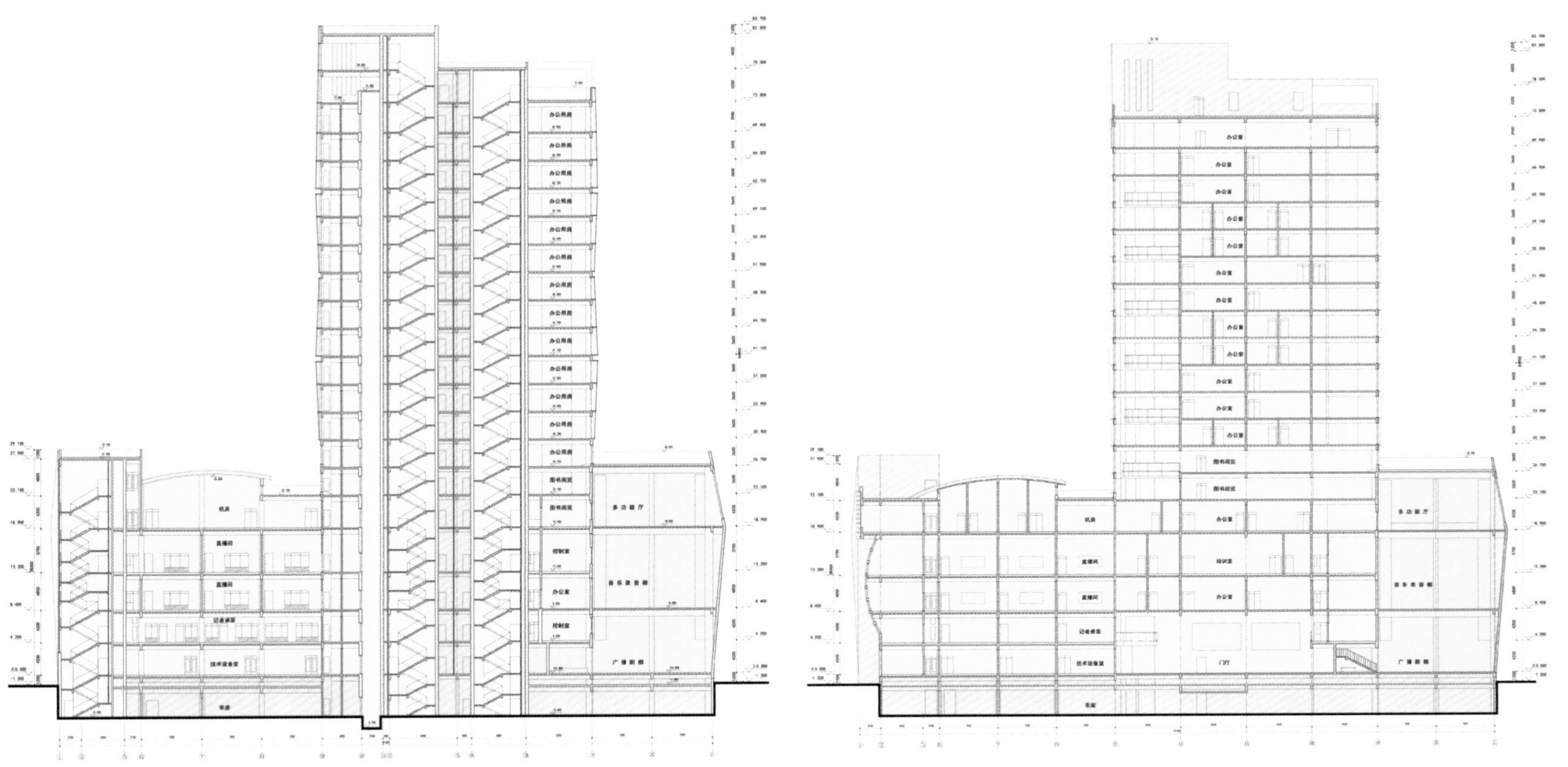

Sectional Drawing 剖面图

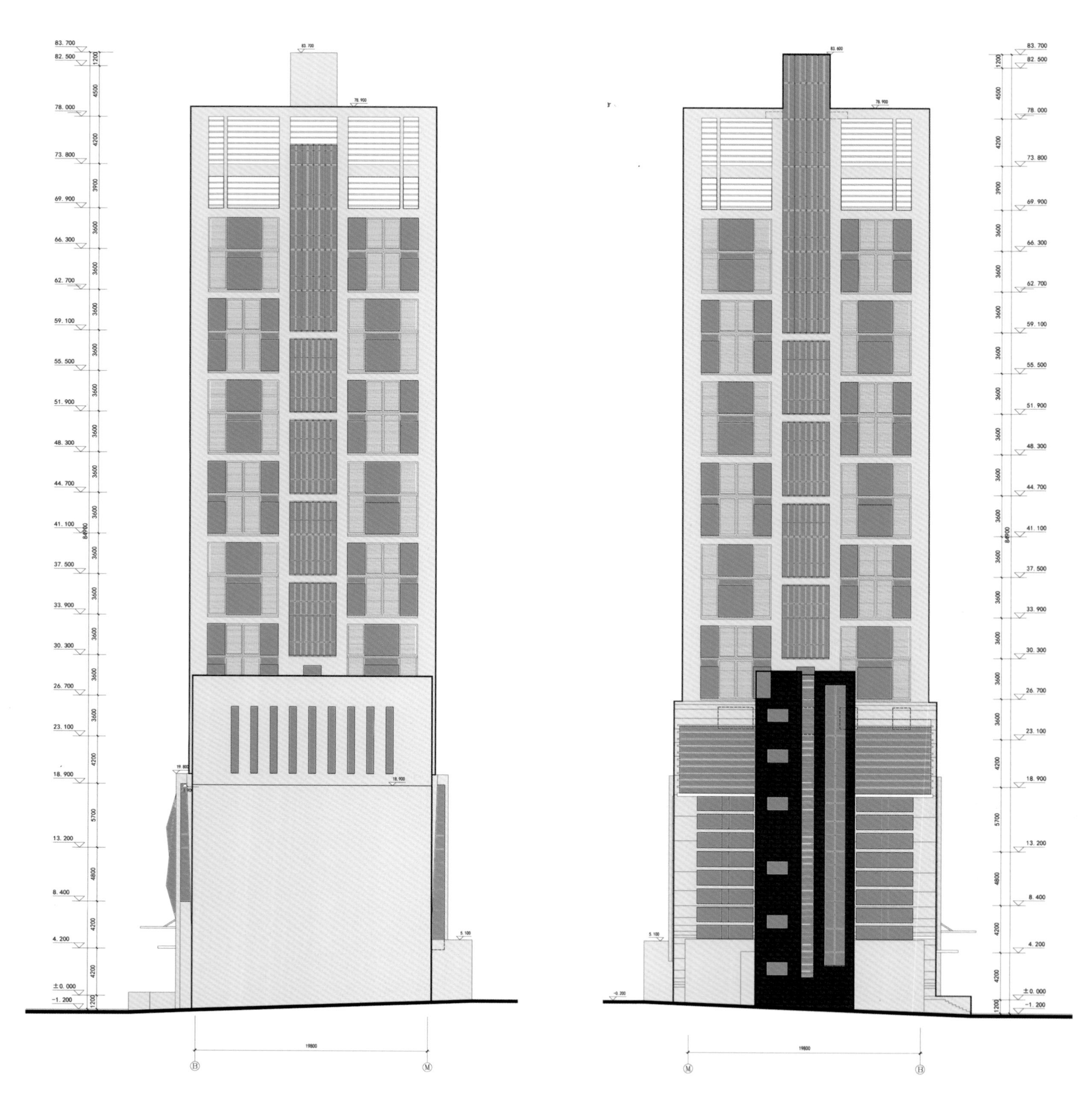

Elevation 立面图

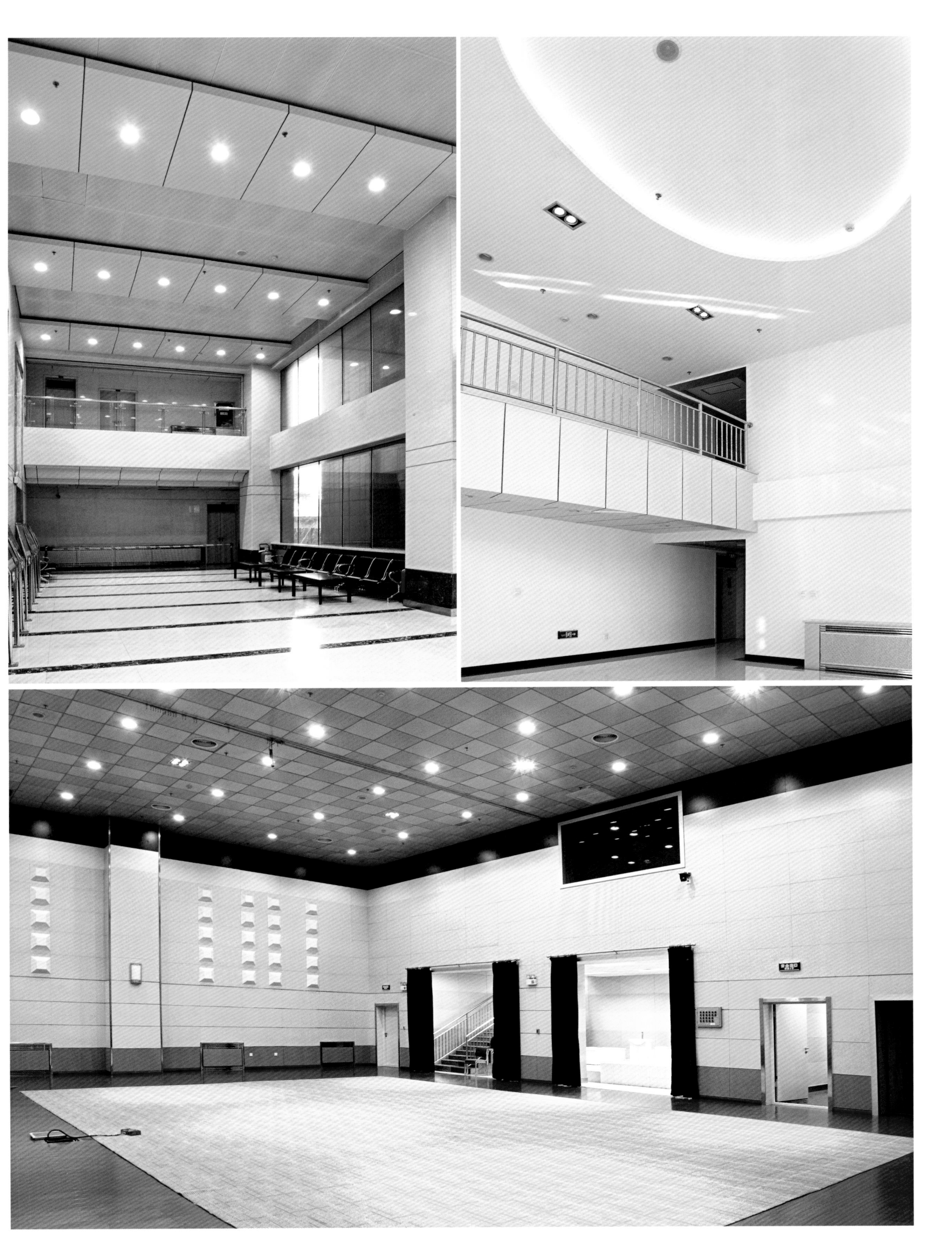

KEY WORDS
关键词

Regional Characteristic 地域特色

Space Layout 空间布局

Facade Design 立面设计

The Office Building of Yanbian Administrative Center
延边州行政中心办公楼

FEATURES
项目亮点

The designers blend the local ethnic characteristics into space layout and facade design, of the project, so as to create a government office building in new era with distinctive national characteristics.

设计师将当地的民族特色融入项目的空间布局以及立面设计等方面，从而打造出具有鲜明民族特色的新时期政府办公建筑。

Location: Yanbian Korean Autonomous Prefecture, Jilin, China
Architectural Design: Jilin Architectural Design Institute Co., Ltd.
Total Land Area: 129,360 m^2
Total Floor Area: 96,164.56 m^2
Building Height: 54 m
Plot Ratio: 1.11
Completion: 2011

项目地点：中国吉林省延边朝鲜族自治州
设计单位：吉林省建筑设计院有限责任公司
总用地面积：129 360 m^2
总建筑面积：96 164.56 m^2
建筑高度：54 m
容积率：1.11
竣工时间：2011 年

Overview

The project fully shows respect for the Korean people's habits and customs on the space layout, creating a flexible space style harmonious with the nature and returning to nature. The general layout makes full use of the indoor, outdoor and grey space combined in different forms, so as to create a government office building in new era with distinctive national characteristics.

项目概况

本项目在空间布局上充分尊重朝鲜族人民的生活习惯，充分融入自然，回归自然，创造灵活的空间样式。总平面图中充分利用室内、室外、灰空间等不同形式的空间组合，从而打造出具有鲜明民族特色的新时期政府办公建筑。

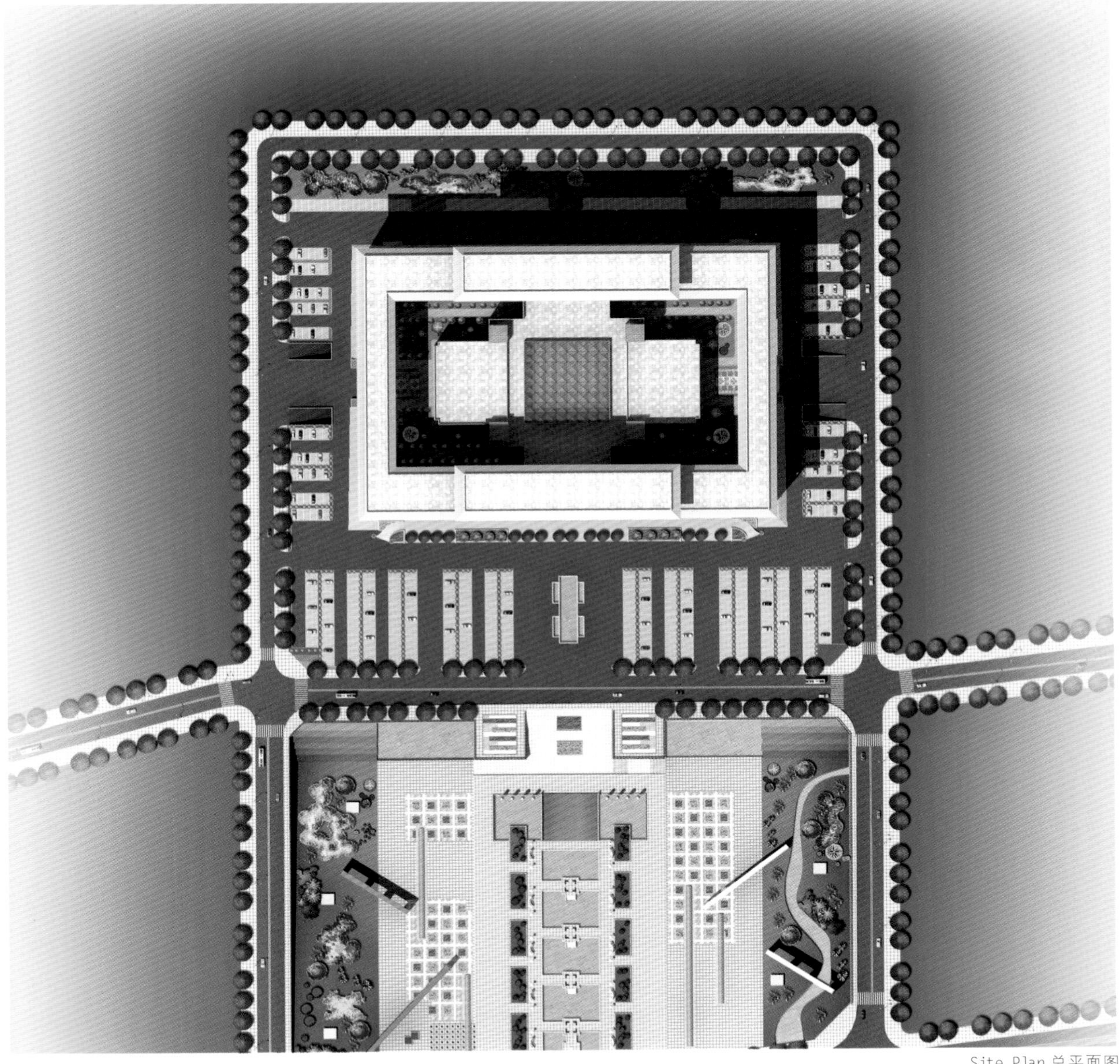

Site Plan 总平面图

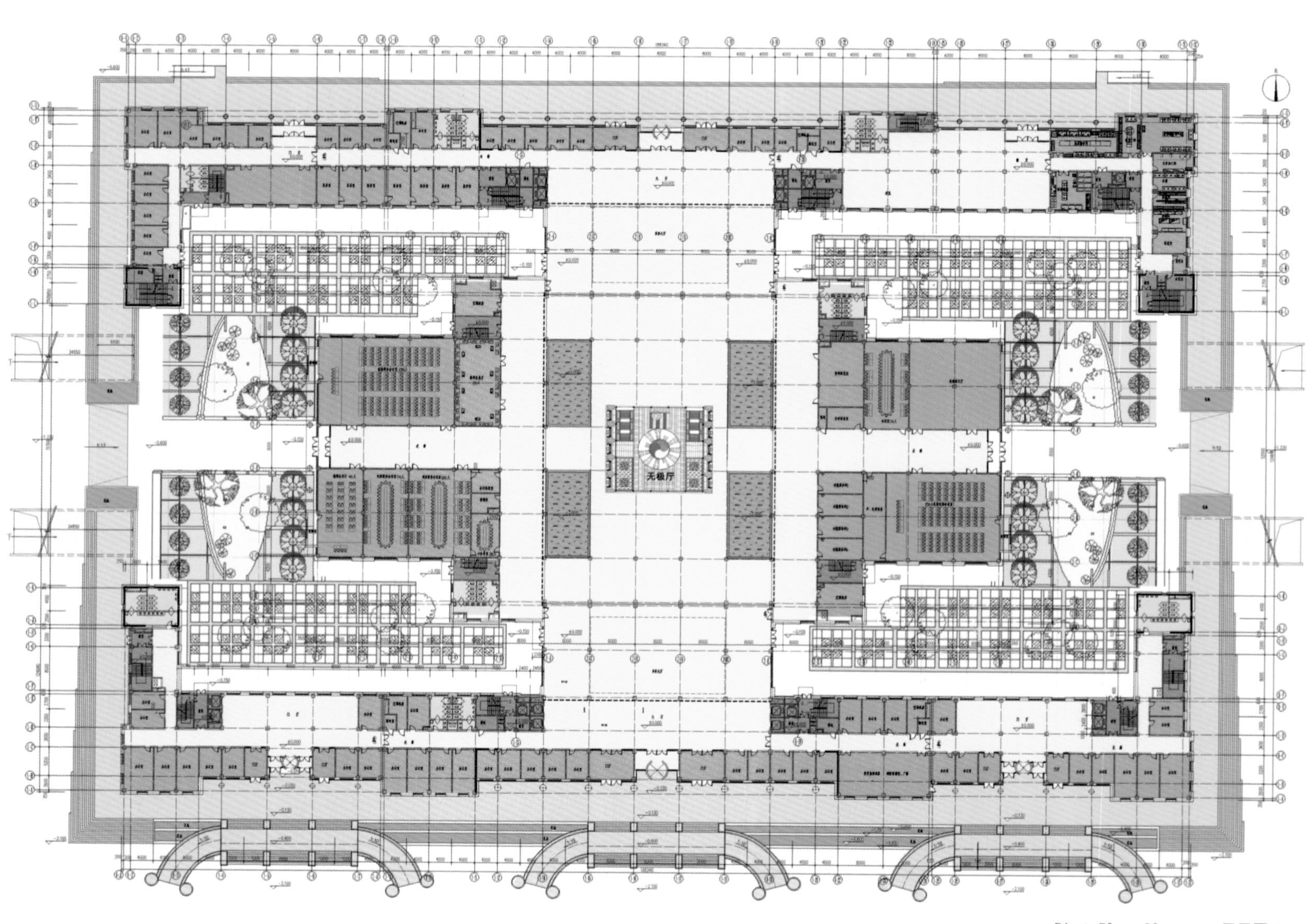

First Floor Plan　一层平面图

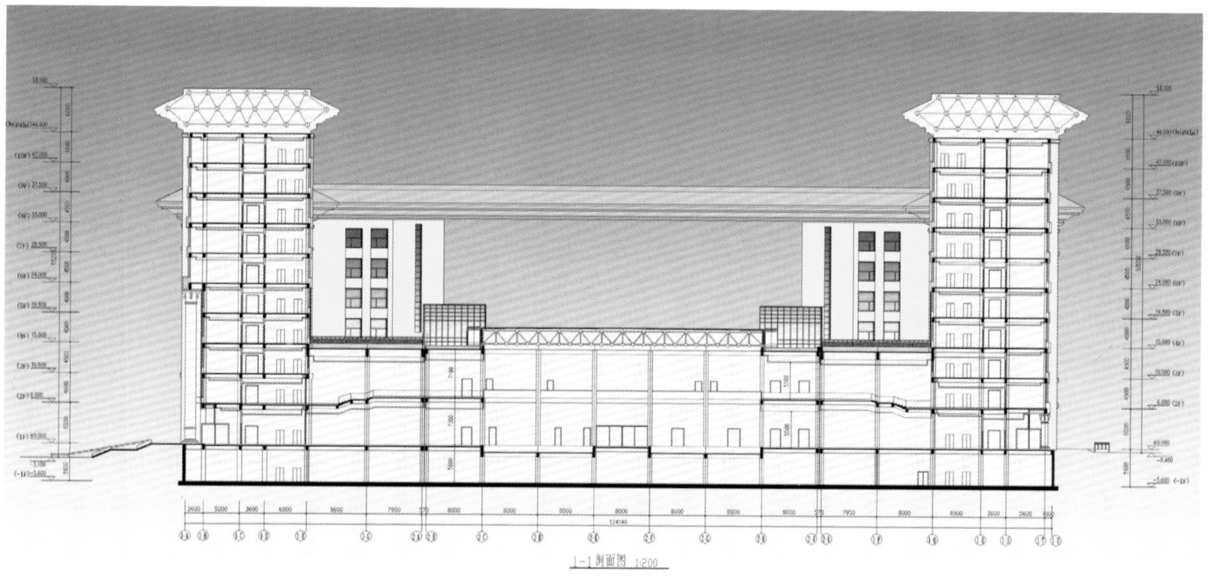

Sectional Drawing 剖面图

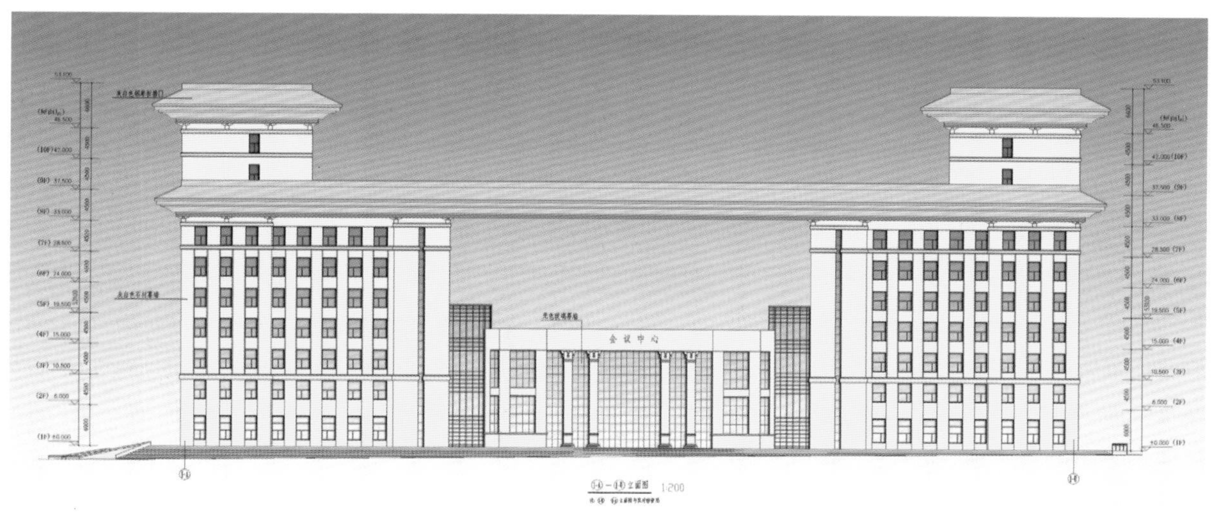

Sectional Drawing 剖面图

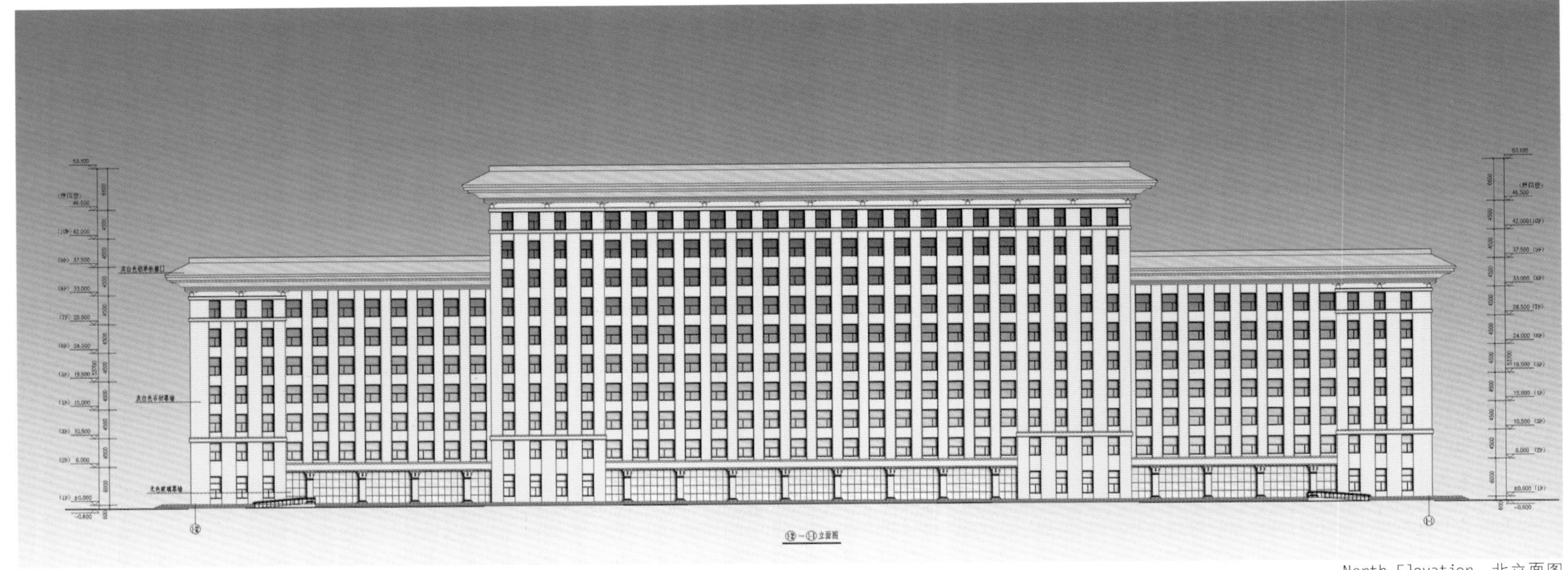

North Elevation 北立面图

Facade Design

Based on many years of work experience and keen insight on the special buildings, the designers make comprehensive use of the unique ethnic characteristics symbols in the facade design, not pursuing blundering and false decoration but fully considering the characteristics of regional culture and national identity. The project scores a success by its size and characteristic space, showing people solemn and vigorous sense from various angles of the city landscape.

立面设计

立面设计中，设计师利用多年的工作经验及对特殊建筑的敏锐洞察力，综合运用民族特有的符号特征，不追求浮躁的表现和虚假的装饰，充分考虑地域文化特点及民族认同感，大象无形、大音希声，以体量及特色空间取胜，使建筑在城市景观的各个角度上均给人以庄重浑厚、大气磅礴之感。

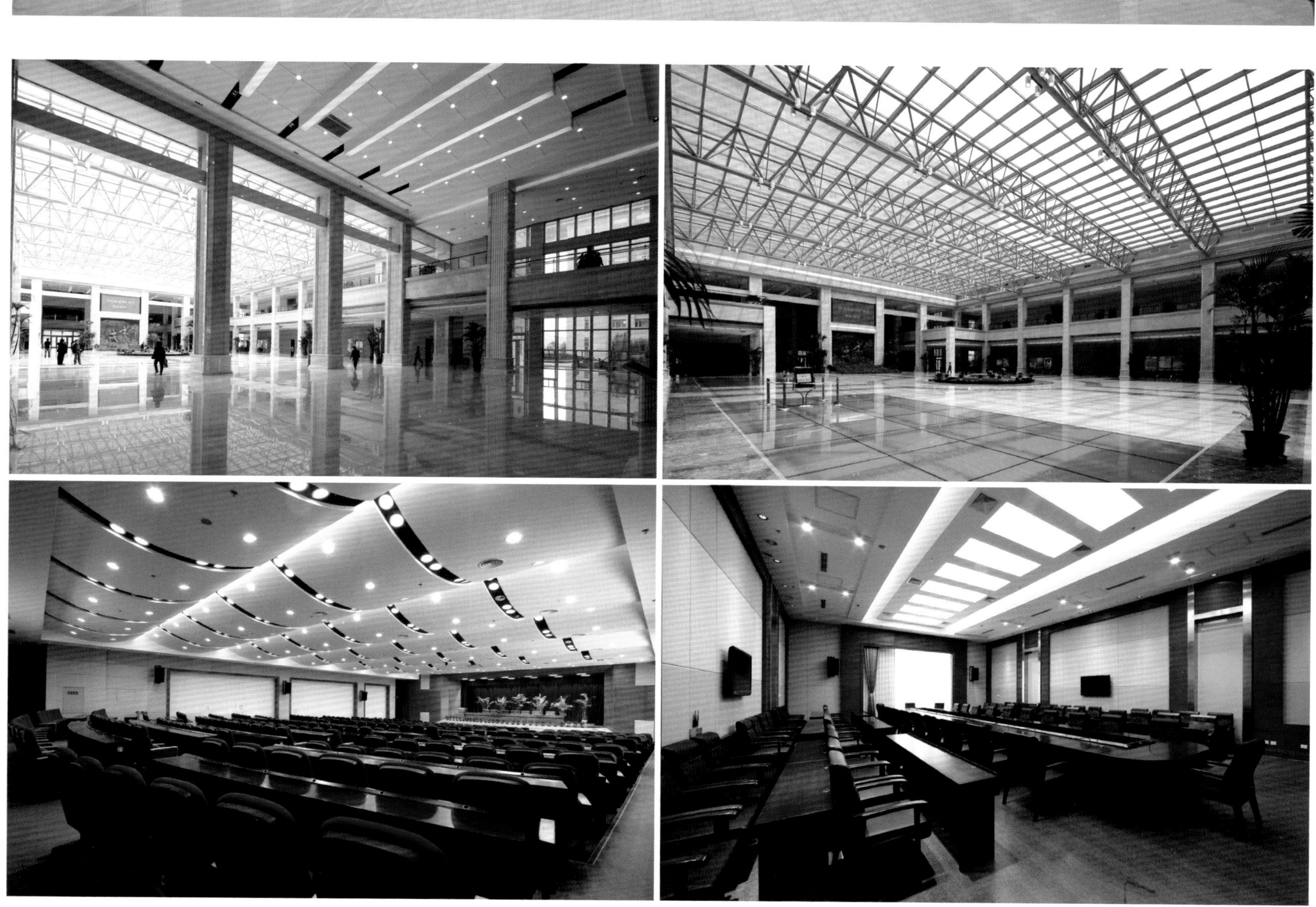

KEY WORDS
关键词

Roof Design 屋顶设计

Open 开放性

Public Space 公共空间

Changxing Radio and Television Bureau
长兴广播电视局

FEATURES
项目亮点

This project puts focus on the study of openness in urban space; it adopts three main design strategies to bring back the construction land to the city and give back the space to the urban residents.

这是一个关注城市空间开放性的实践，设计采用了三个主要设计策略，将建筑用地还给城市，将空间还给城市市民。

Location: ChangXing, Zhejiang, China
Architectural Design: Southeast University Architectural Design Institute Co.,Ltd
Total Land Area: 18,676 m^2
Total Floor Area: 23,660 m^2 (14,625 m^2 above the ground, 9,035 m^2 under the ground)
Plot Ratio: 1.09
Green Coverage Ratio: 28.4%
Completion: 2009

项目地点：中国浙江省长兴县
设计单位：东南大学建筑设计研究院有限公司
总用地面积：18 676 m^2
总建筑面积：23 660 m^2（地上部分 14 625 m^2，地下部分 9 035 m^2）
容积率：1.09
绿化率：28.4%
竣工时间：2009 年

Overview

Changxing Radio and Television Bureau puts focus on the study of openness in urban space, which brings back the construction land to the city and gives back the space to the urban residents; the building should be a place that gives people freedom and restores the original free activity routes.

项目概况

长兴广播电视局是一个关注城市空间开放性的实践，将建筑用地再还给城市，将空间还给城市市民，建筑旨在成为一个可以让人自由前往的场所，并恢复原来的自由活动路线。

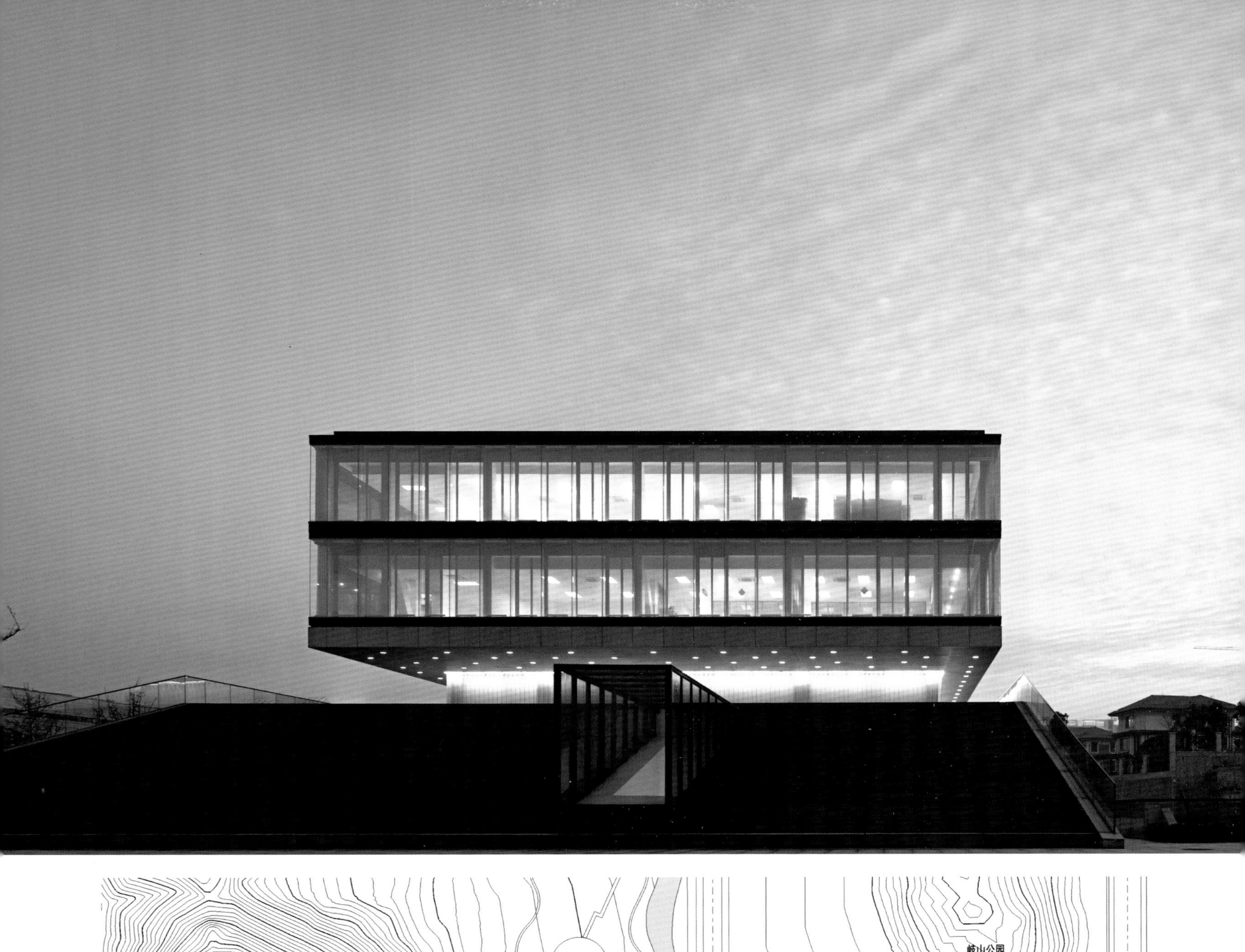

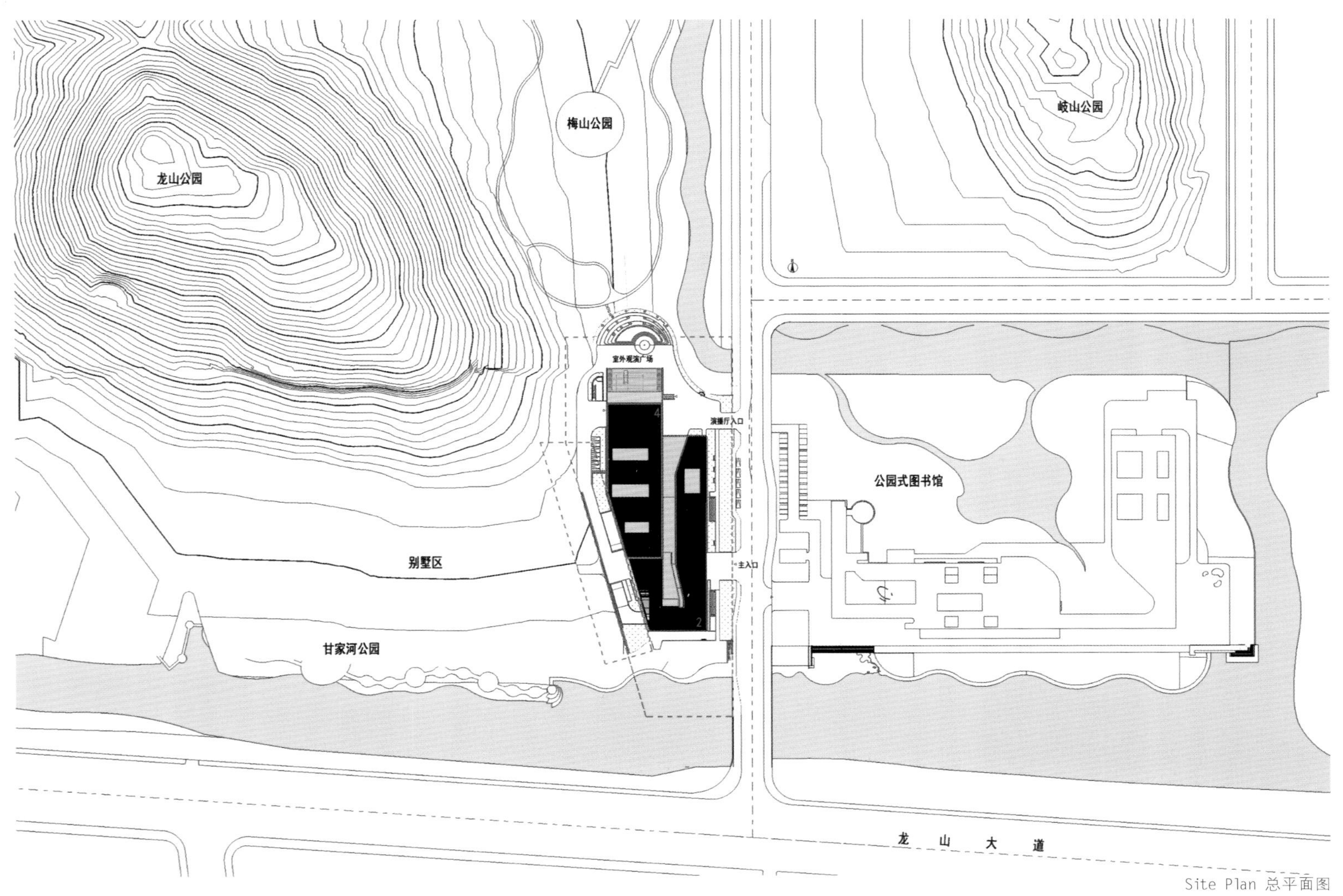

Site Plan 总平面图

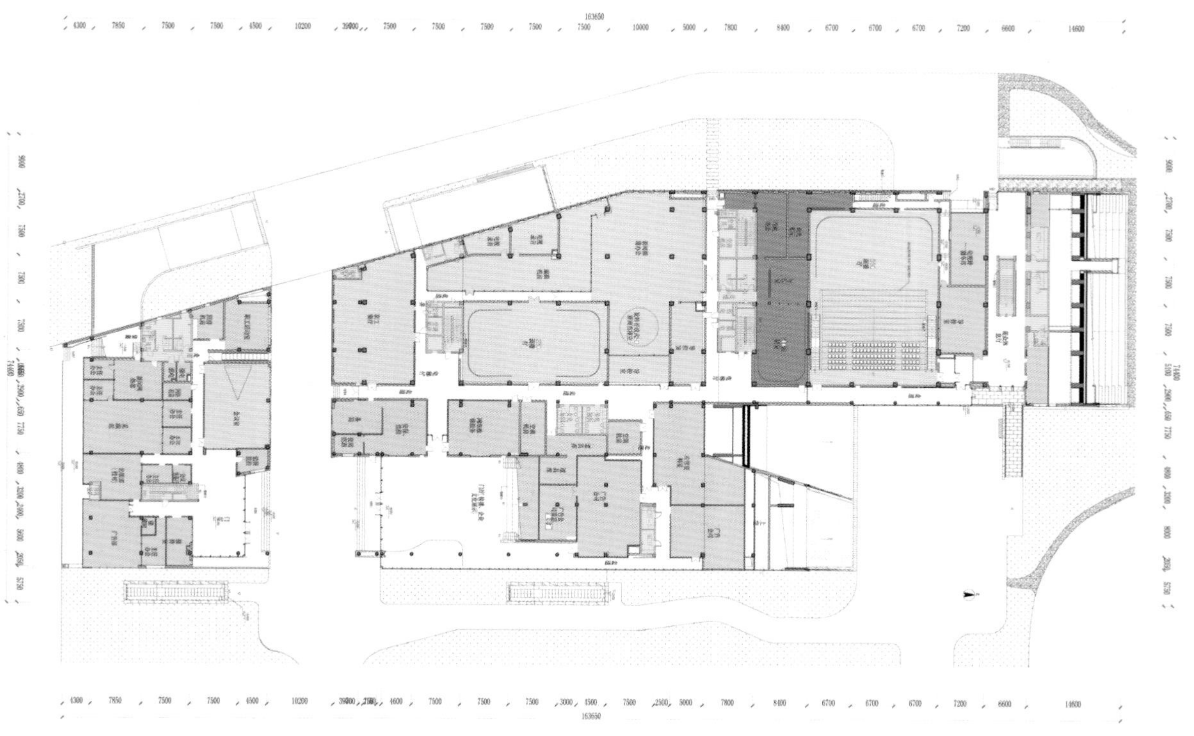

First Floor Plan　一层平面图

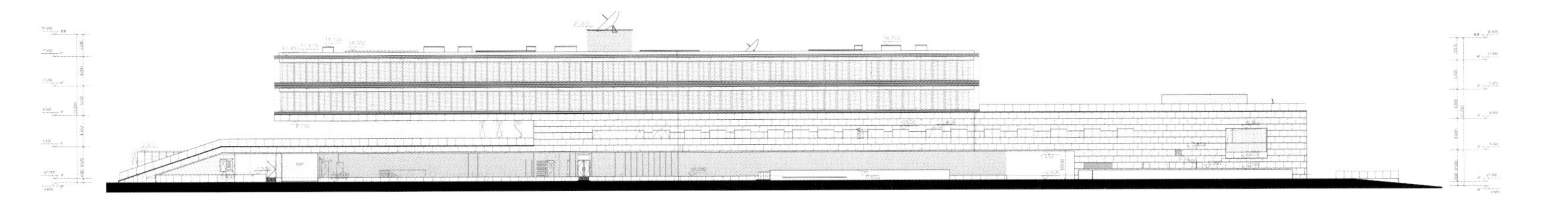

East Elevation 东立面图

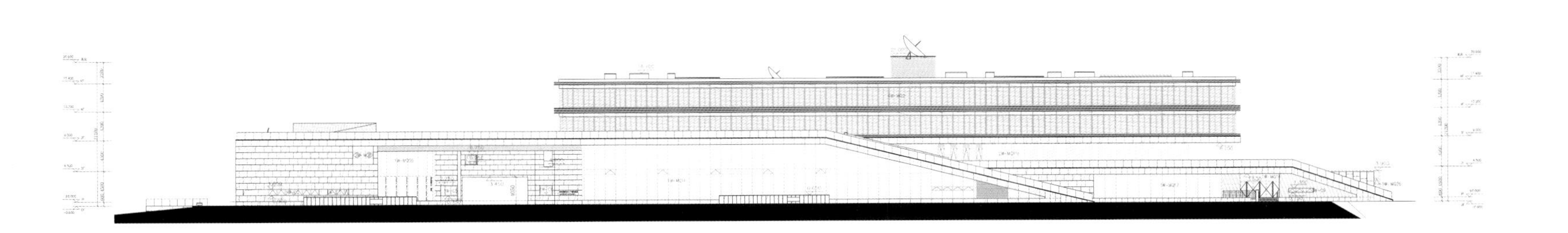

West Elevation 西立面图

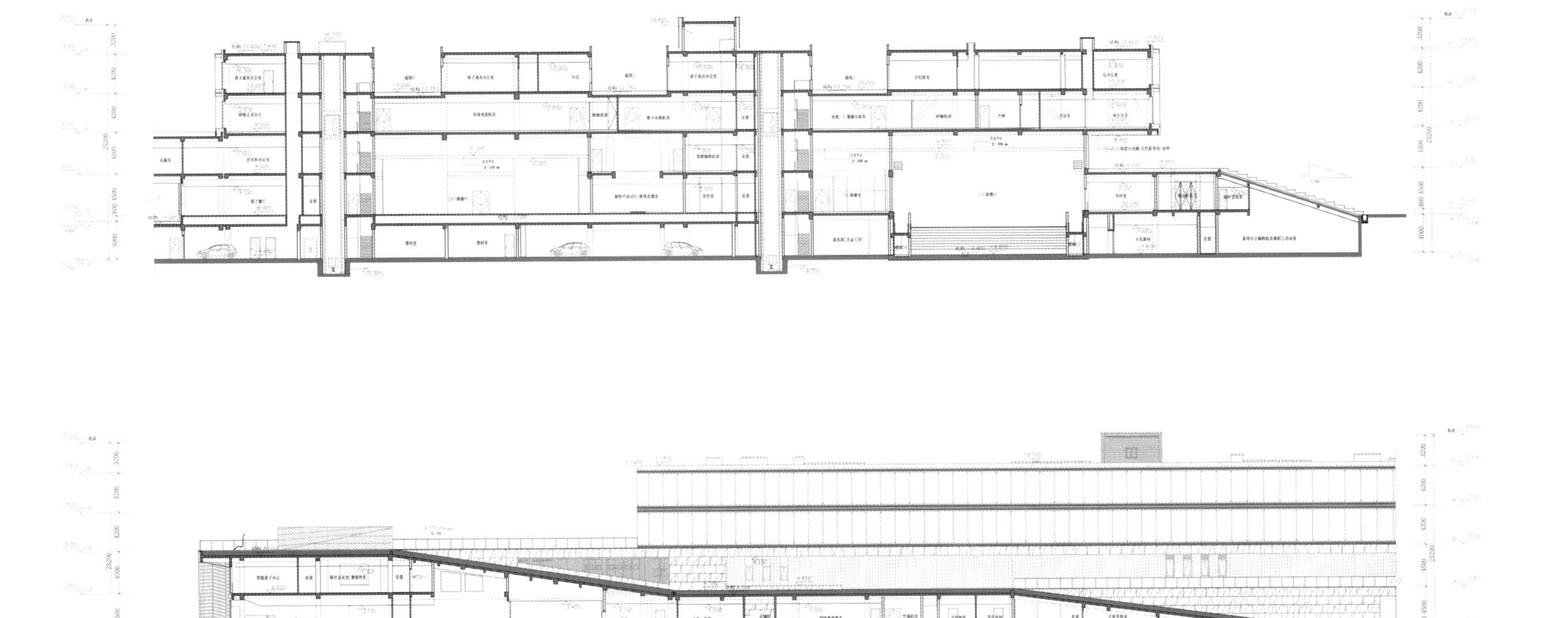

Sectional Drawing 剖面图

Architectural Design

This project is a multistory building with a total of 4 layers and building height of 18.475 m. It adopts three main design strategies to make the public roof become a free walking trail leading people slowly from the northern end of the base and the trail connects together the parks around and recovers the citizens' activity routes; its roof is designed to a planted roof for people to overlook the surrounding scenery. The north of the building together with the roof design forms a outdoor performing plaza, which links the building closely with the Meishan Park; the external hall is multi-functional and people can go there free to drink coffee, surf the Internet, view exhibitions and watch shows, etc.

建筑设计

本项目为多层建筑，共4层，建筑高度为18.475 m。设计采用了三个主要设计策略，将公共部分的屋顶设计成自由漫步道，从基地北端引导人们缓缓而上，步道串联了周边的公园，并恢复了市民的活动路线；而屋顶设计成种植屋面，在这里人们可以俯瞰周边的风景。建筑北面结合建筑屋顶设计了室外观演广场，将建筑与梅山公园紧密相连；对外大厅具有多种用途，人们可以自由前往，在这里喝咖啡、上网、观看展览、观演等。

KEY WORDS
关键词

Ecological Sustainability 生态可持续
Characteristics of the Times 时代特征
Environment Relationship 环境关系

Zhejiang Daily Press Group Editing Building
浙江日报报业集团采编大楼

FEATURES
项目亮点

The project design brings up an efficient, ecological and sustainably developing editorial office, and it also deals well the spatial figure-ground relationship with the old editorial building and relation with the surrounding environment.

设计在打造高效、生态、可持续发展的新采编大楼的同时，很好地处理了与老采编大楼的空间图底关系，以及与周边的环境关系。

Location: Hangzhou, Zhejiang, China
Architectural Design: Architectural Design & Research Institute of Zhejiang University
Land Area: 7,525 m^2
Total Floor Area: 65,261 m^2
Plot Ratio: 2.2
Completion: 2010

项目地点：中国浙江省杭州市
设计单位：浙江大学建筑设计研究院
占地面积：7 525 m^2
总建筑面积：65 261 m^2
容积率：2.2
竣工时间：2010 年

Overview

Zhejiang Daily Press Group Editing Building has a total of 23 layers of the main building, 3 layers of annexes, and some part of 5 layers; the main building bears a height of 84.7 m, and the building density (including the old editorial building) is 31.5%.

项目概况

浙江日报报业集团采编大楼主楼共 23 层，裙房 3 层，局部 5 层，主楼高度为 84.7 m，大院内建筑密度（含老采编大楼）为 31.5%。

Architectural Design

Located on the east side of Zhejiang Daily yard as a mosaic expansion project in the old city, the Zhejiang Daily Press Group Editing Building tries to find a balance between the old and new building construction. The project design tries to bring up an efficient, ecological and sustainably developing editorial office to reshape the Zhejiang Daily Press Group as a leading media agency in a new era; at the same time, it also deals friendly the spatial figure-ground relationship with the old editorial building, the relationship with the original stream and greening vegetation, and relationship with the surrounding environment.

建筑设计

坐落于浙江日报大院东区的新采编大楼，作为镶嵌式的老城区扩建项目，试图寻找出新老营造之间的一个平衡点。设计在努力打造高效、生态、可持续发展的新采编大楼，重塑浙江日报报业集团作为传媒龙头机构的新时代特质的同时，很好地处理了与老采编大楼的空间图底关系，与原有小河和植被绿化的关系，以及与东、北侧居住小区的环境关系。

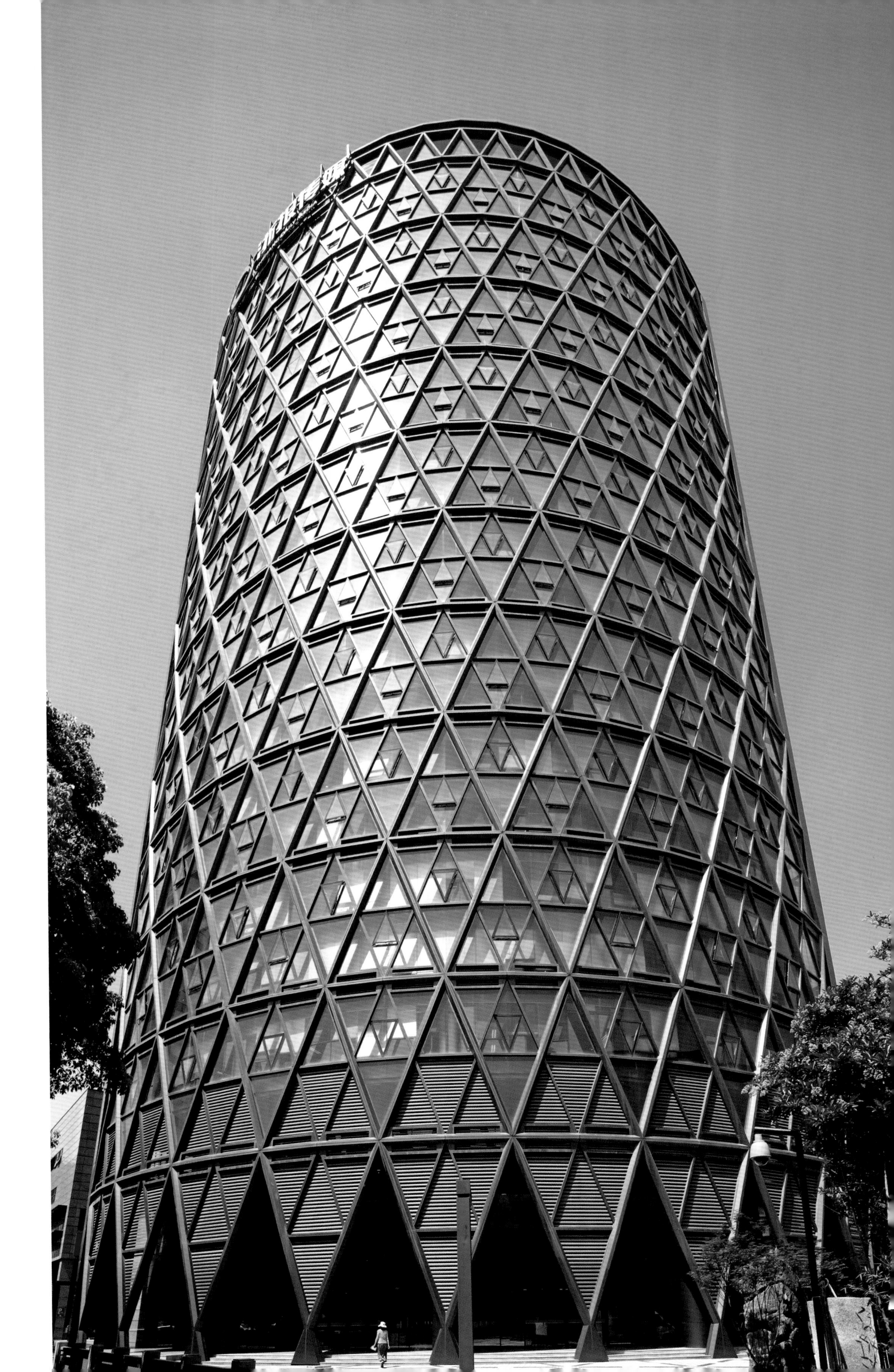

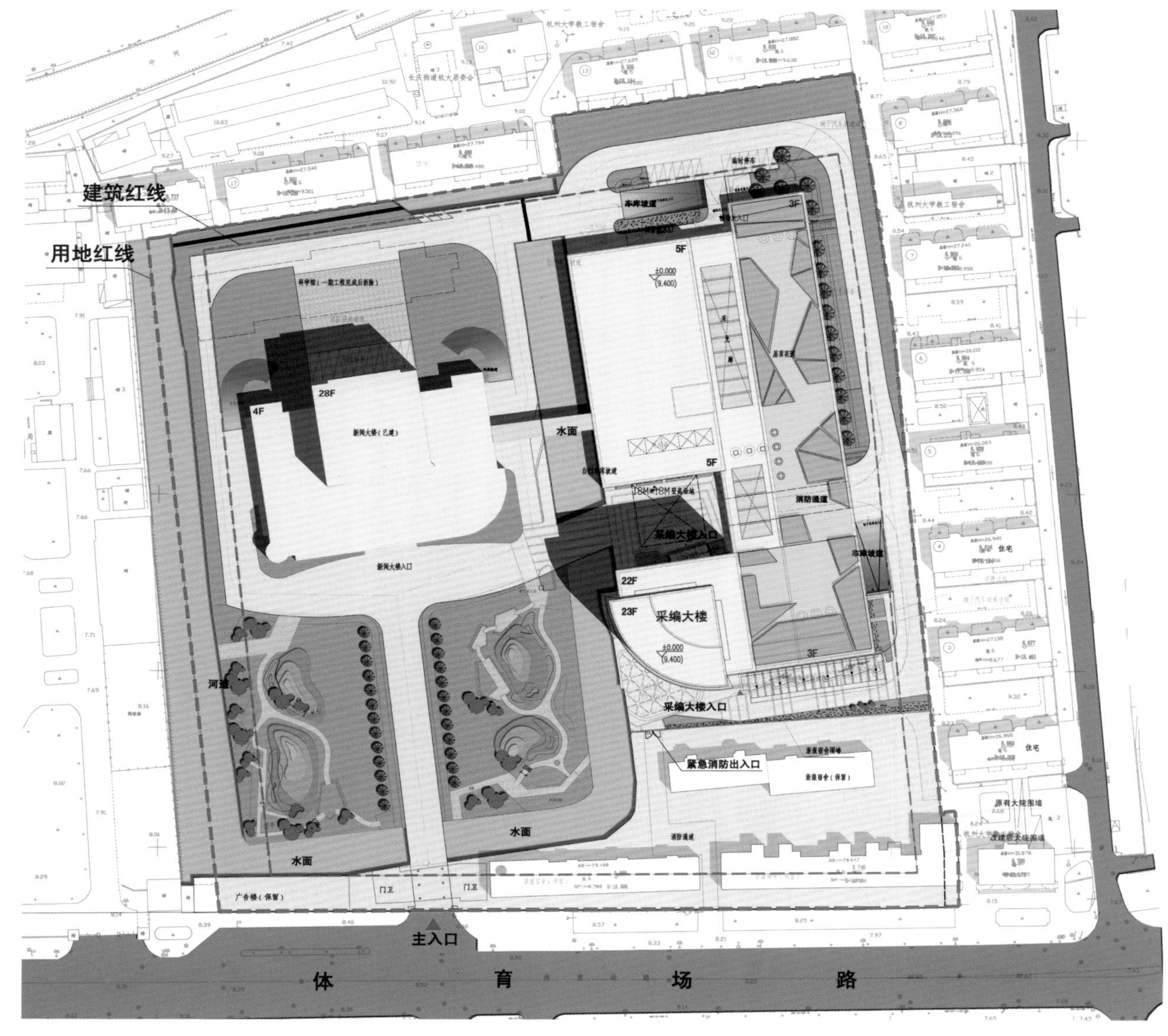

Site Plan 总平面图

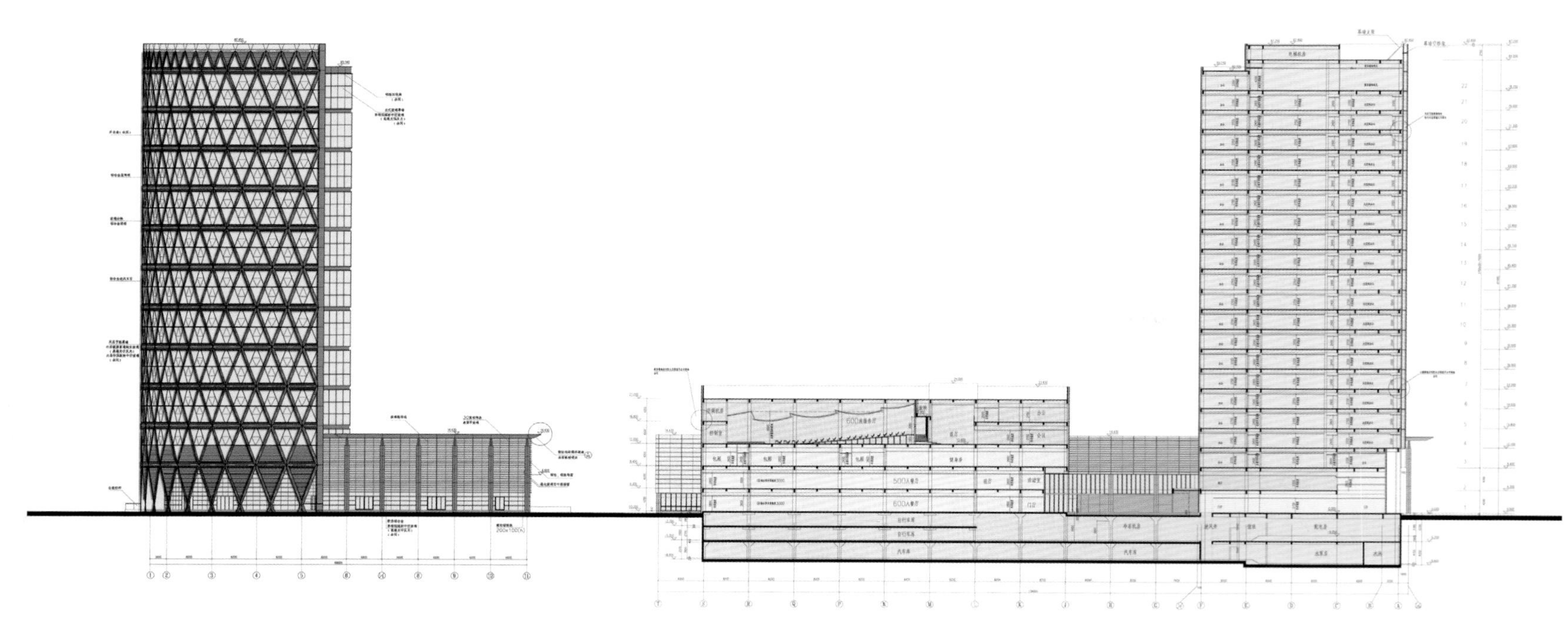

Sectional Drawing 剖面图

浙报传媒
CATHAYA

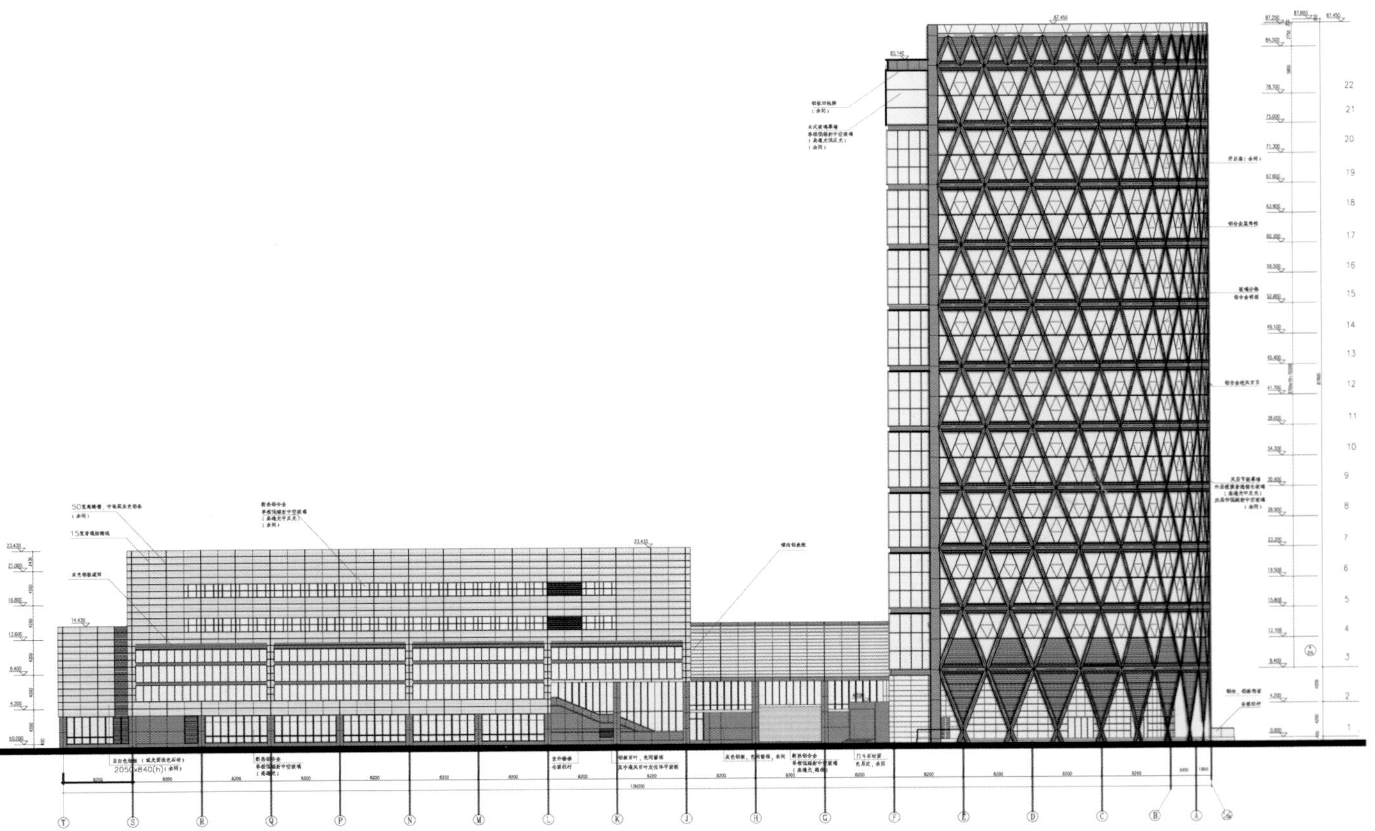

West Elevation 西立面图

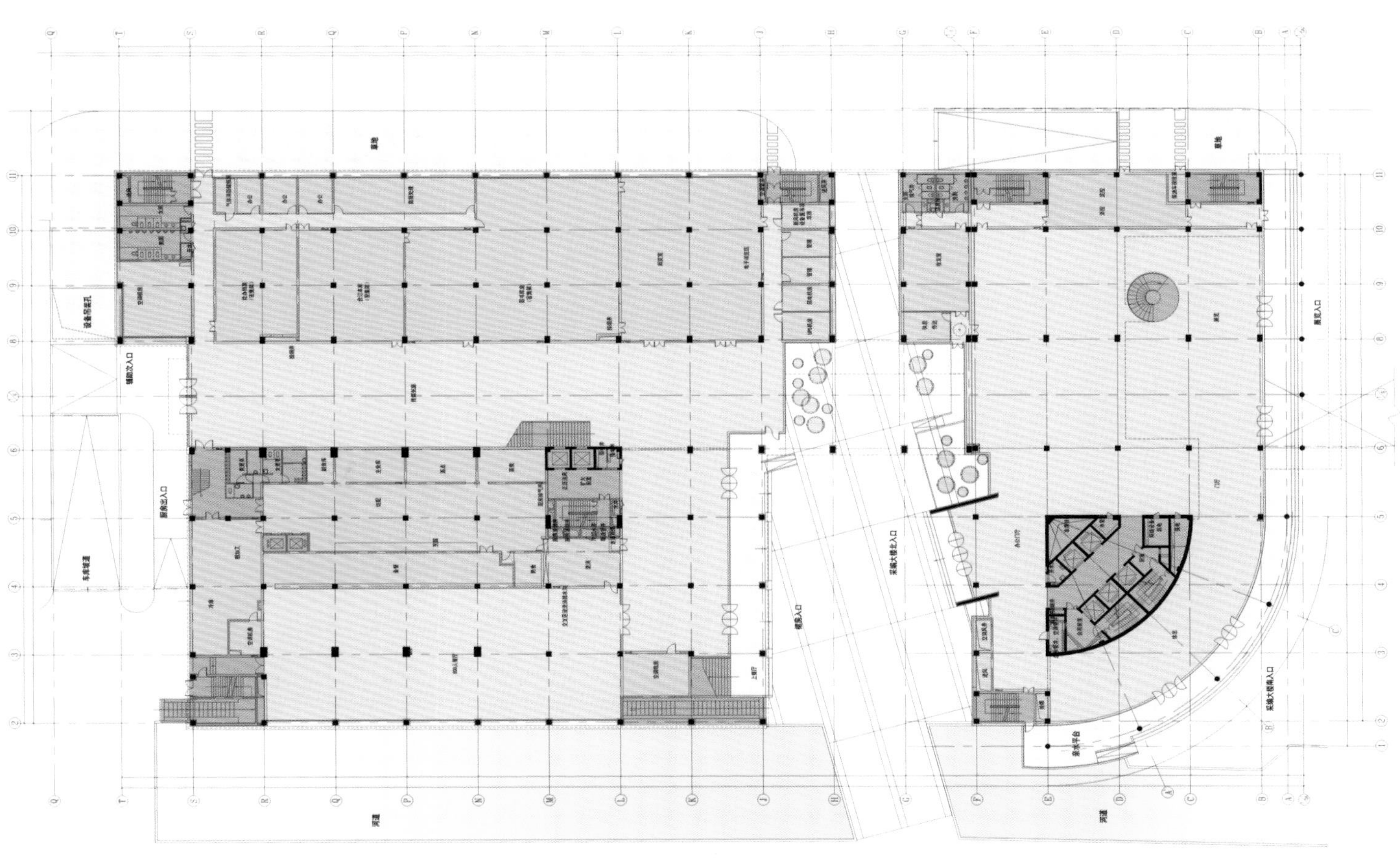

First Floor Plan　一层平面图

KEY WORDS
关键词

Spatial Regulation 空间调节
Green Building 绿色建筑
Low Energy Consumption 低能耗

China Potevio Shanghai Industrial Park HQ Scientific Research Building
中国普天信息产业上海工业园总部科研楼

FEATURES
项目亮点

The project adopts passive design strategy of "spatial regulation", which is through effective spatial organization, reasonable size and structure design to achieve performance-based regulation of the indoor and outdoor environment with its own space form and organizational structure, and effectively reduce the energy consumption as well.

项目采用“空间调节”的被动式设计策略，即通过有效的空间组织、合理的体型和构造设计，以空间本身的形态和组织结构来实现对室内外环境的性能化调节，同时有效降低能耗。

Location: Fengxian District, Shanghai, China
Architectural Design: Southeast University Architectural Design Institute Co.,Ltd.
Land Area: 10,057 m^2
Construction Scale: 4,369 m^2
Plot Ratio: 0.43
Green Coverage Ratio: 45.7%
Completion: 2009

项目地点：中国上海市奉贤区
设计单位：东南大学建筑设计研究院
用地面积：10 057 m^2
建筑规模：4 369 m^2
容积率：0.43
绿化率：45.7%
竣工时间：2009 年

Overview

China Potevio Shanghai Industrial Park HQ Scientific Research Building is located in Shanghai Fengxian Industrial Park, part of the Potevio Shanghai A1 (including A2) Project Plot Phase I. This building has a total of five layers and a building height of 23.1 m. The project construction aims to integrate and optimize the green architectural design strategies and technical equipments with sustainability principles, conduct the research and development of intelligent building environment control system and industrialized experiment, so as to create a green building demonstration project in the area with hot summer and cold winter in China.

项目概况

中国普天信息产业上海工业园总部科研楼位于上海市奉贤区，是普天上海工业园 A1（含 A2）地块一期工程的组成项目，该大楼共 5 层，建筑高度为 23.1m。项目的建设旨在集成与优化体现可持续性原则的绿色建筑设计策略与技术设备，开展建筑环境控制智能化系统的研发与产业化实验，创建我国冬冷夏热地区绿色建筑的示范工程。

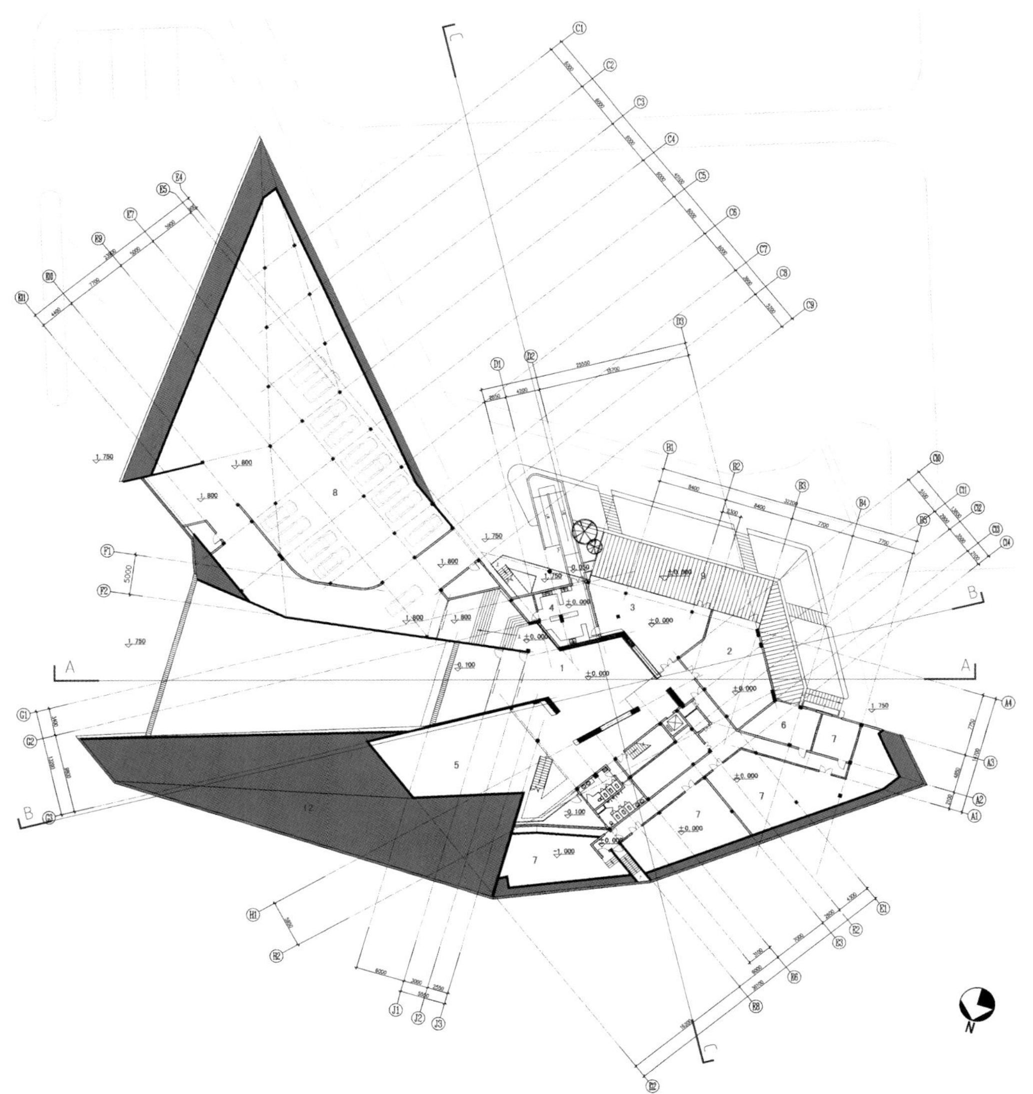

1. 中庭
2. 会议室
3. 餐厅
4. 厨房
5. 展厅
6. 消防控制室
7. 设备用房
8. 车库
9. 庭院
10. 科研办公用房
11. 设备夹层
12. 草坡
13. 植草屋面

Site Plan 总平面图

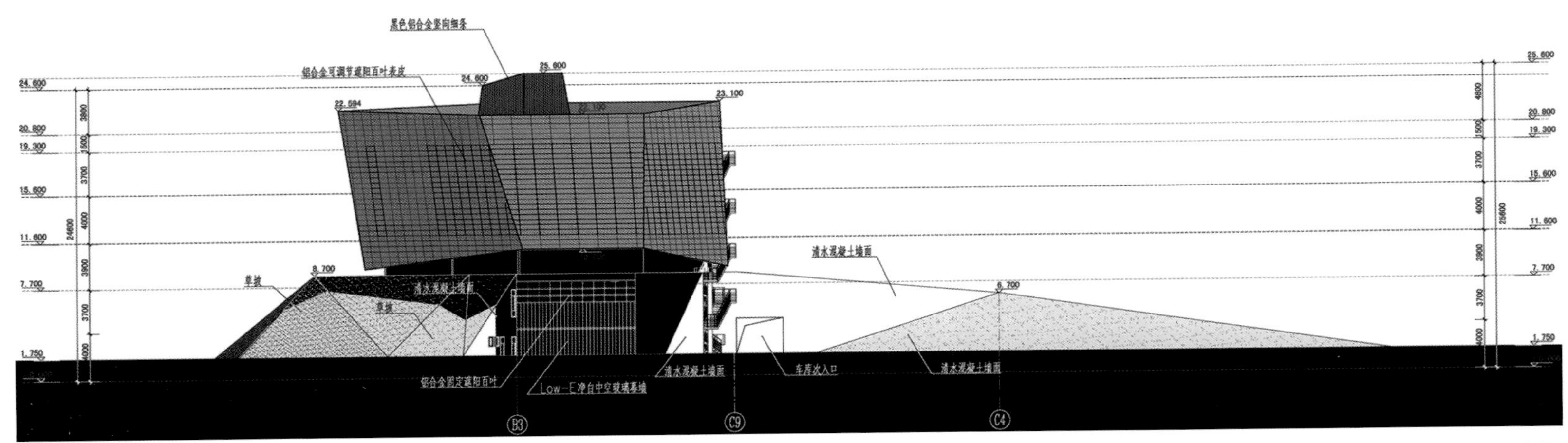

West Elevation 西立面图

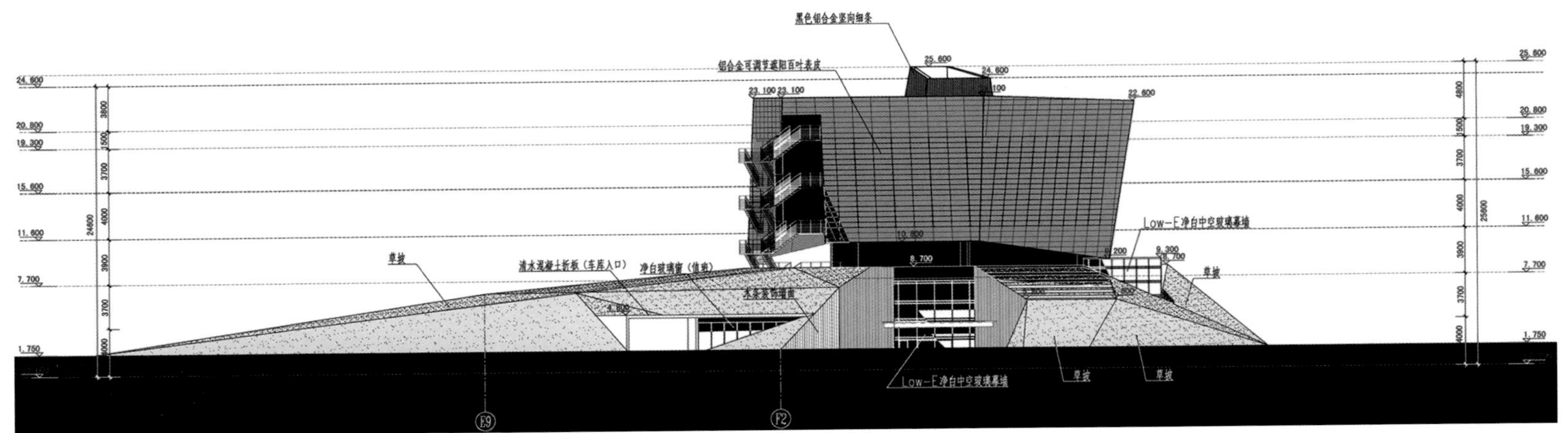

East Elevation 东立面图

Architectural Design

The main characteristics of this project design is through effective spatial organization, reasonable size and structure design to achieve performance-based regulation of the indoor and outdoor environment with its own space form and organizational structure, and effectively reduce the energy consumption as well. The so-called passive design strategy of "spatial regulation" is mainly embodied in the aspects of low shape coefficient, self sunshade, interactive regulated envelop enclosure, natural ventilation and natural lighting, etc.

建筑设计

本项目建筑设计的主要特点在于通过有效的空间组织、合理的体型和构造设计，以空间本身的形态和组织结构来实现对室内外环境的性能化调节，同时有效降低能耗。这种被称为"空间调节"的被动式设计策略主要表现在低体型系数、形体自遮阳、交互式可调控围护结构、自然通风、天然采光等方面。

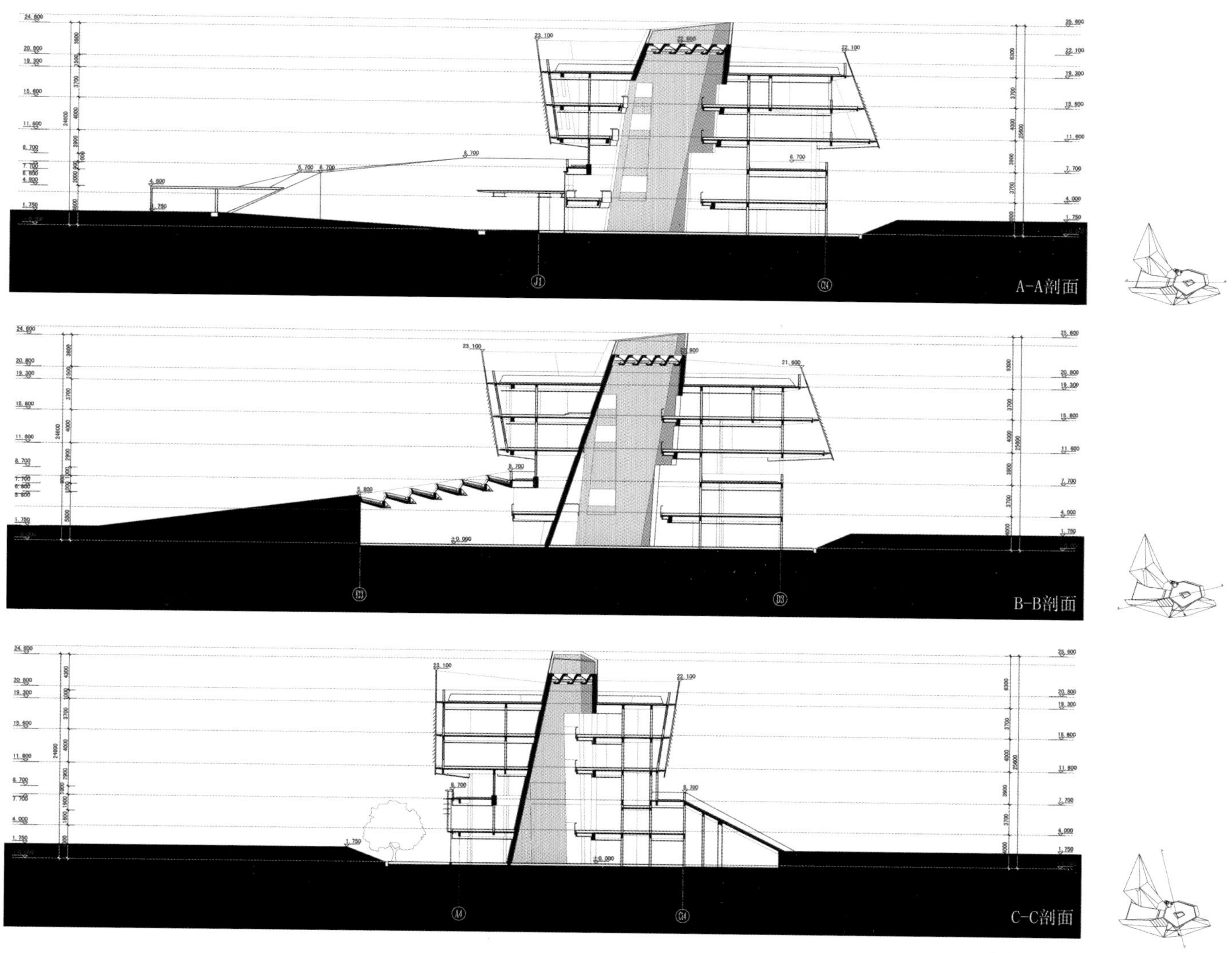

Sectional Drawing 剖面图

KEY WORDS
关键词

Stone-like Material 仿石面材

Characteristic of Qiang Nationality 羌族特色

Sense of Depth 层次感

Beichuan Administrative Center
北川行政中心

FEATURES
项目亮点

The design fully reflects the idea of "serving the people", and respects the terrain and makes the facades have characteristics of both Qiang nationality and modern times.

整个设计充分体现了便捷服务、庄重亲民的新理念，尊重山地地形，立面造型既富有羌族特色，又具有现代建筑特质。

Location: Beichuan, Mianyang, Sichuan, China
Architectural Design: HongKong HUAYI Design Consultants.Ltd.
Total Land Area: 69,329.22 m^2
Total Floor Area: 58,856 m^2
Plot Ratio: 0.75
Green Coverage Ratio: 35%

项目地点：中国四川省绵阳市北川县
设计单位：香港华艺设计顾问（深圳）有限公司
总占地面积：69 329.22 m^2
总建筑面积：58 856 m^2
容积率：0.75
绿化率：35%

Overview

Located in the northern end of central axis of New Beichuan, this project is the largest public building group that accommodates 30 government agencies such as Party committee, NPC, government & CPPCC (the "four groups"), public security organs and other units in different levels. It is the hub to operate public affairs and the core guarantee and strong supporter to help Beichuan people to rebuild their homes.

项目概况

项目位于北川新县城中轴线北端，是北川新县城规模最大的一个公共建筑群，包括党委、政协、人大、政府（"四套班子"）以及其他办公机关等30余个单位。建成后它将成为北川各项公共事务运转的中枢，以及北川人民重建家园的核心保障和坚强后盾。

Site Plan 总平面图

Architectural Design

Inherited characteristics of local residence, the buildings are constructed with stones and along the mountain. The front and the back of the buildings are set in different heights which create a sense of depth. The roof is designed in flat and pitch, interpreting the traditional Qiang architecture in a modern way.

The "four groups" are arranged in the core area and accompanied by the other agencies around, meaning that Beichuan people can make joint efforts to rebuild Beichuan under the party's leadership. The "four groups" embrace Maanshan inwardly and opposite the urban landscape axis, forming a visual focal point for the north and playing as the center of the building group as well. North-south administrative axis and east-west landscape axis constitute an extraordinary space, where the intersection is exactly the public square "the Heart of Beichuan". Starting from Maanshan, from north to south, three terraces are developed along the central axis gradually.

Since there are too many governmental departments, designers tried to make hard things simple, making the buildings scattered without chaos and identifiable in the crowd. Local element "white stone" combined modern and tradition subtly, and the stone-like materials saved costs without losing the solemn of administrative buildings.

建筑设计

设计传承羌寨民居“依山而建，垒石为室”的山地关系，整合办公楼前后不同的标高体系，略微收分，形成倚山而立的层次感。采用平坡结合的屋顶造型，彰显出传统羌族建筑的现代化表现形式。

本案核心部分设立“四套班子”建筑，青山环抱左右拱卫，各局级办公以“四套班子”中心轴线东西侧布置，形成簇拥核心的建筑群体感，寓意在党的领导下，北川人民“万众一心”重建北川的力量与信心。“四套班子”以内凹环抱马鞍山的建筑姿态与城市景观轴相对，形成城市的底景与视觉焦点，是建筑群的中心；南北向的行政轴线与东西向的景观轴线构成了气势不凡的尺度空间，两轴的交汇处恰好是市民广场“北川之心”；从马鞍山由北至南，三个台地沿中轴线逐步展开。

行政中心部门众多，设计师采取“化繁为简”的设计思路，使得建筑散而不乱，聚而可识。建筑适当收分，局部退台以及运用羌族白石头元素等处理手法，巧妙地将现代建筑与传统羌族建筑相结合。采用仿石面材的处理，既节约了成本，又不失行政类建筑的大气与庄严。

KEY WORDS 关键词	
	Vertical Line 竖向线条
	Fold Line Form 折线形式
	Color Aesthetic Appreciation 色彩审美

Electric Power Distribution and Trade Center of Northeast Grid

东北电网电力调度交易中心大楼

FEATURES

项目亮点

The treatment of elevation skin adopts two methods of "irrespective of inside and outside" that two different textured skins and materials interpenetrate to concisely express the modern aesthetic temperament and interest, and reveal the enterprise industry characteristics and personality attributes.

立面表皮的处理采用两种“内外有别”的方式，两种肌理的表皮和不同材质体块的穿插，简洁地彰显出当代的审美情趣，展现了企业的行业特征与个体属性。

Location: Shenyang, Liaoning, China
Architectural Design: Liaoning Architectural Design and Research Institute
Floor Area: 85,713 m^2

项目地点：中国辽宁省沈阳市
设计单位：辽宁省建筑设计研究院
建筑面积：85 713 m^2

Overview

Electric Power Distribution and Trade Center of Northeast Grid is located in Shenyang, which is the window of the Northeast Grid in the northeast area, and also the landmark building developed in Hunnan new area.

项目概况

东北电网电力调度交易中心大楼位于沈阳市，是东北电网在东北地区的窗口，也是新浑南开发的标志性建筑。

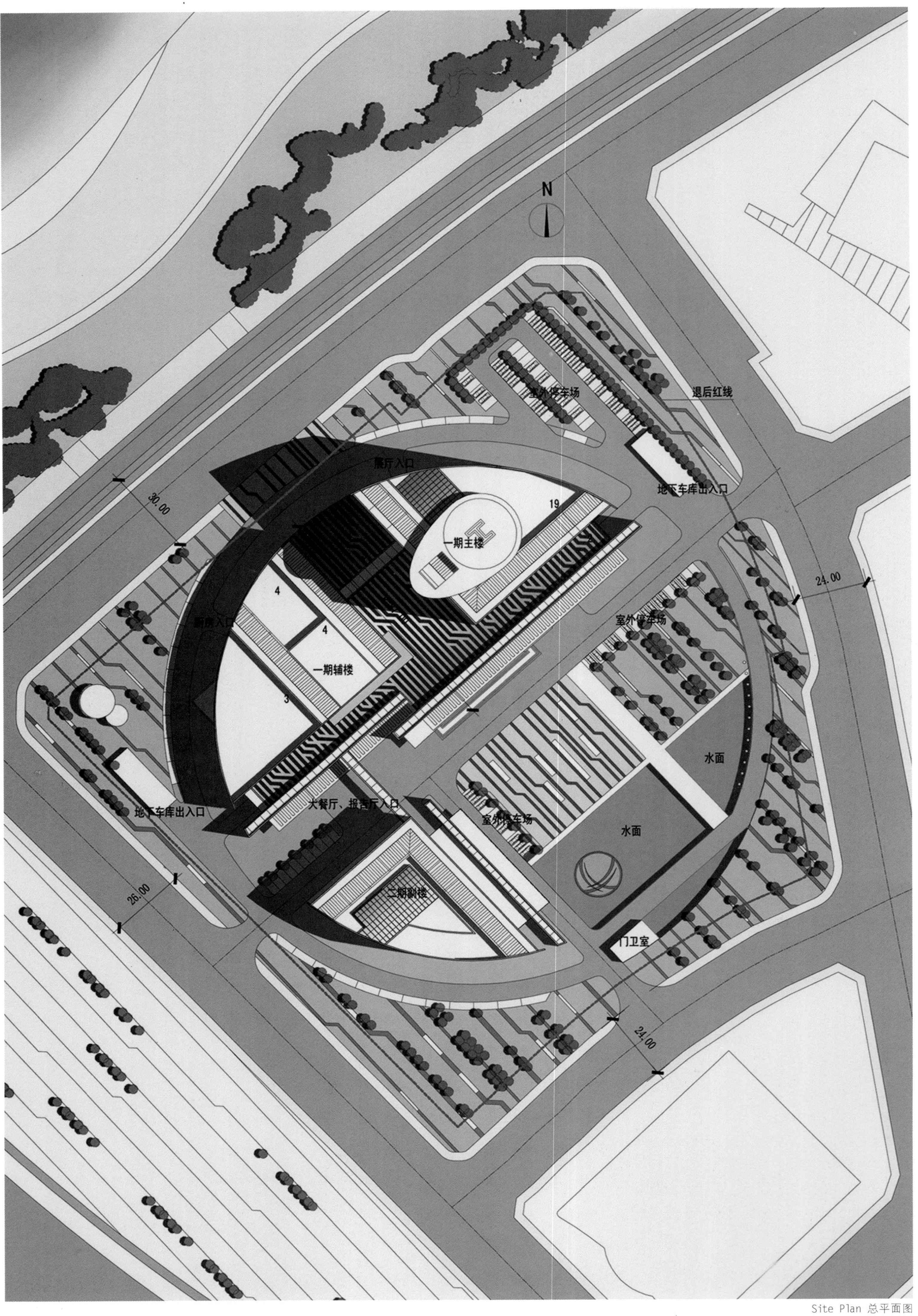

Site Plan 总平面图

Planning and Layout

The overall layout of the building is based on the base environment around; there are large-scale highways and rivers, large area of vacant space, but only circle plane of Xinhua News Agency and square plane of automobile exhibition room, so adopting the form of circle plane will form friendly dialogue uncertainty surrounding buildings and highways, rivers in the future; at the same time, this kind of form should be identical with the identification of Northeast Grid and China's traditional value of round heaven and square earth.

According to the analysis of the whole plane design, the building is divided into the main building, auxiliary building and annex building, which situate three orientations from the southwest to the northeast of the base to become the supporting part. The entrance square in the southeast corner, as a cushion space to the surrounding environment, becomes the city's living room, which forms a complete circle together with the arc water surface. The main building and auxiliary building are freely divided and combined by a three-layer shared hall, and this spacious shared space can also hold all kinds of gatherings and celebrations as well. Through the shared hall, there is the graceful HunHe Park, which introduce landscape into the interior; the urban constructions at the north shore of HunHe River become the natural background of this project, and achieve some kind of spiritual relationship with the original east power building on the same orientation extending line.

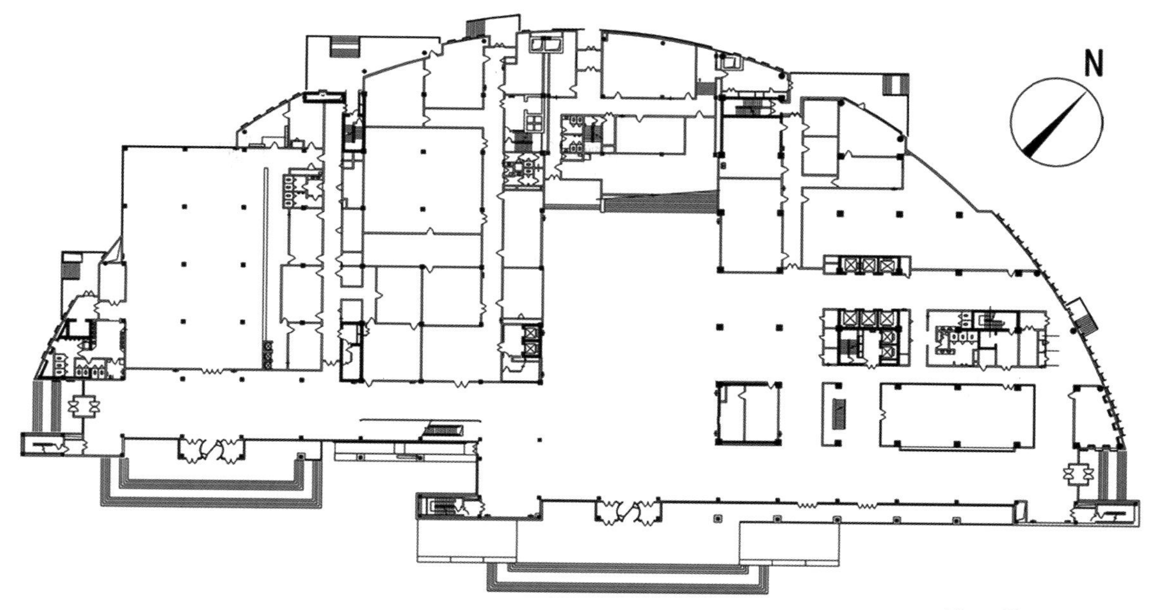

First Floor Plan 一层平面图

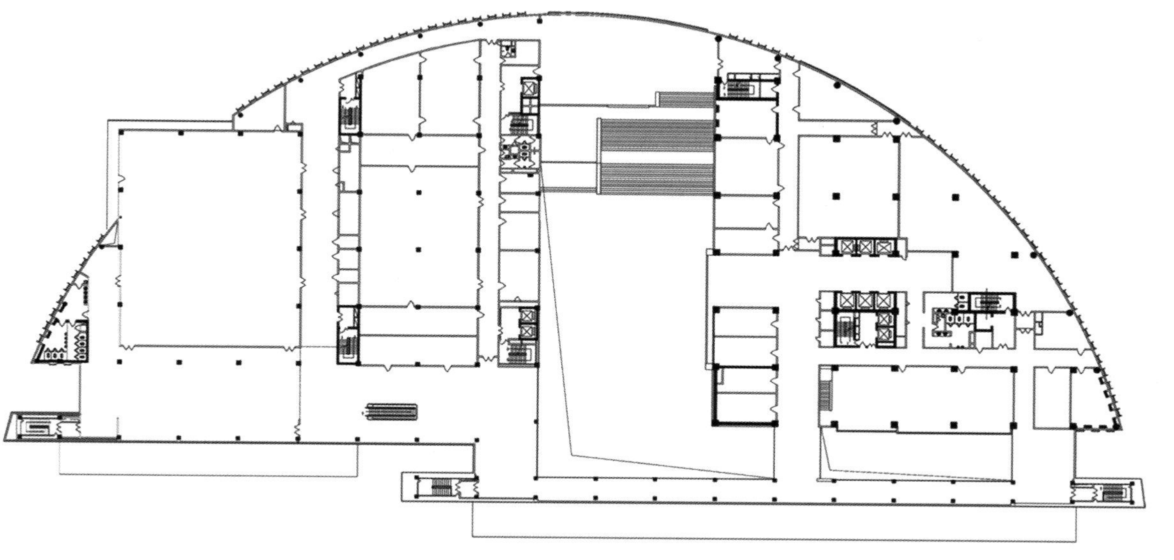

Second Floor Plan 二层平面图

规划布局

建筑的总体布局首先来自基地周边环境，基地周边有大尺度的高速路与河流，周边空地面积较大，只有圆形平面的新华社和方形平面的汽车展厅，因此采用圆形平面与未来不确定的周边建筑和高速路、河流形成友好对话；与此同时，在形式上与东北电网有限公司的标识和“天圆地方”的中国传统价值观相契合。

根据总平面设计的分析，把建筑划分为主楼、辅楼、副楼三个部分。主楼、辅楼和副楼分别占据基地的西南到东北三个朝向，形成了基地的倚靠。东南角的入口广场作为与周边环境之间的缓冲空间，成为城市的“客厅”，与弧形水面构成总平面完整的圆形。主楼与辅楼通过3层高的共享大厅分合自如，宽敞的共享空间，同时也可以举行各种集会和庆典活动；透过共享大厅可见秀美的浑河公园，景致被引入建筑内部，浑河北岸的城市建筑成为本项目的天然背景，同时也与这一方位延长线上的原东电大楼取得了某种精神上的联系。

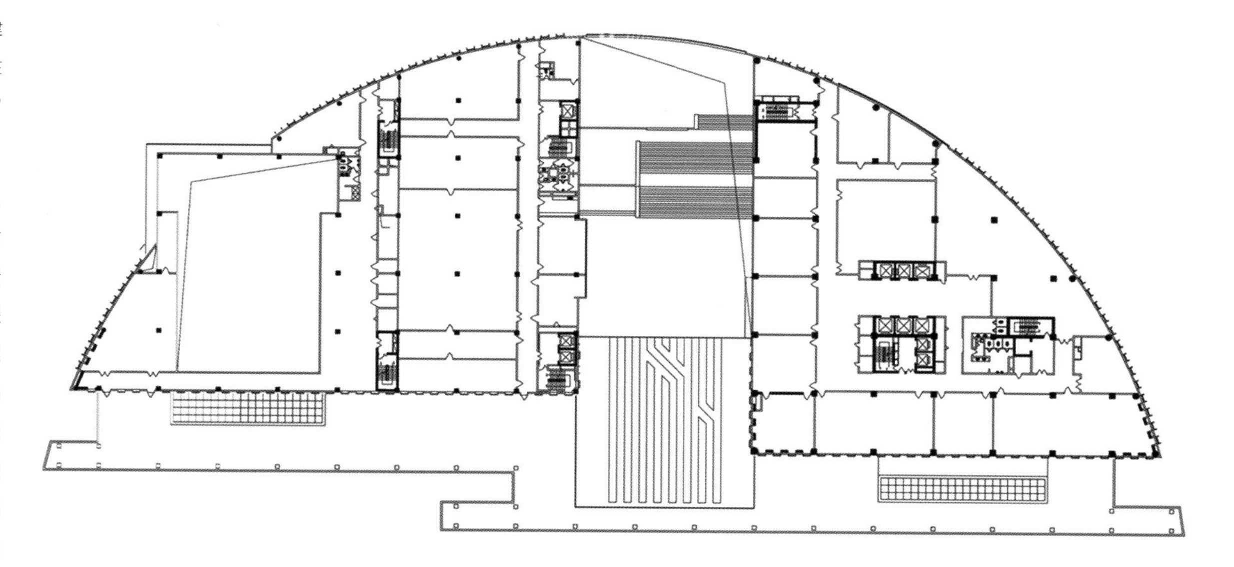

Third Floor Plan 三层平面图

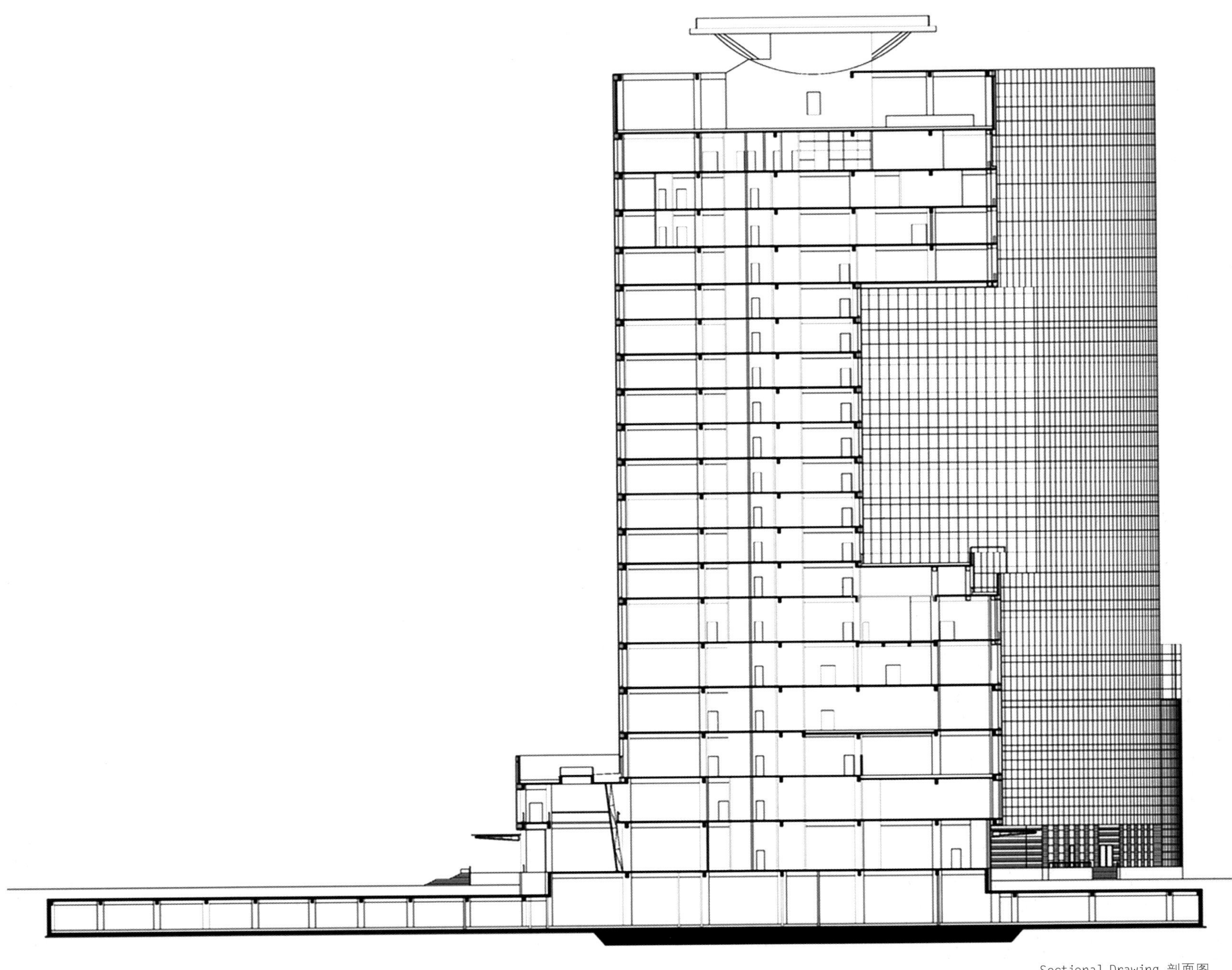

Sectional Drawing 剖面图

Architectural Design

The treatment of elevation skin adopts two methods of "irrespective of inside and outside" that three buildings' elevations along the lateral arc boundary of the project base adopt unified form of vertical lines to express the obedience and common attributes between the construction and city, and the inside facades adopt modular lines to symbolize the fold line form of the electric power, revealing the enterprise industry characteristics and personality attributes. The architectural color is rooted in the understanding of northern regional climate and color aesthetic habits that it uses heavy color tone as the fundamental key of buildings and adds warm color clay board to reveal the change out of the unity; two different textured skins and materials interpenetrate to concisely express the modern aesthetic temperament and interest, while the detail and scale is supposed to have eternal charm.

建筑设计

立面表皮的处理采用两种“内外有别”的方式，三栋建筑沿基地外侧弧形边界的立面均采用统一的竖向线条形式，彰显建筑与城市的服从关系和共性属性，内向基地的各立面采用模数化的象征电力的折线形式，展现企业的行业特征与个体属性。建筑的色彩来源于对北方地域气候和色彩审美习惯的理解，以重色作为建筑的基调，穿插暖色的陶土板，希望流露其统一之中的变化；两种肌理的表皮和不同材质体块的穿插，简洁地彰显出当代的审美情趣，而对细节和比例尺度的关注则是希望建筑能够魅力永驻。

KEY WORDS
关键词

Image Element 意象元素

Innovative Modeling 创新造型

Detail Treatment 细部处理

Guangdong Xinghai Performance Art Group (New Address)

广东星海演艺集团（新址）

FEATURES
项目亮点

Based on the perspective of urban design, the architectural modeling and facade design seek changes in harmony and look for innovation in the unity to create an elegant, pure, fresh and active performing art culture enterprise image.

建筑造型和立面设计从城市设计的角度出发，在和谐中求变化，于统一中求创新，创造出优雅、清新又积极向上的演艺文化企业形象。

Location: Guangzhou, Guangdong, China
Architectural Design: Guangzhou Urban Planning & Design Survey Research Institute
Total Land Area: 9,042 m^2
Building Base Area: 2,713 m^2
Total Floor Area: 13,953 m^2
Plot Ratio: 1.4
Completion: 2009

项目地点：中国广东省广州市
设计单位：广州市城市规划勘测设计研究院
总用地面积：9 042 m^2
建筑基底面积：2 713 m^2
总建筑面积：13 953 m^2
容积率：1.4
竣工时间：2009 年

Overview

Guangdong Xinghai Performance Art Group office building is the landmark annex building of Xinghai Concert Hall in Ersha Island, which bears the appearance characteristics of public cultural construction. It has 6 layers above the ground, 1 layer under the ground, and a building height of 23.2 m. There is a total construction area of 13,953 m^2, of which 10,824 m^2 above the ground, and 3,129 m^2 under the ground.

项目概况

广东星海演艺集团办公楼是二沙岛上的标志性建筑星海音乐厅的配楼，具有公共文化建筑的形态特征。建筑地上 6 层，地下 1 层，建筑高度为 23.2 m。总建筑面积为 13 953 m^2，其中地上部分 10 824 m^2，地下部分 3 129 m^2。

Design Concept

The artistic idea of the construction scheme comes from the "bamboo" image in traditional Chinese literati landscape paintings; "bamboo" is the symbol of Chinese living environment, the symbol of Chinese traditional cultural spirit, and also the symbol of Chinese traditional music artistic conception.

设计构思

建筑方案的艺术构思源于中国传统文人山水画中“竹子”的形态，“竹子”是中国人居环境的象征，也是中国传统文化精神的象征，更是中国传统音乐意境的象征。

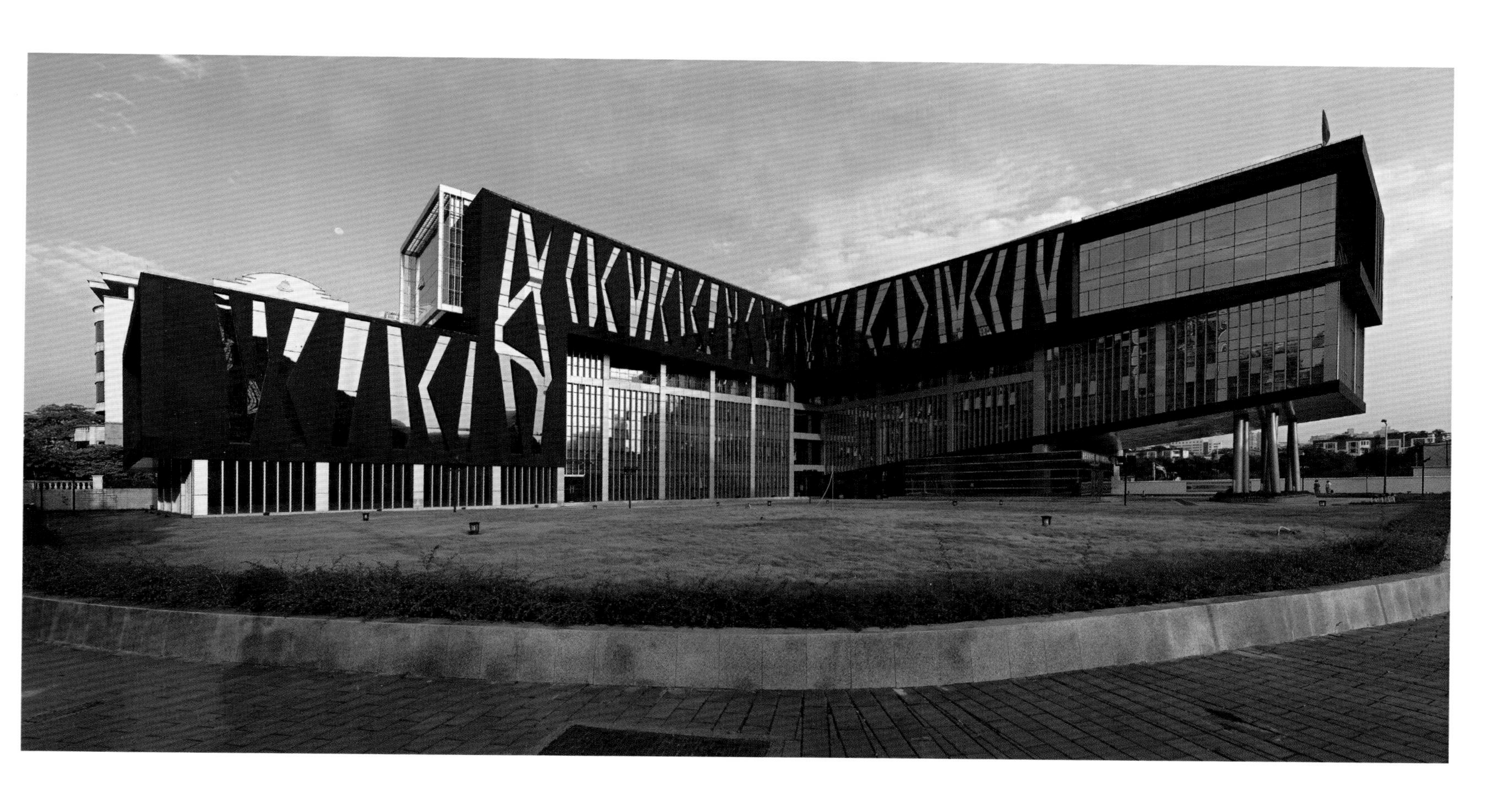

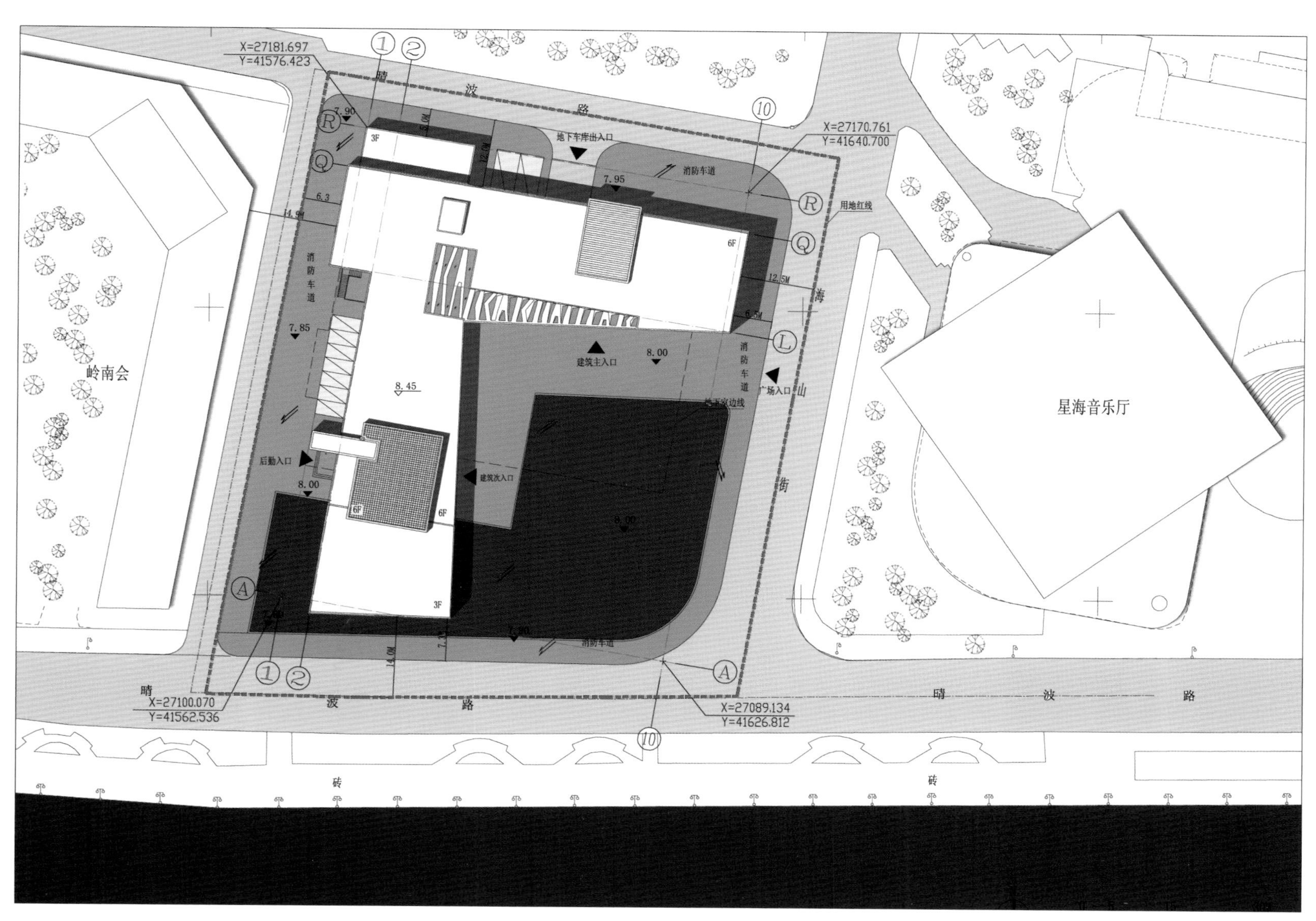

Site Plan 总平面图

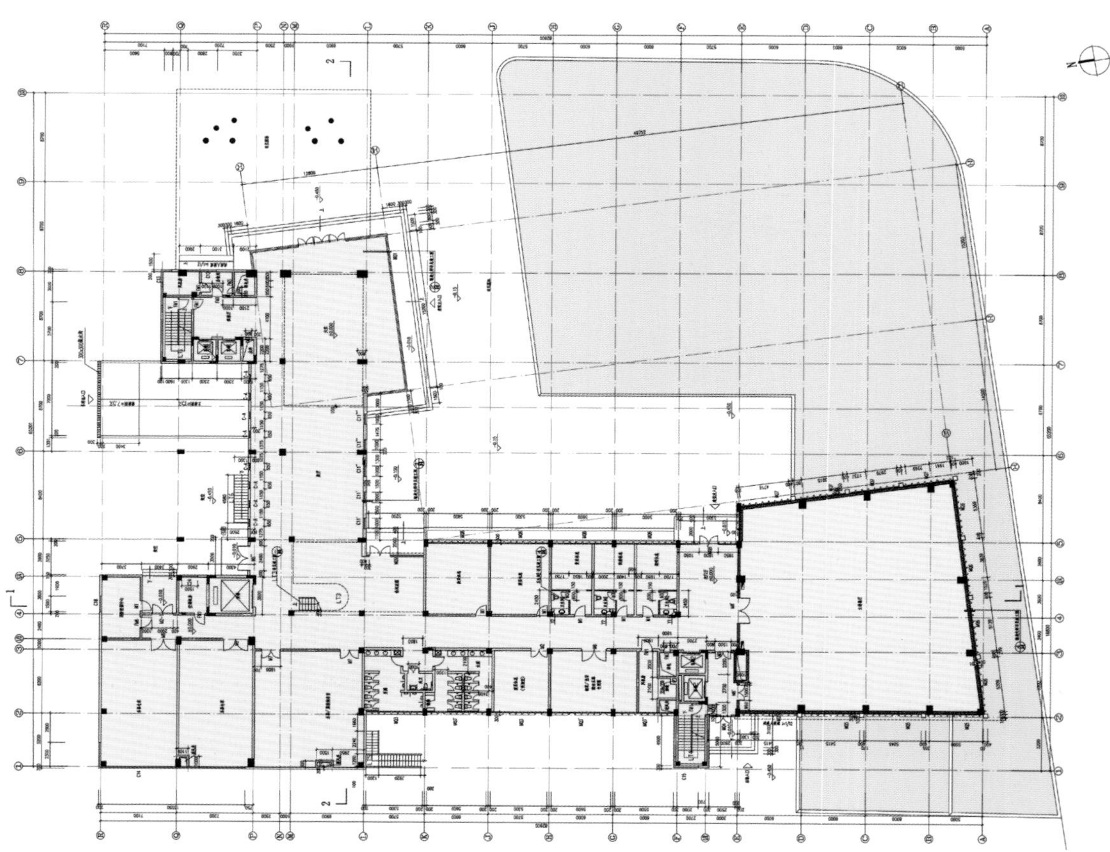

First Floor Plan 一层平面图

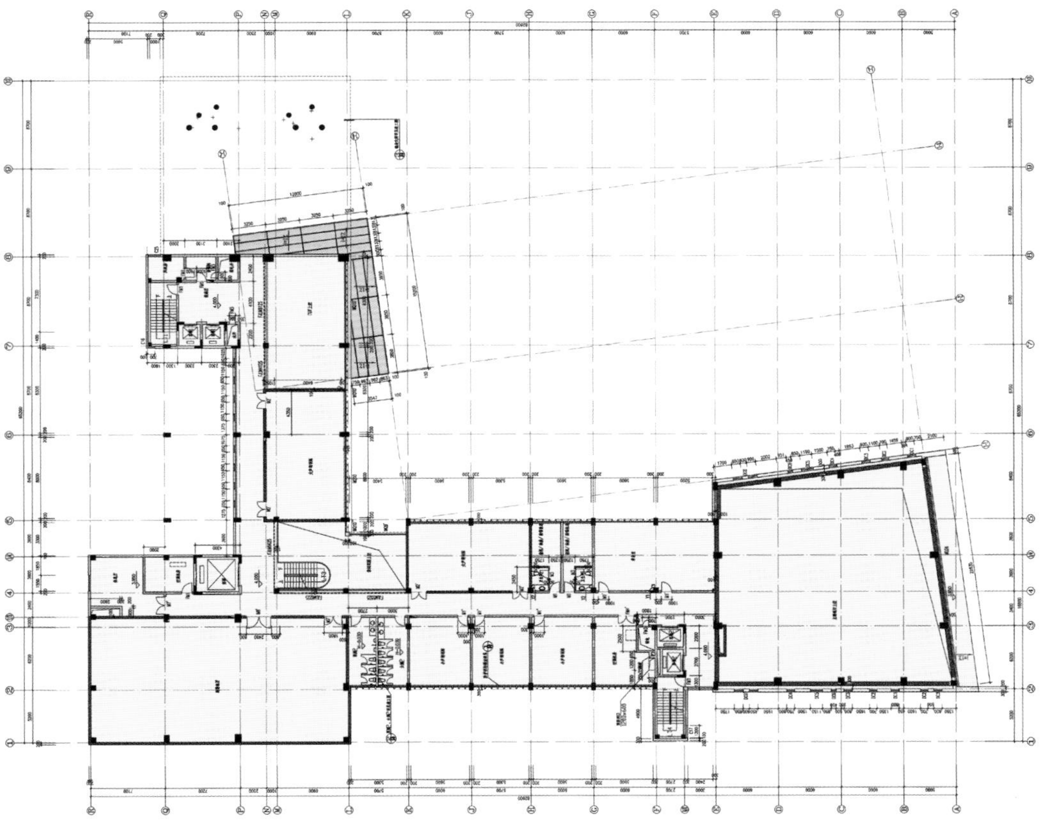

Second Floor Plan 二层平面图

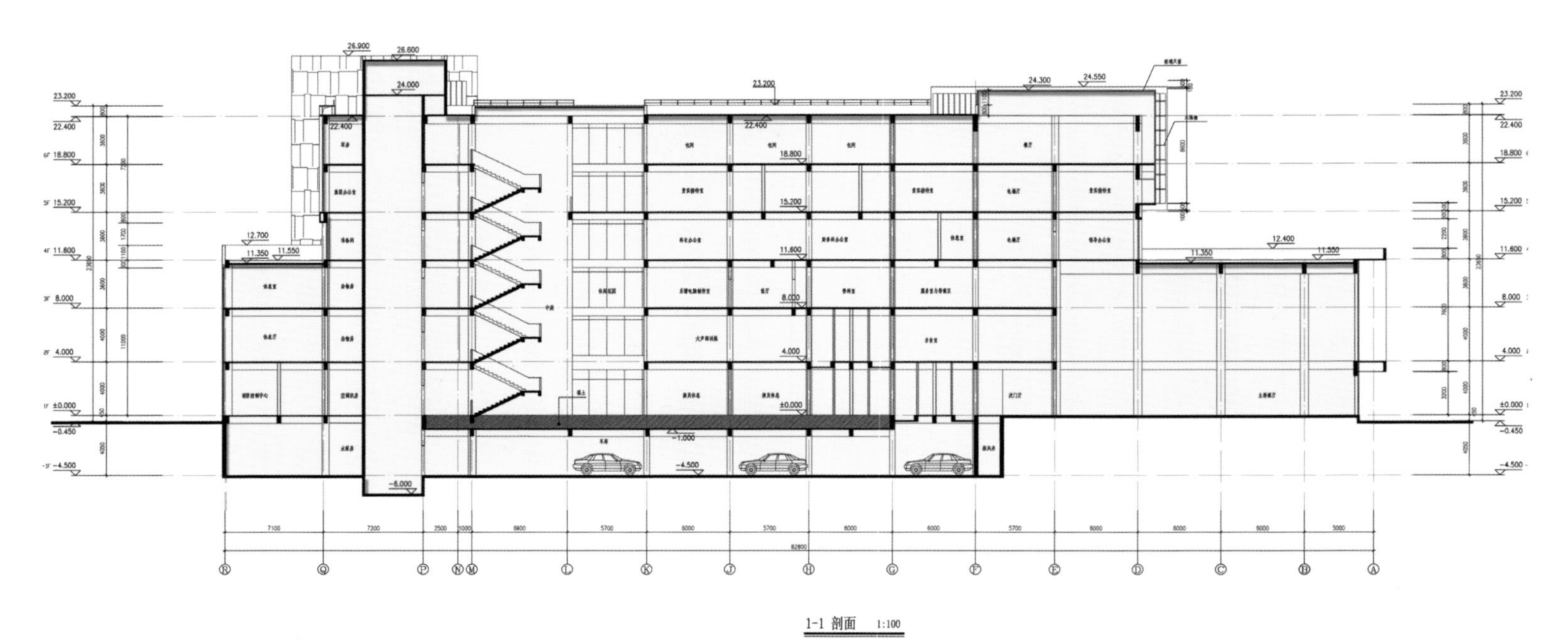

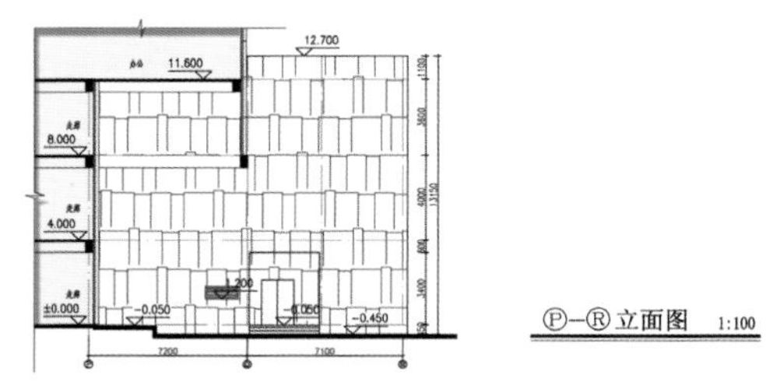

Sectional Drawing 剖面图

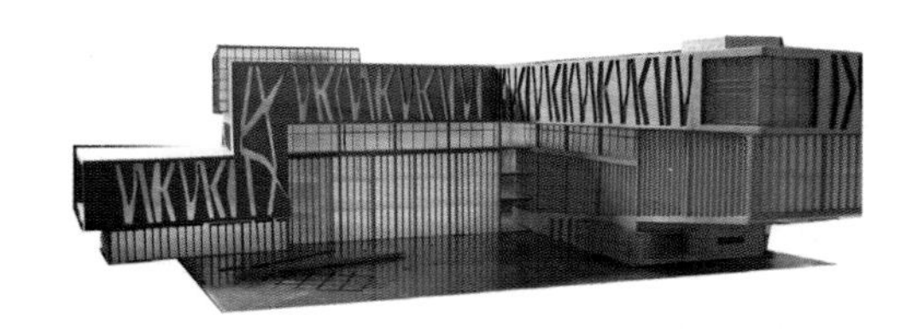

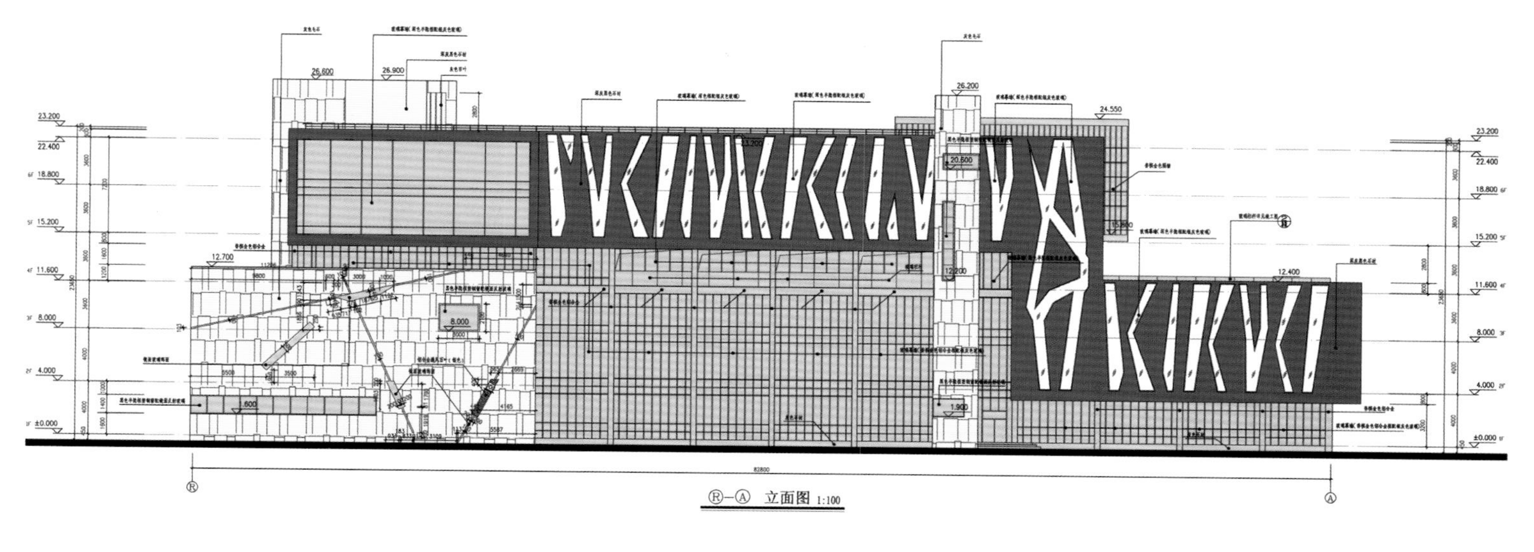

Ⓡ—Ⓐ 立面图 1:100

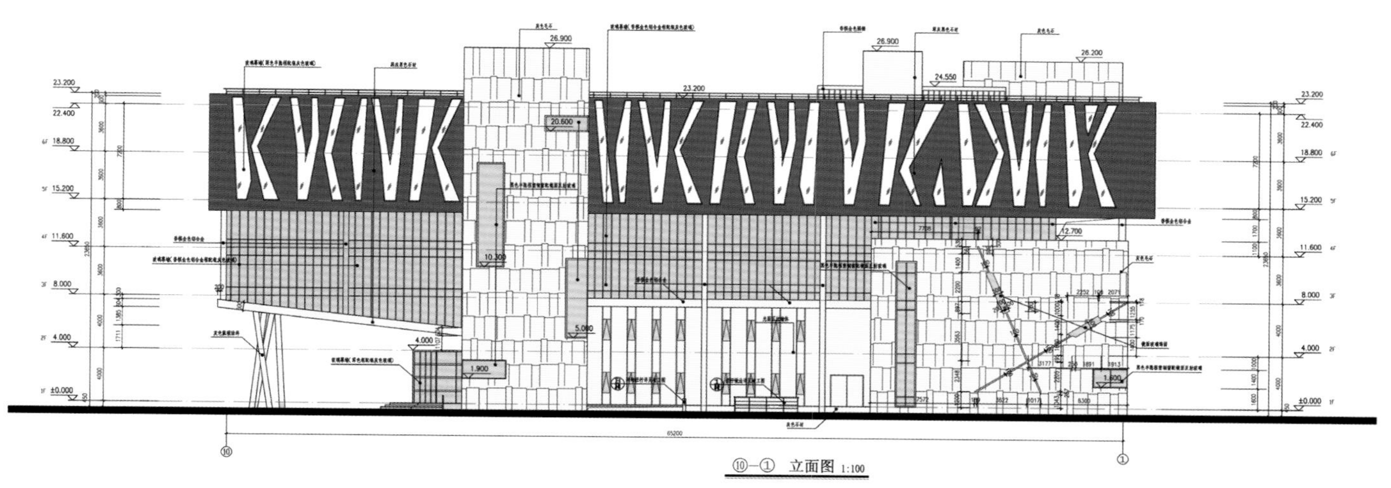

⑩—① 立面图 1:100

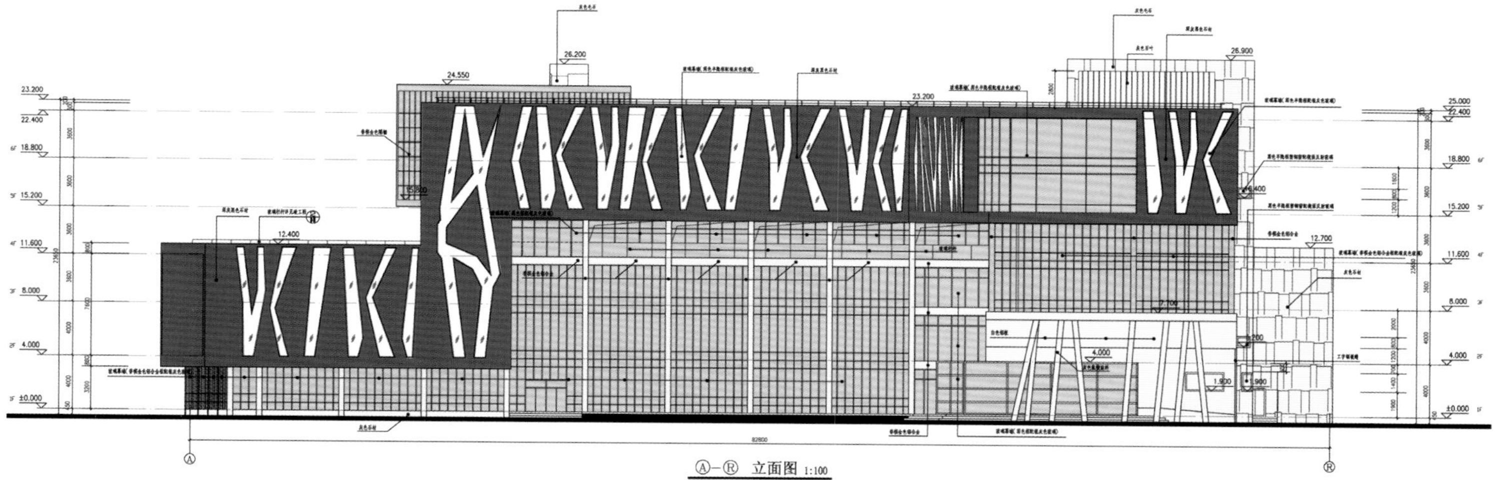

Ⓐ—Ⓡ 立面图 1:100

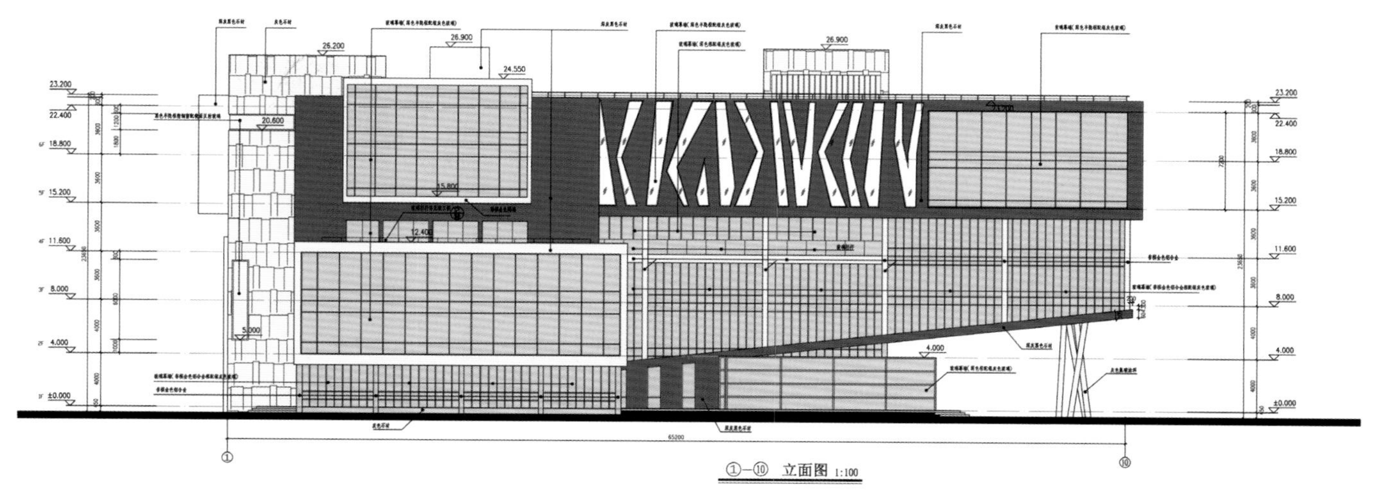

①—⑩ 立面图 1:100

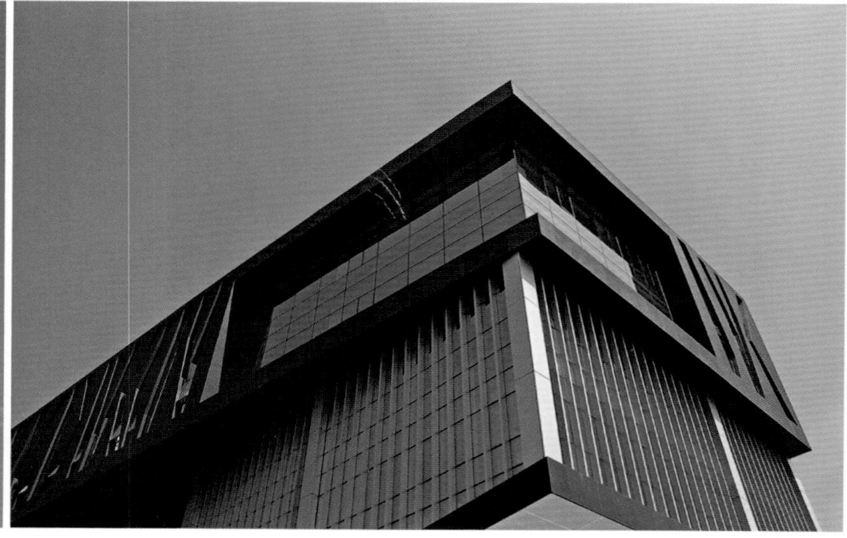

Architectural Design

Based on the perspective of urban design, the architectural modeling and facade design take the surrounding cultural environment as the background, and lay focus on the relationship of the new buildings with Xinghai Concert Hall and the construction contours of the Pearl River north shore. It seeks changes in harmony and looks for innovation in the unity to make the new building and the concert hall rely on each other and enhance each other. For the material selection, it prefers to the simple, economic and appropriate material to achieve good effect, trying to create an elegant, pure, fresh and active performing art culture enterprise image.

建筑设计

建筑造型和立面设计从城市设计的角度出发，以周边文化建筑区的环境为背景，重点考虑新建筑与星海音乐厅及珠江北岸建筑轮廓线的关系。在和谐中求变化，于统一中求创新，使新建筑与音乐厅相互依托、交相辉映。在材料选用上力求朴素、经济，以适宜的选材，达到良好的效果，努力创造优雅、清新又积极向上的演艺文化企业形象。

KEY WORDS
关键词

Breathing Curtain Wall 呼吸幕墙

Concise and Atmosphere 简洁大气

Shading Syestem 遮阳系统

The French Embassy in Beijing
法国驻华大使馆新馆

FEATURES
项目亮点

Combining the technics of three-phase in French classical architecture and city desk design and pavilion in Chinese classical architecture, based on the traditional idea of French and Chinese architecture, meanwhile, applying modern design technics and materials, expresses unique noble and elegant architectural disposition.

建筑形态通过结合法国古典建筑三段式及中国古典建筑中城台与楼阁的手法，基于法、中建筑传统理念的运用，同时加以现代的设计手法和材料，彰显出高贵、优雅、独特的建筑气质。

Location：Chaoyang District, Beijing, China
Architectural Design: Beijing Architecture Design and Research Institute

项目地点：中国北京市朝阳区
设计单位：北京市建筑设计研究院

Overview

The French Embassy in Beijing is located in the third embassy area, northern side of Liangmaqiao Road, Chaoyang District, Beijing. The new embassy building is constituted by foreign business area, embassy office and ambassador residence.

项目概况

法国驻华大使馆新馆位于北京朝阳区，亮马桥路北第三使馆区。新使馆建筑主要由使馆对外公务区、大使办公区和大使官邸三部分组成。

Planning and Layout

The third floor to the fourth floor on the southern, western and northern side develop into podium part, stretching and grave, an equilateral triangle office tower is arranged in the acute angle corner of Liangmaqiao Road and Tianze Road, which is concise but great, full of sculptural sensibility, meanwhile, it fully integrates with the surrounding city skyline.

规划布局

建筑主体南、西、北三面以三至四层形成裙房部分，舒展、庄重，仅在亮马桥路与天泽路的锐角转角处布置等边三角形办公高塔，简洁、大气，充满雕塑感，同时充分呼应周边城市天际线。

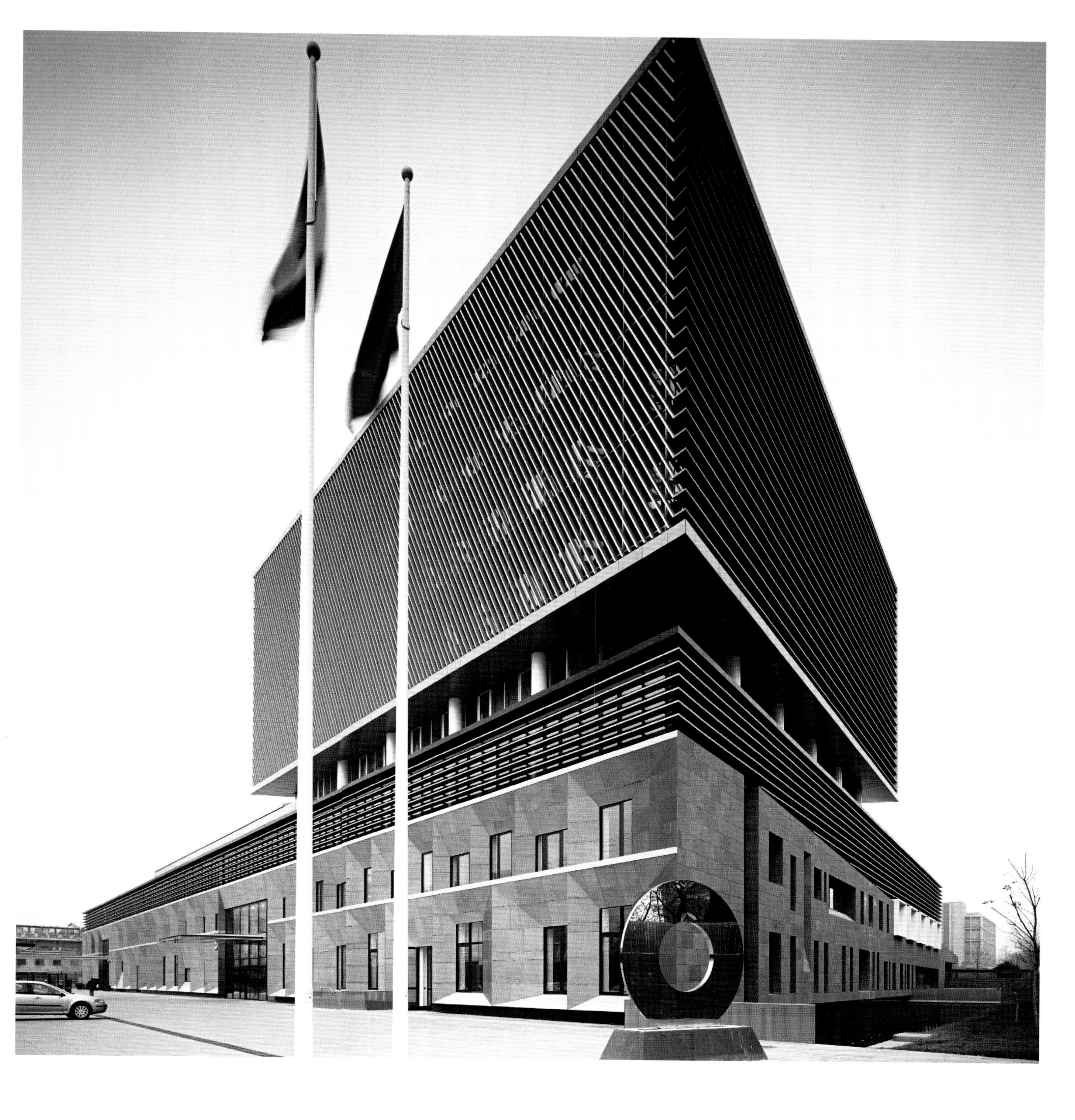

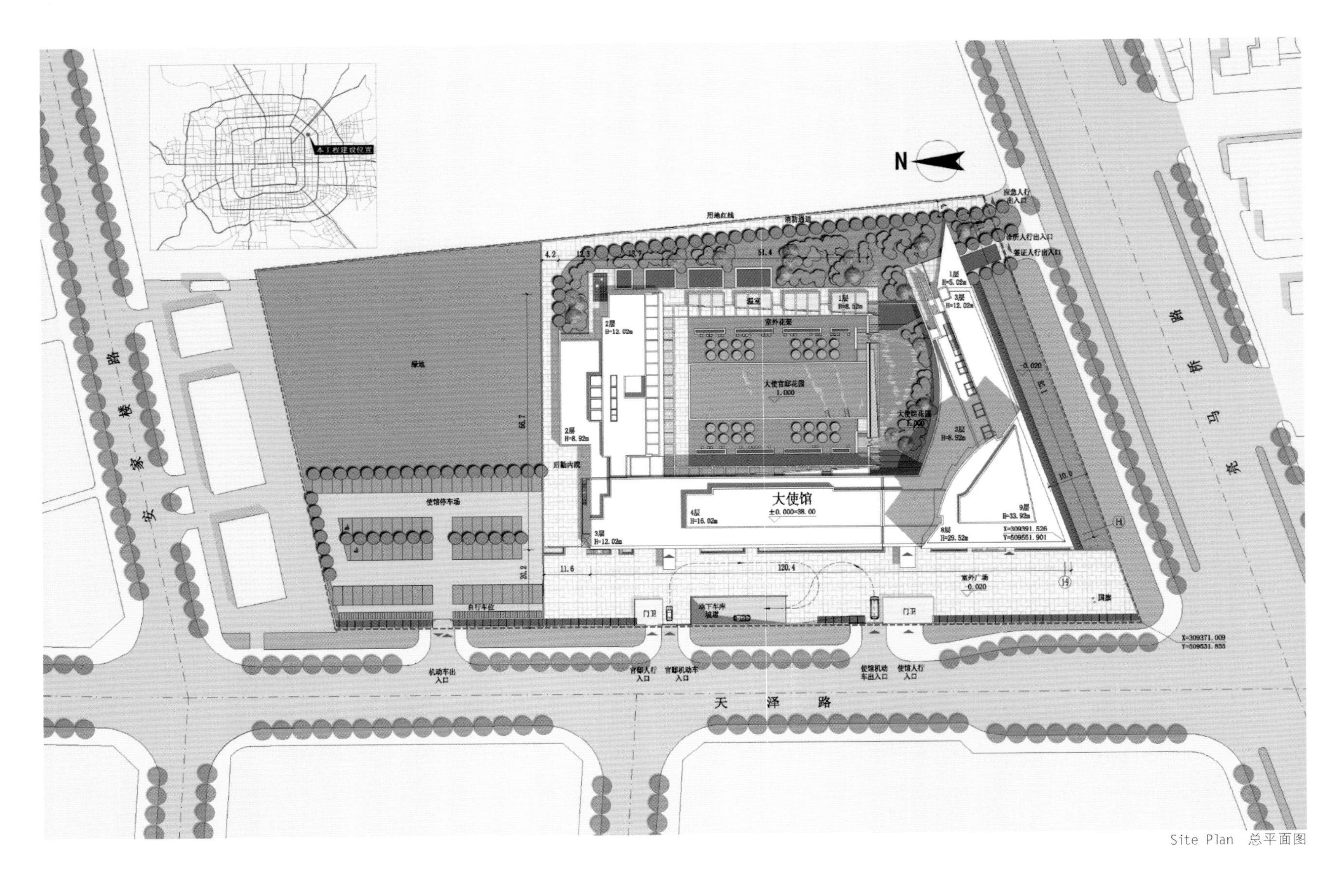

Site Plan 总平面图

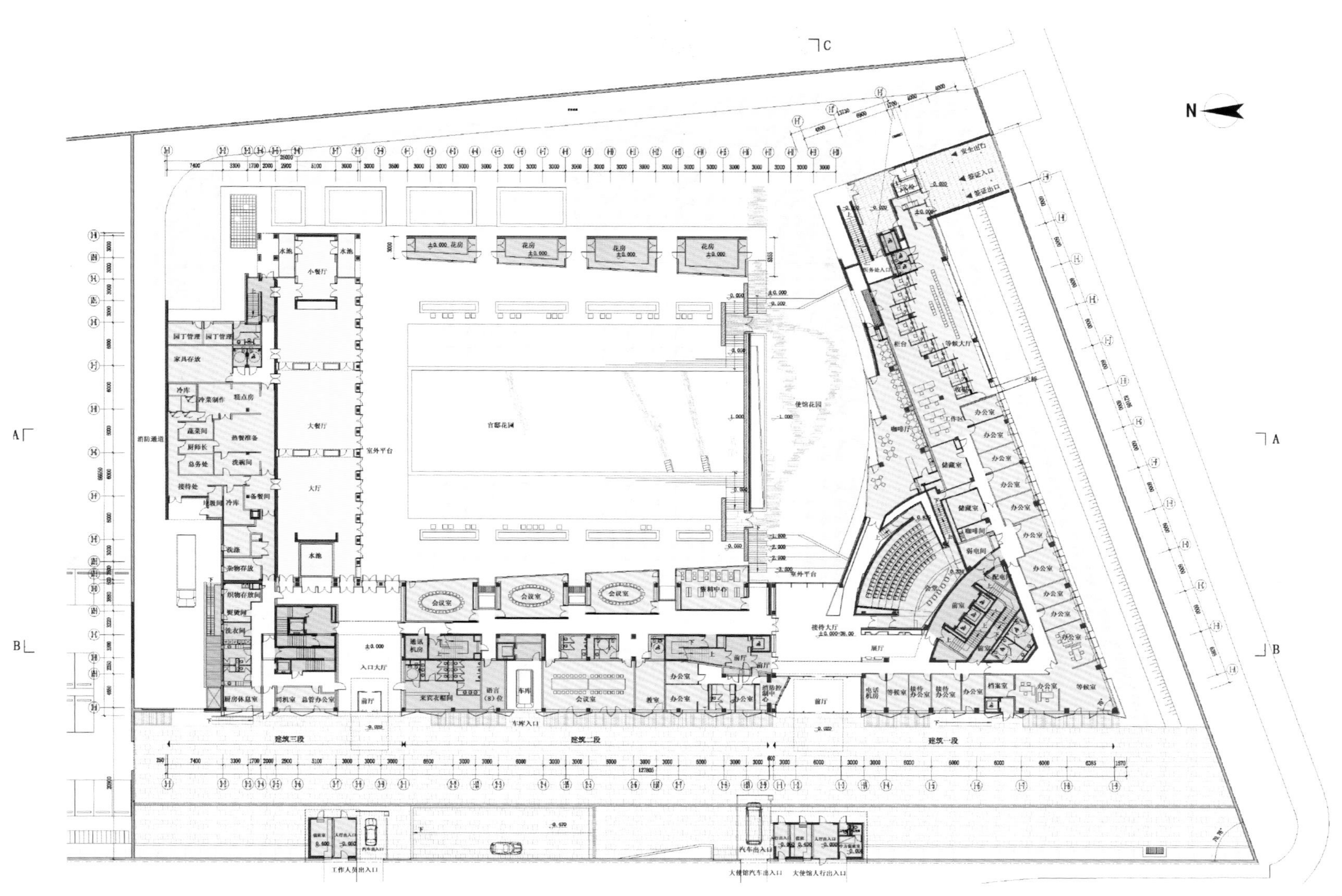

First Floor Plan 一层平面图

Architectural Design

Combining the technics of three-phase in French classical architecture and city desk design and pavilion in Chinese classical architecture, based on the traditional idea of French and Chinese architecture, meanwhile, applying modern design technics and materials, expresses unique noble and elegant architectural disposition. The podium as city desk with brunet Mongolia stone, the exterior wall stretches along southern side, continually lasts to the northern and eastern side, gives person strong and majestic sensibility; the pavilion in the high-level employs golden aluminum alloy blinds and glass, the shutter on the southern side with the aid of ingenious contrast of transition details in the corner, changed into the shading system in the western side. The courtyard facade of the pavilion in the higher floor employs double printing glass breathing curtain wall system, breathing, it can solve the problem of energy saving, also provides a whirling gauze clothing for the high-rise loft. Epitaxial sunshade, breathing curtain wall, printing glass veranda in the inner courtyard, greenhouse, a variety of skin facade lighting appears with rich variation, but also corresponds to ecological design concept perfectly.

建筑设计

建筑形态通过结合法国古典建筑三段式及中国古典建筑中城台与楼阁的手法，基于法、中建筑传统理念的运用，同时加以现代的设计手法和材料，彰显出建筑高贵、优雅的独特气质。作为城台的裙房采用深色蒙古石材，外墙沿南侧展开，并向北面和东面延续，给人坚固、厚重的感觉；高层楼阁运用金色铝合金遮阳百叶和玻璃，南向水平百叶借助转角处交接细节的巧妙对比，在西向转换为斜向遮阳系统。高层楼阁内庭院立面采用双层印花玻璃呼吸幕墙系统，在解决了节能问题的同时，也为高层阁楼披上了一层婆娑的纱衣。外延遮阳、呼吸幕墙，内院的印刷玻璃游廊、温室，多样的表皮在呈现立面丰富的光影变化的同时，又完美地契合了不同朝向的生态设计理念。

KEY WORDS
关键词

Dome **穹顶**

Regional Culture **地域文化**

Steady Image **沉稳形象**

Harbin International Communication Platform
哈尔滨市国际交流平台

FEATURES
项目亮点

The project design adopts the combination strategy of European classic symbol of dome and the membrane structure, which not only shows the Harbin region culture characteristic, but also well meets the functional requirements for the steady image of the building.

设计采用欧式经典标志的穹顶与膜结构相结合的策略，既展示了哈尔滨的地域文化特色，也很好地满足了功能对建筑庄重、沉稳的形象要求。

Location: Harbin, Heilongjiang, China
Architectural Design: Architectural Design and Research Institute of HIT
Total Floor Area: 5,195 m^2
Completion: 2011

项目地点：中国黑龙江省哈尔滨市
设计单位：哈尔滨工业大学建筑设计院
总建筑面积：5 195 m^2
竣工时间：2011 年

Overview

The project land is situated inside Harbin Sun Island State Guesthouse, facing Songhua River in the south, near the No.1 building of State Guesthouse with beautiful landscape around; it has a planning area of 18,000 m^2 and a total construction area of about 5,195 m^2.

项目概况

本工程用地位于哈尔滨市太阳岛国宾馆内，南临松花江，位于国宾馆一号楼附近，周边景观优美，规划总占地 18 000 m^2，总建筑面积约 5 195 m^2。

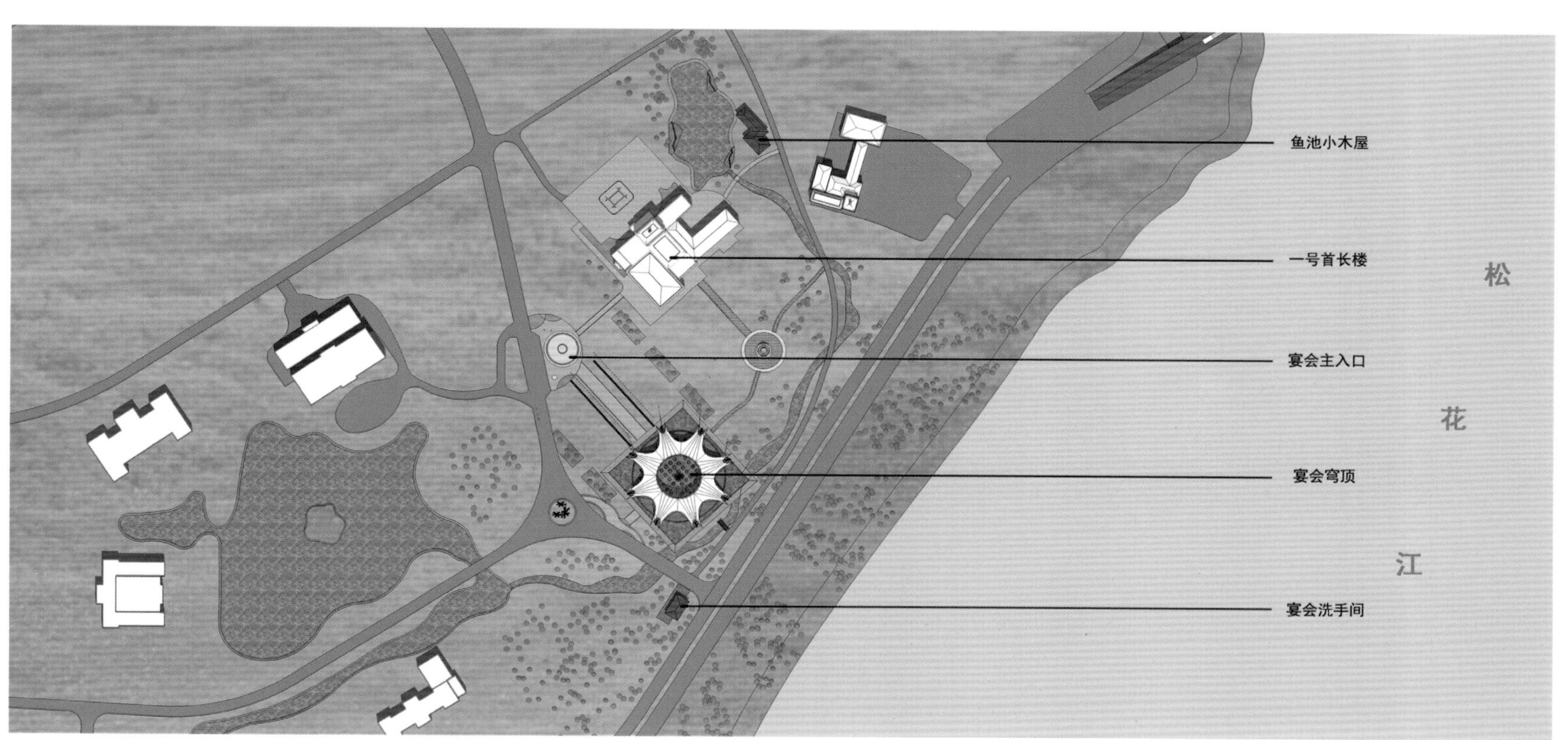

Site Plan 总平面图

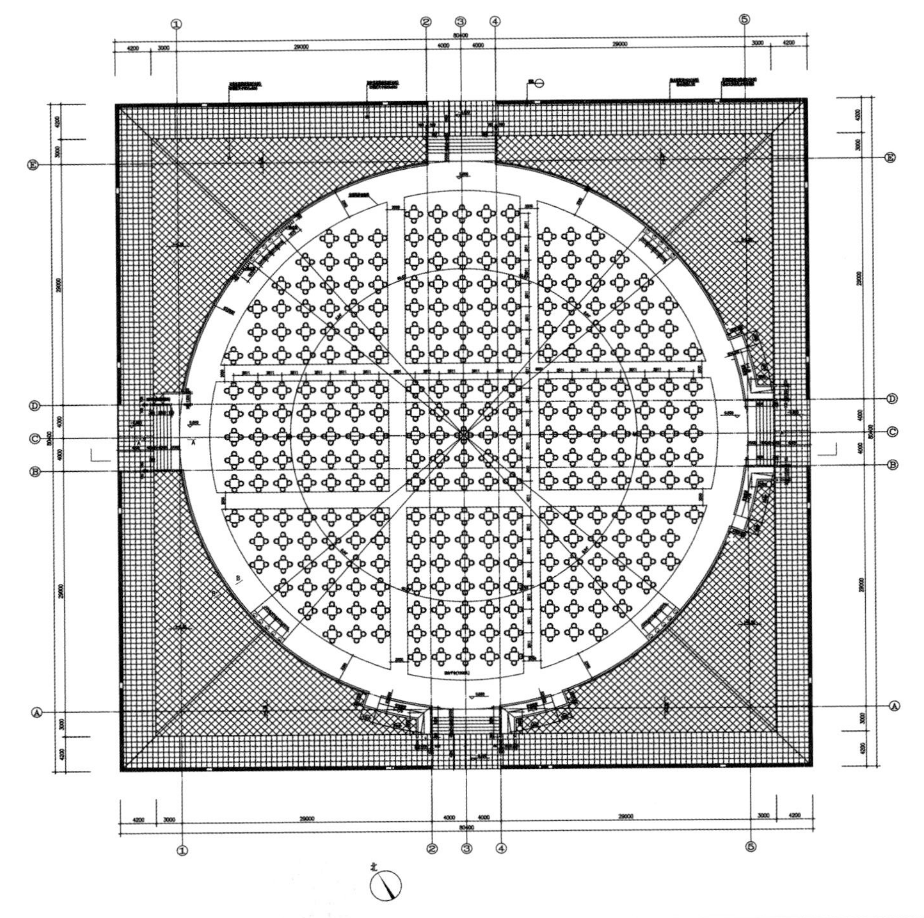

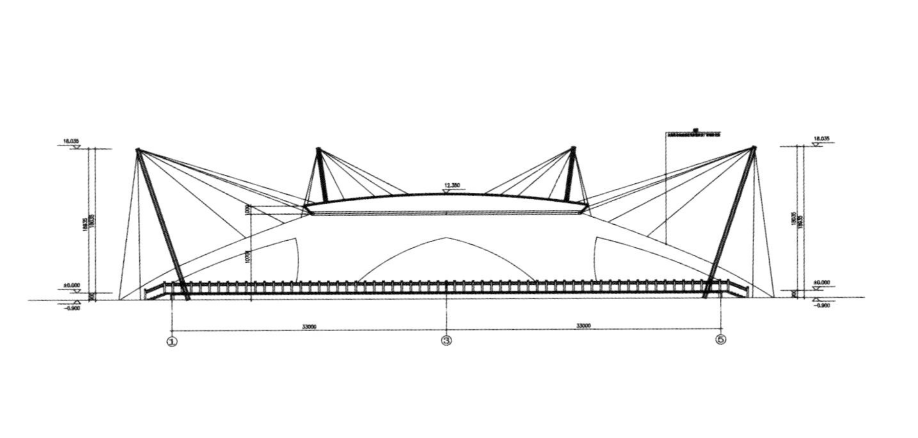

剖面图　1：800

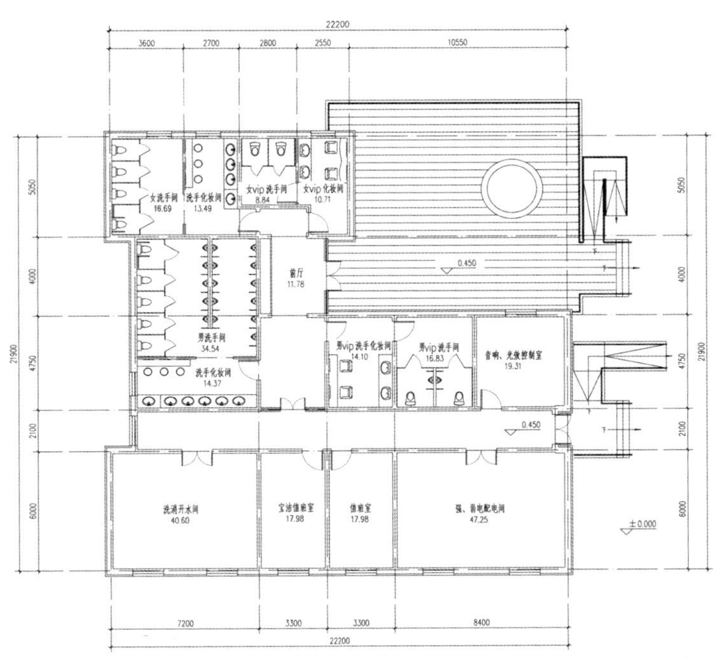

宴会平台辅楼平面图　1：400

Architectural Design

The construction target of Harbin International Exchange Platform is positioned to meet the chief leader reception function of China Harbin International Economic and Trade Fair, demonstrating Harbin city characteristics and the reality of Harbin's rapid expansion and development. The project design adopts the combination strategy of European classic symbol of dome and the membrane structure, which not only shows the Harbin region culture characteristic, but also well meets the functional requirements for the steady image of the building. And the surrounding vegetation makes the building more close to the earth, look more sedate and profound and integrate more harmoniously into the environment without being too abrupt, which embodies peace and noble beauty of the building.

建筑设计

哈尔滨市国际交流平台建筑目标定位为能够满足中国哈尔滨国际经济贸易洽谈会（哈洽会）首长接待功能，展示哈尔滨城市特色并彰显哈尔滨城市快速发展崛起的现实。设计采用欧式经典标志的穹顶与膜结构相结合的策略，用欧式建筑中的穹顶为形象基础，既展示了哈尔滨的地域文化特色，也很好地满足了功能对建筑庄重、沉稳的形象要求。而周边的植被使得建筑更亲近大地，更加稳重、大方，更能融入环境而不显突兀，体现了建筑的宁静、高贵之美。

Commercial Building
商业建筑

Fashionable Appearance
外观时尚

Luxury Facade
立面豪华

Mixed Use
功能多样

Leisure Atmosphere
休闲氛围

KEY WORDS
关键词

Complete Function 功能齐全

Unique Shape 造型独特

Graceful Appearance 外观优美

Chengdu Suning Plaza
成都苏宁广场

FEATURES
项目亮点

The buildings are centrally arranged on the flat and square site, and surrounded by circular passages which are not only used as fire lanes, but also outdoor parking lot. Entrance plaza is placed near the main entrance of the building to meet the requirement of evacuation and commercial activities.

项目基地地块方正、平整，在充分利用基地覆盖率的基础上，将建筑集中布置在基地中央，四周留有环形通道，既是消防车道，又能满足室外场地停车的需求。

Location: Chengdu, Sichuan, China
Architectural Design: Nanjing Yangtze River Urban Architectural Design Co., Ltd.
Developer: Chengdu Hongye Real Estate Co., Ltd.

项目地点：中国四川省成都市
设计单位：南京长江都市建筑设计股份有限公司
开发商：成都鸿业置业有限公司

Overview

This project covers an area of 26,736 m^2, and the base area of the building is 16,041 m^2, plot ratio is 3.5, total floor area is 125,355 m^2, in which the underground floor area is 31,769 m^2 and the overground floor area is 93,566 m^2. The 31.15 m high project has two floors underground and six floors overground. It was designed to be a recreational plaza for the middle-class families in Chengdu urban area.

项目概况

本项目用地面积 26 736 m^2，建筑基底面积 16 041 m^2，容积率为 3.5，总建筑面积 125 335 m^2，其中地下建筑面积 31 769 m^2，地上建筑面积 93 566 m^2。项目地下二层，地上六层，建筑高度 39.15 m。项目整体定位为满足成都市区中产家庭消费为主的休闲娱乐广场。

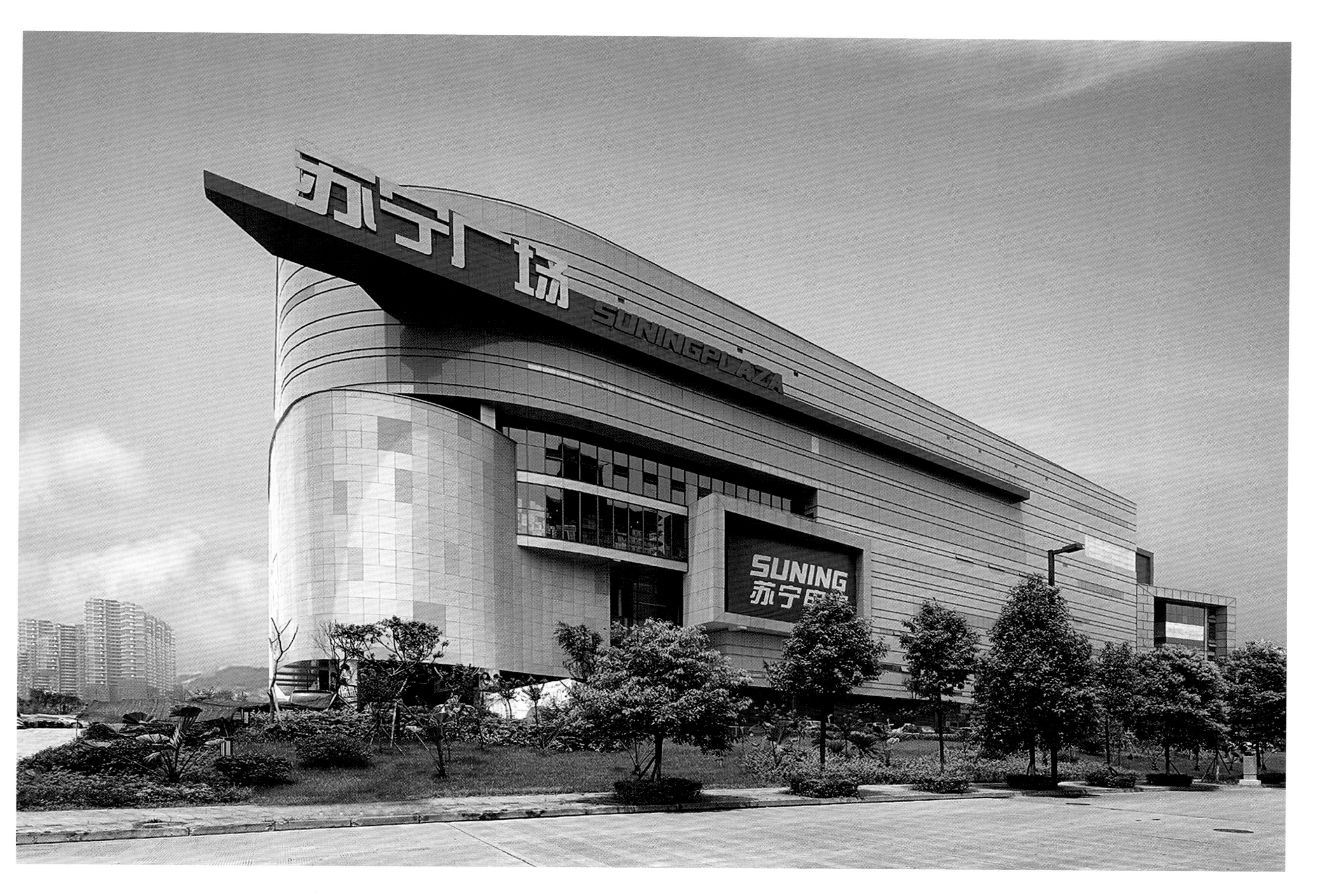

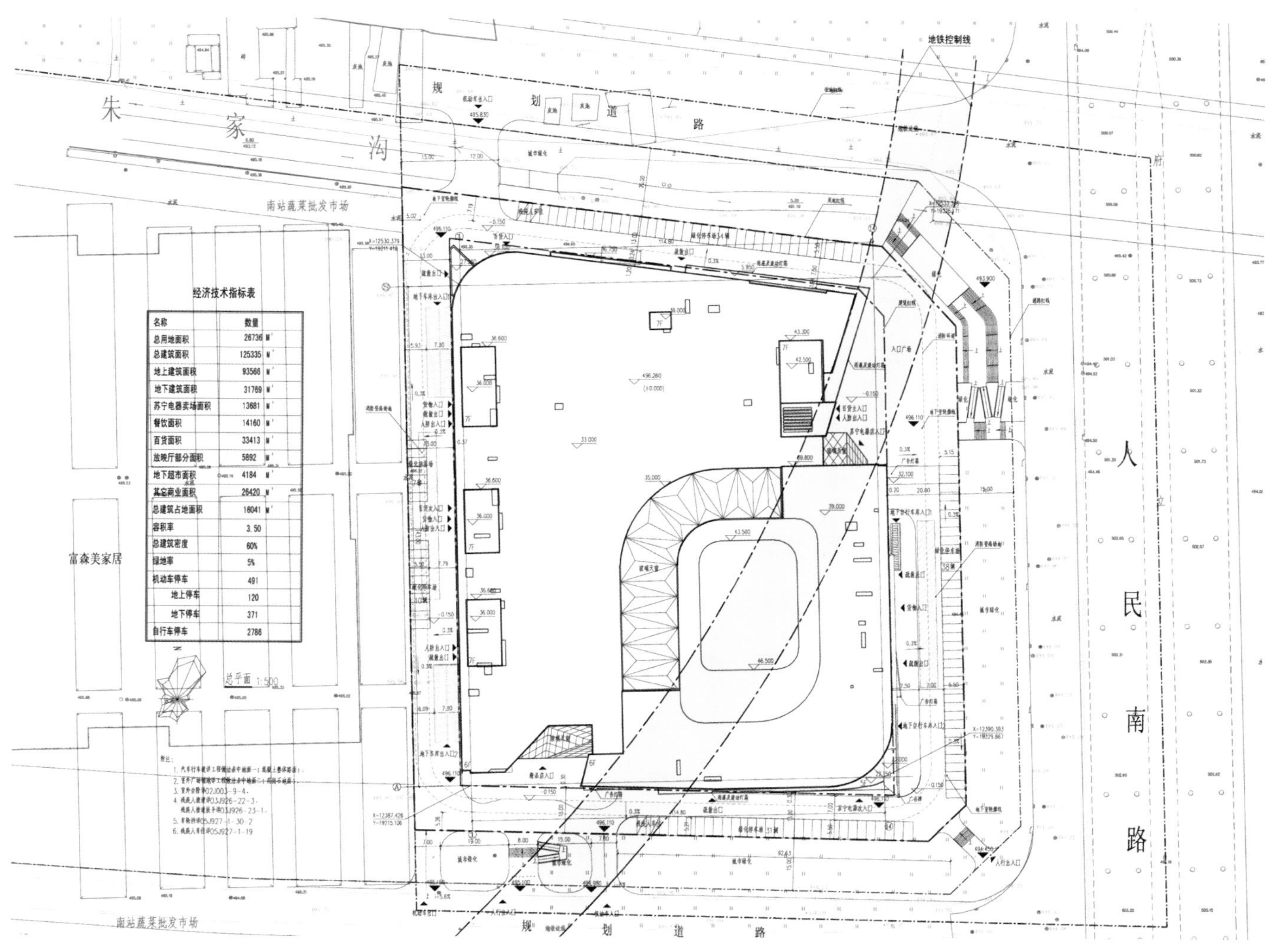

经济技术指标表

名称	数量
总用地面积	26736 M'
总建筑面积	125335 M'
地上建筑面积	93566 M'
地下建筑面积	31769 M'
苏宁电器卖场面积	13681 M'
餐饮面积	14160 M'
百货面积	33413 M'
放映厅部分面积	5892 M'
地下超市面积	4184 M'
其它商业面积	26420 M'
总建筑占地面积	16041 M'
容积率	3.50
总建筑密度	60%
绿地率	5%
机动车停车	491
地上停车	120
地下停车	371
自行车停车	2786

Site Plan 总平面图

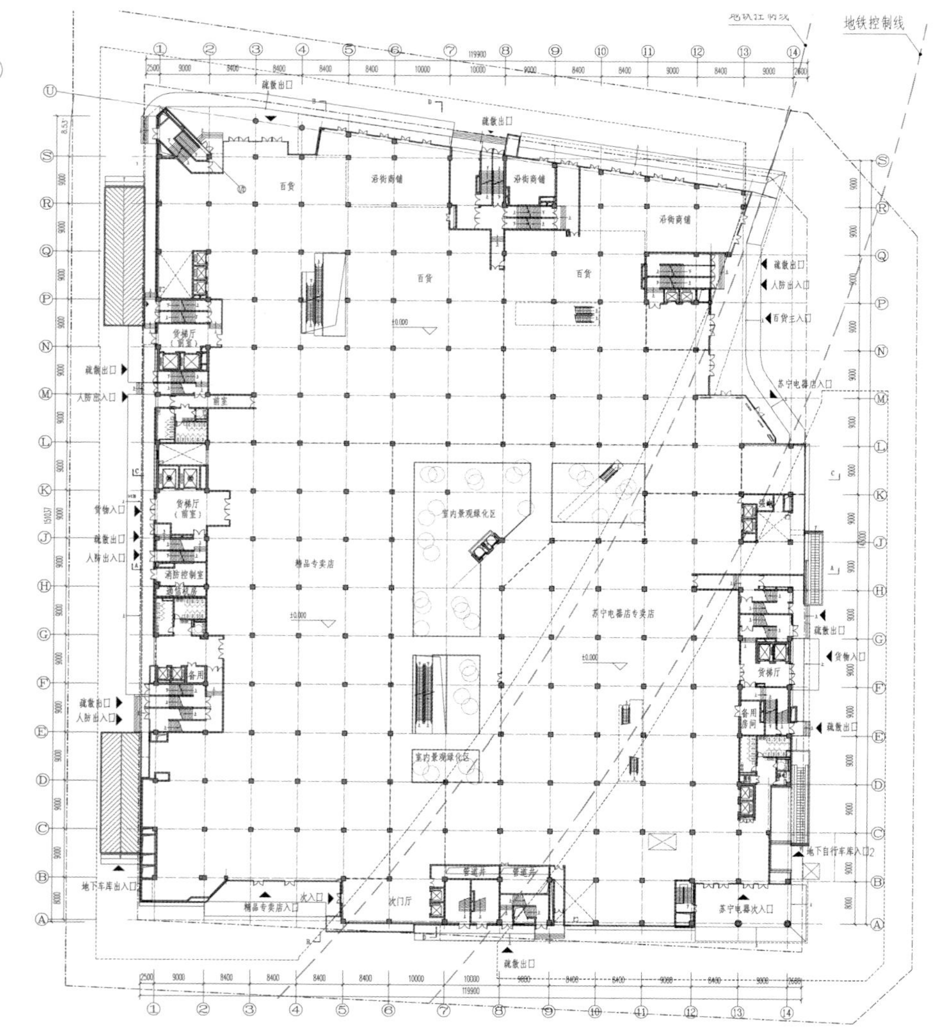

First Floor Plan 一层平面图

Planning

The buildings are centrally arranged on the flat and square site, and surrounded by circular passages which are not only used as fire lanes, but also outdoor parking lot. Entrance plaza is placed near the main entrance of the building to meet the requirement of evacuation and commercial activities. The surrounding roads have different altitudes: the west is higher than the east, and the site shares a height difference of 2 m with the road on the east side. As for planning, vehicle entrances are arranged in north-south direction on the west side of the site and a 15 m wide greenbelt is situated between the east side, the south side and the city road. In addition, outdoor step ramp is utilized to connect the site to the main entrance direction with the city road.

规划布局

项目基地地块方正、平整，在充分利用基地覆盖率的基础上，将建筑集中布置在基地中央，四周留有环形通道，既是消防车道，又能满足室外场地停车的需求。建筑的主要入口附近均设计有入口广场，可以满足人员疏散及商业活动的要求。基地周边道路整体标高西高东低，基地和东侧道路之间约有 2 m 高差。在规划设计上，将机动车出入口布置在基地西侧的南北两个方向，建筑东侧、南侧和城市道路之间有 15 m 宽的城市绿化带，在建筑主要入口的方向，通过室外台阶坡道等将基地和城市道路连接起来。

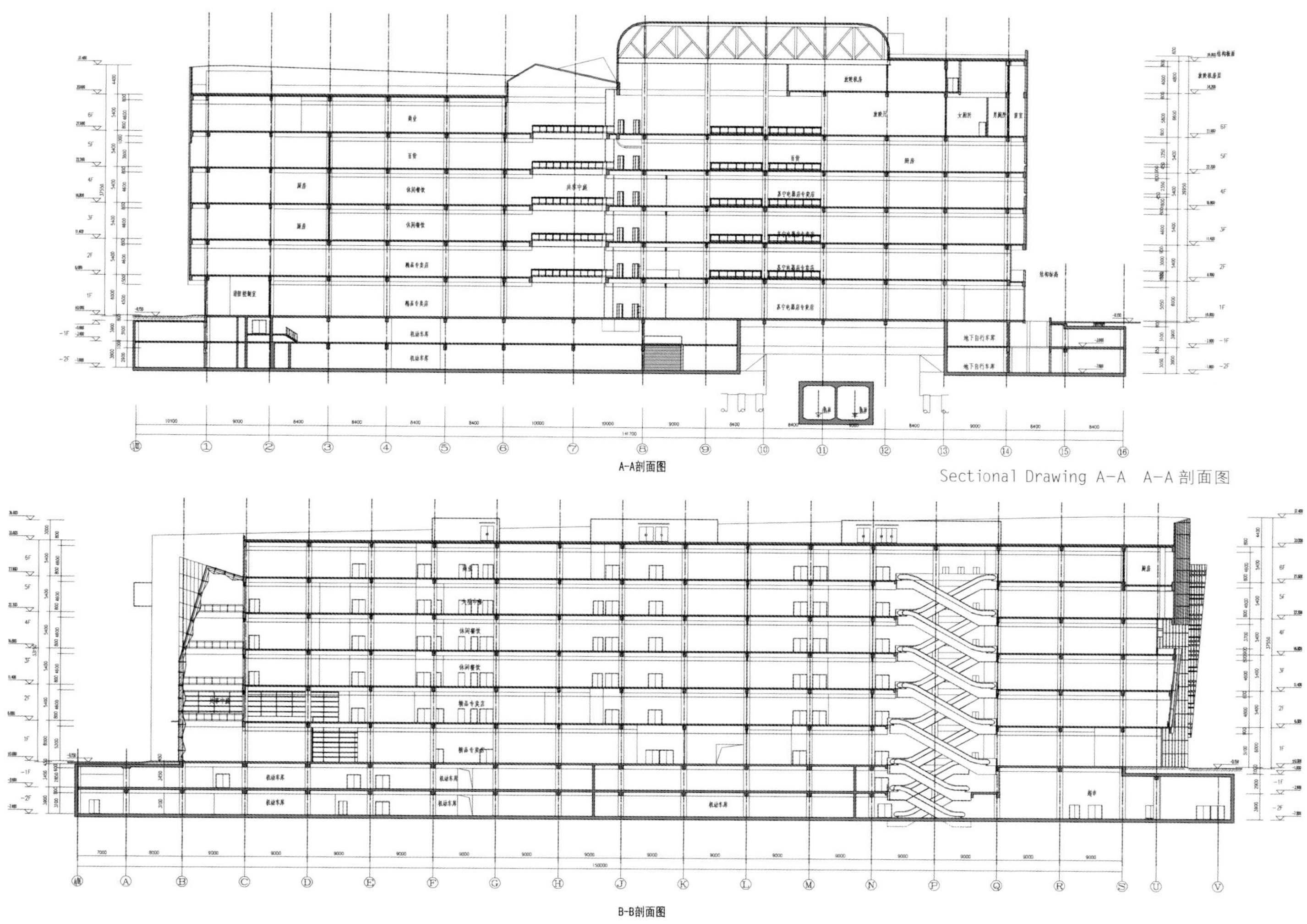

Sectional Drawing A-A A-A 剖面图

Sectional Drawing B-B B-B 剖面图

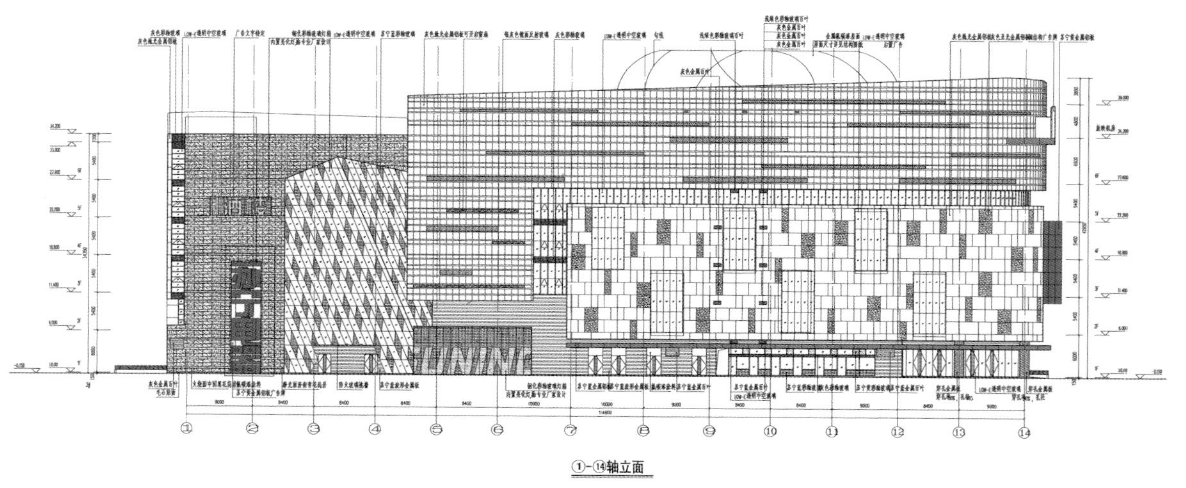

①-⑭轴立面

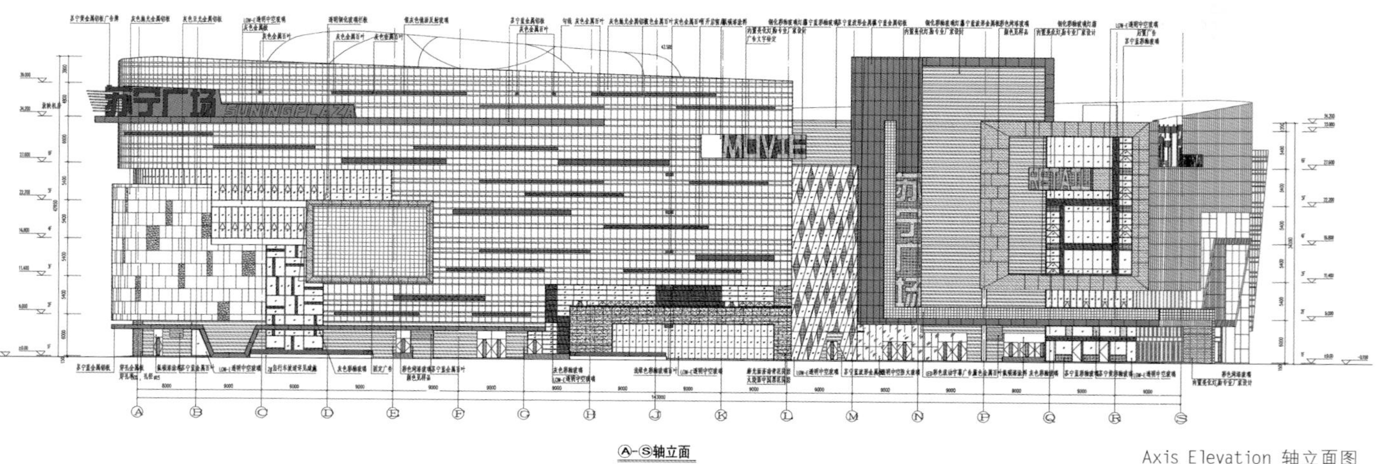

Ⓐ-Ⓢ轴立面

Axis Elevation 轴立面图

SUNING PLAZA

享美食
首站，
广阔儿童天地 复苏城南欢乐童趣梦想

KEY WORDS
关键词

Unique Design 造型新颖

Simple Shape 形体简洁

Abundant Details 细部丰富

B06 Plot Project in Donggang District (Wanda Center)

东港区 B06 地块项目（万达中心）

FEATURES
项目亮点

Wanda Center owns unique and modern spatial shape. Viewing it from far away, two tower buildings stand upright in this urban space, and the light and the shadow create the "X" shape textures of the facade wall makes the buildings to be one of the landmark constructions in Donggang CBD.

万达中心具有新颖、现代的空间造型。远观，两座塔楼一高一矮矗立在城市空间里，立面幕墙 "X" 形肌理的光与影，使建筑宛如破土而出的竹笋，成为东港商务区的标志性建筑之一。

Location: Dalian, Liaoning, China
Architectural Design: Dalian Architectural Design and Research Institute Co.,Ltd
Total Land Area: 23,200 m^2
Total Floor Area: 208,754 m^2
Plot Ratio: 7.05

项目地点：中国辽宁省大连市
设计单位：大连市建筑设计研究院有限公司
总用地面积：23 200 m^2
总建筑面积：208 754 m^2
容积率：7.05

Overview

Wanda Center, the international 5A office building and five star and six star hotel, sits in the core location of Donggan CBD in Zhongshan Distrit, Dalian city, which is next to Dalian International conference center and an international cruise terminal that are under planning. This project is a complex skyscraper constituted by a hotel, 149.5 m height and have 36 floors, an office building, 202.4 m height and 44 floors, together with a 4-floor podium and a 3-floor basement.

项目概况

万达中心位于大连市中山区东港商务区核心位置，紧邻大连国际会议中心和规划中的国际邮轮码头，为国际 5A 级写字楼及五星级、六星级酒店。项目是由 149.5 m 高、36 层的酒店和 202.4 m 高、44 层的写字楼以及 4 层裙房、3 层地下室组成的一个复合超高层建筑。

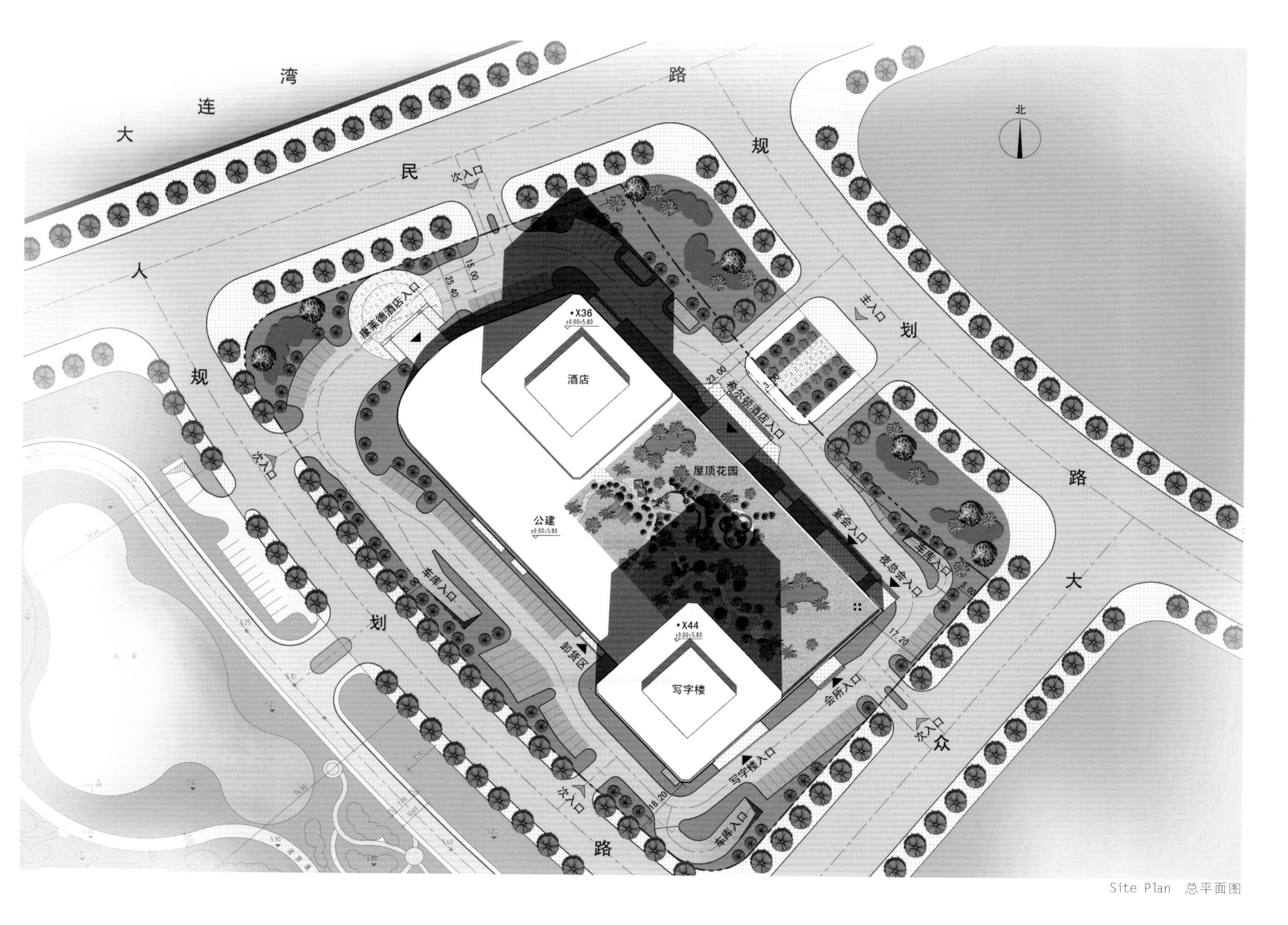

Site Plan 总平面图

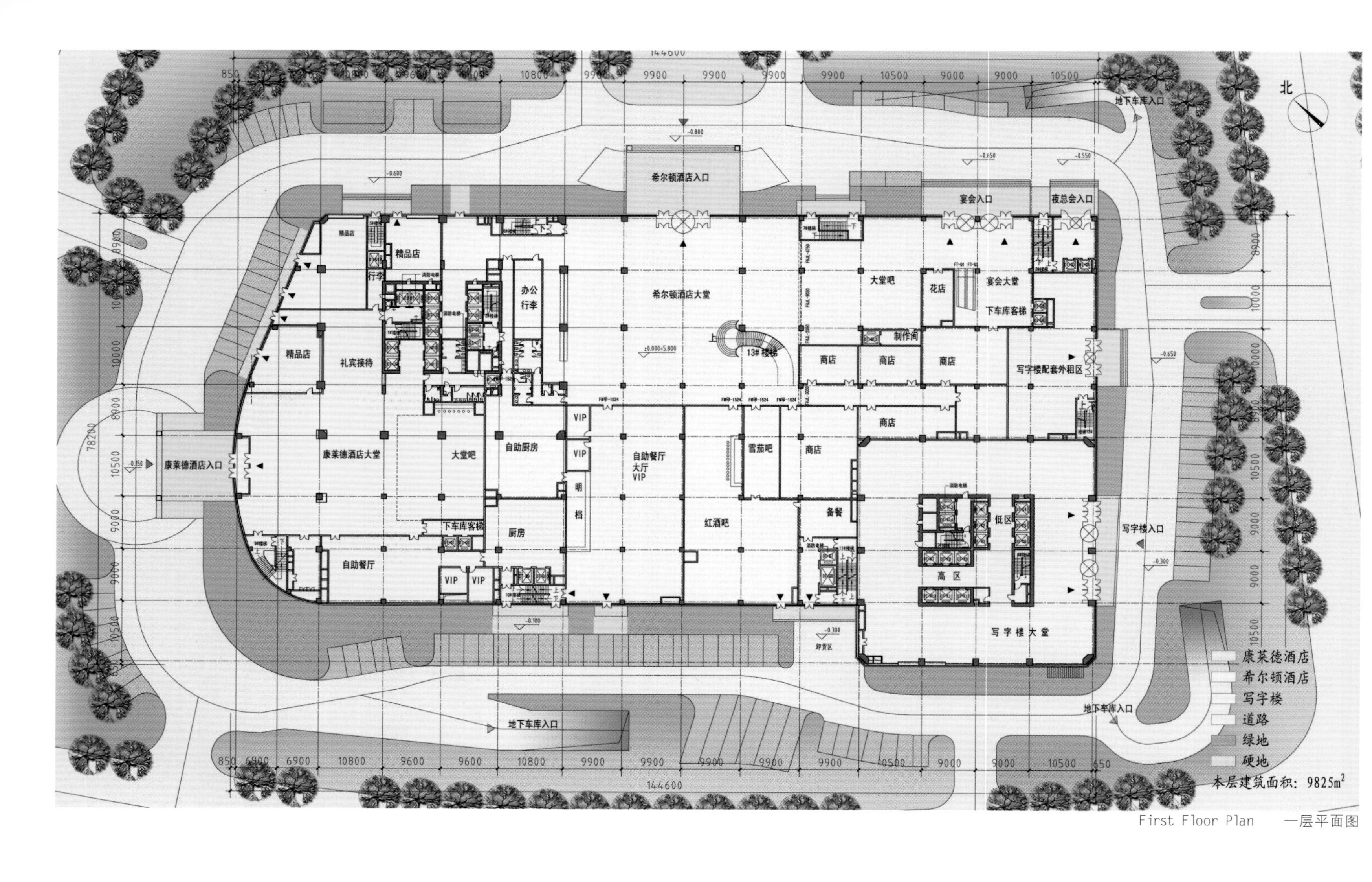

First Floor Plan　一层平面图

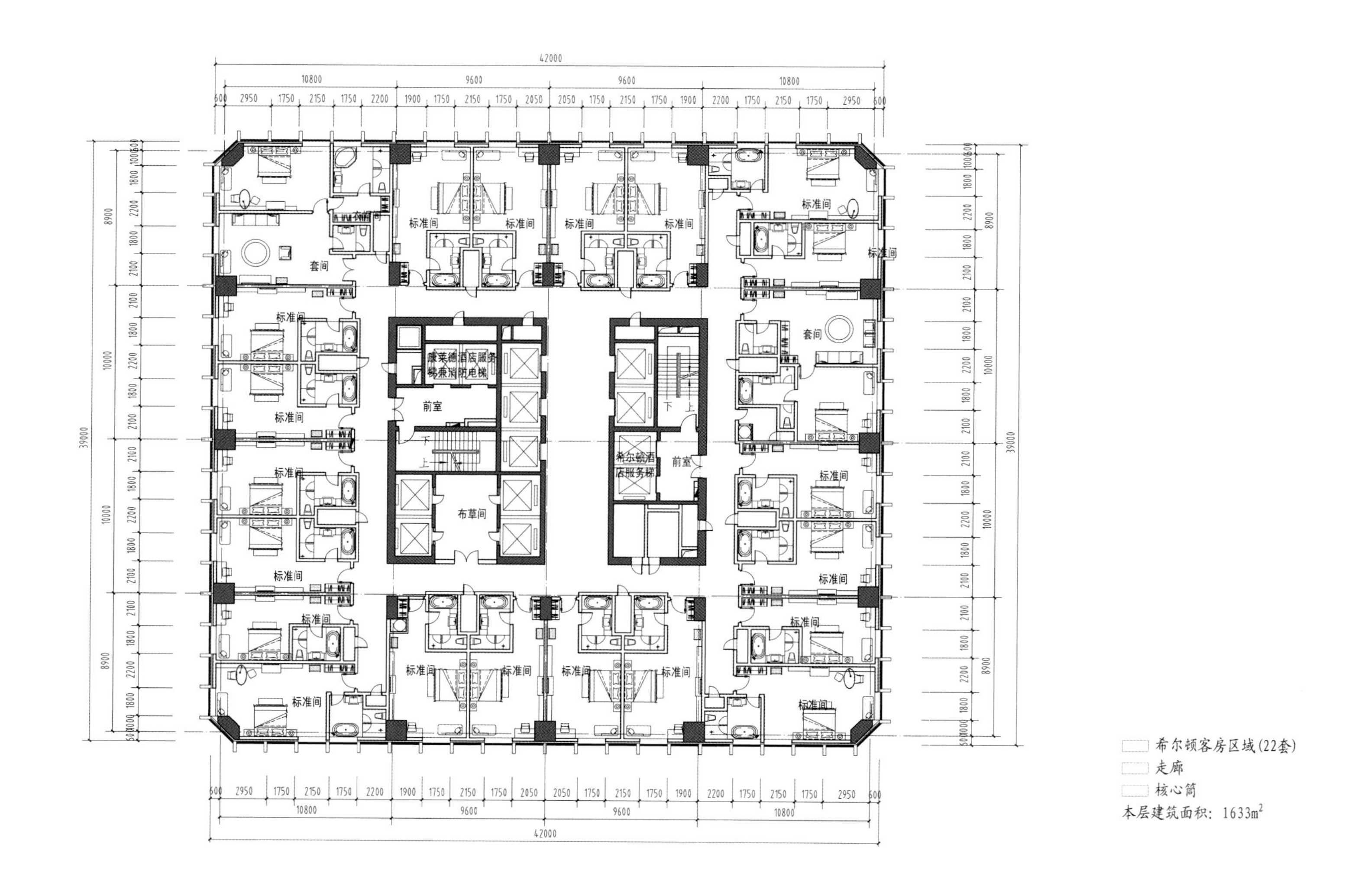

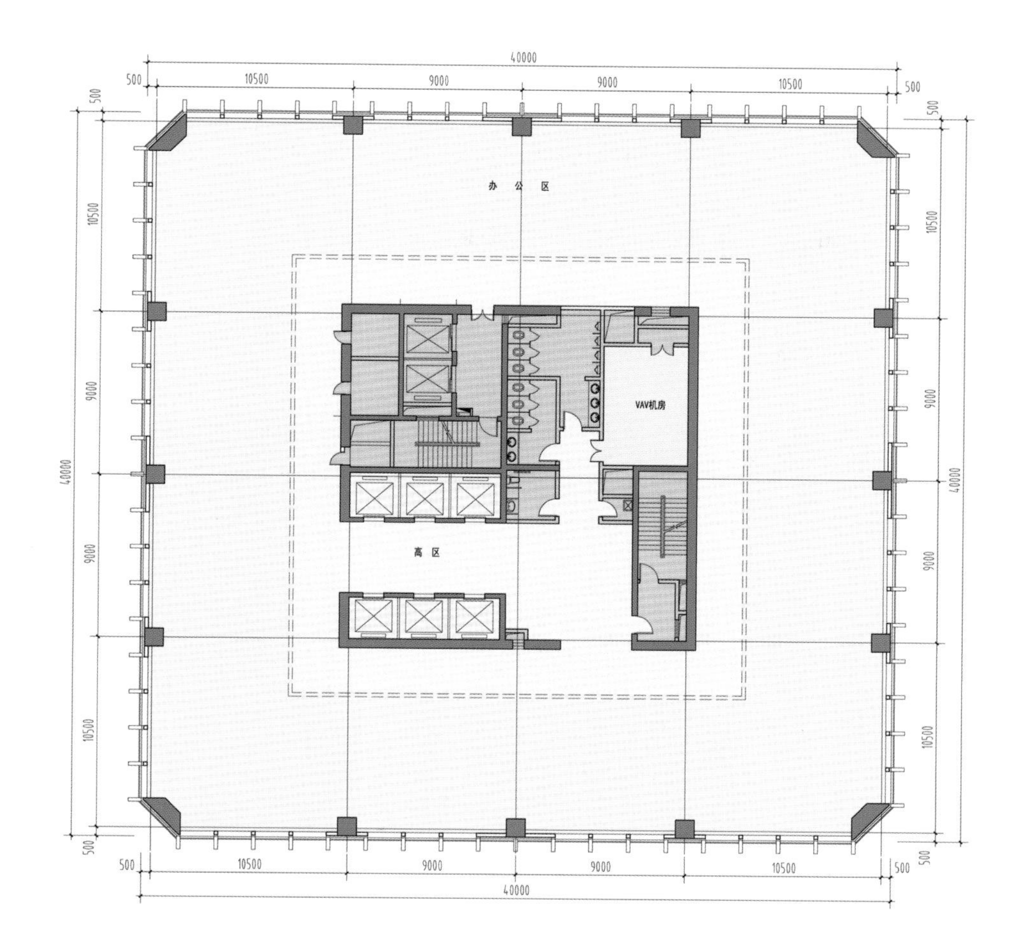
40000
500
10500
9000
9000
10500
办 公 区
VAV机房
高 区

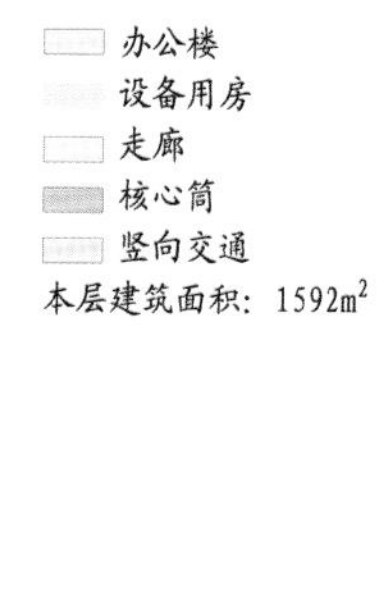
办公楼
设备用房
走廊
核心筒
竖向交通
本层建筑面积：1592m²

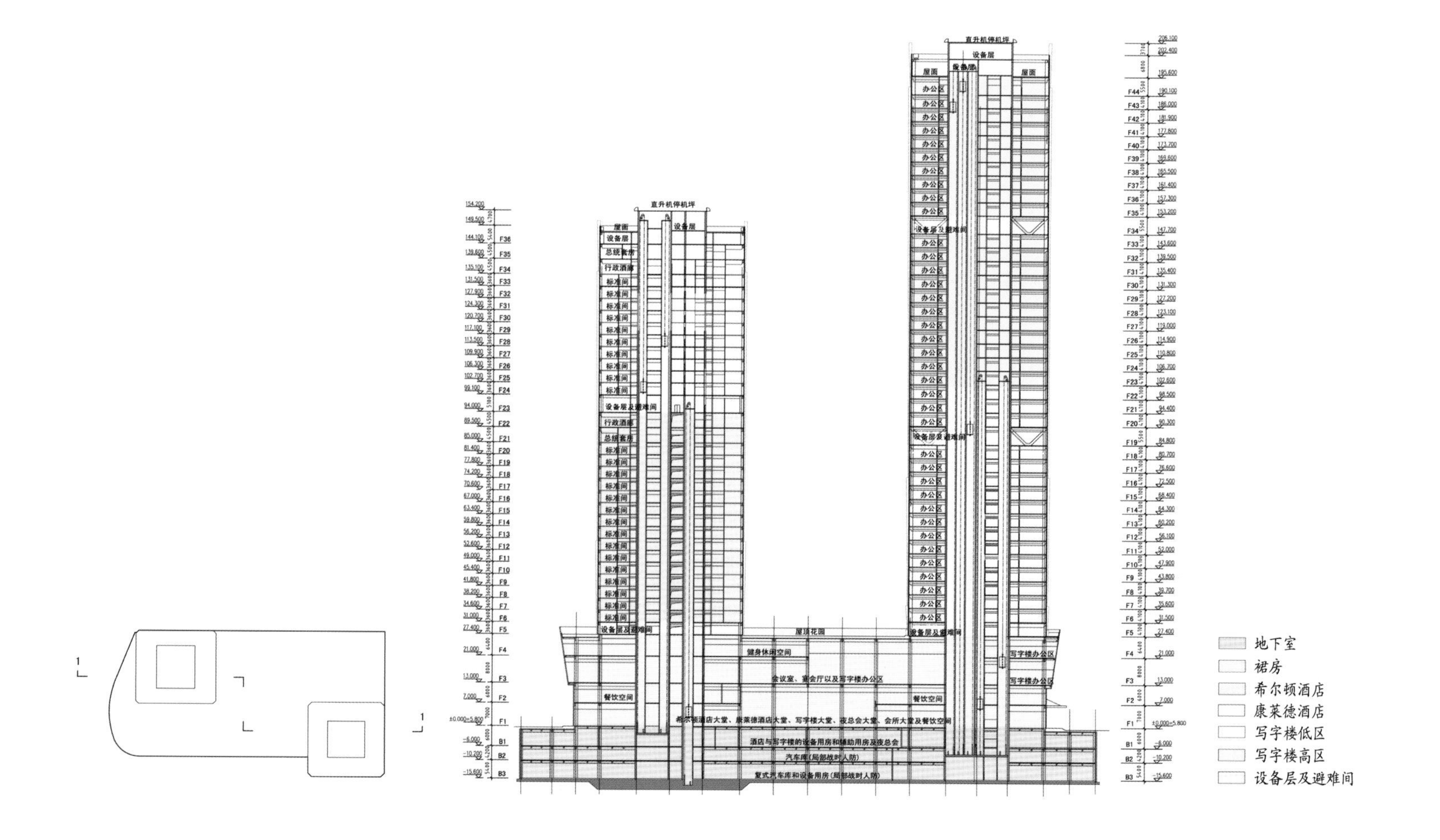
直升机停机坪
设备层
屋面
总统套房
行政酒廊
标准间
设备层及避难间
办公区
屋顶花园
健身休闲空间
会议室、宴会厅以及写字楼办公区
餐饮空间
写字楼办公区
汽车库(局部战时人防)
地下室
裙房
希尔顿酒店
康莱德酒店
写字楼低区
写字楼高区
设备层及避难间

CONRAD

CONRAD
大连康莱德酒店

Hilton

KEY WORDS
关键词

Unified Style 风格统一
Harmonious Facade 立面和谐
Commercial Atmosphere 商业氛围

ASEAN International Business Zone Korean Garden
东盟国际商务区韩国园区

FEATURES
项目亮点

The overall plan of this site revolved around the pattern of the flag of South Korea (Taegukgi). Various independent yet connected exterior open spaces are shaped via the single building. Footpath is arranged along the south-north axis which runs through the elevated bottom of residences. Large open space is situated in the center that acts as the community center as well.

地块总体规划以韩国太极旗格局为布置基础。以建筑单体形塑各个独立又相连的外部开放空间，中轴以人行步道及住宅底部架空南北连贯，中心布置大型开放空间，形成小区生活的中心。

Location: Nanning, Guangxi, China
Architectural Design: Guangxi Hualan Group
Total Land Area: 39,365.65 m²
Total Floor Area: 61,840 m²
Overground Floor Area: 3,137.5 m²
Underground Floor Area: 13,714.28 m²
Plot Ratio: 1.169

项目地点：中国广西壮族自治区南宁市
设计单位：广西华蓝设计（集团）有限公司
总用地面积：39 356.65 m²
总建筑面积：61 840 m²
地上建筑面积：3 137.5 m²
地下建筑面积：13 714.28 m²
容积率：1.169

Overview

The overall plan of this site revolved around the pattern of the flag of South Korea (Taegukgi). Various independent yet connected exterior open spaces are shaped via the single building. Footpath is arranged along the south-north axis which runs through the elevated bottom of residences. Large open space is situated in the center that acts as the community center as well. The liaison office is established at the intersection of Fengling Interchange and Zhujing Road to highlight the importance and representativeness of this building. Commercial and office area runs along Minzu Road, the former is on the east side and the latter on the west. Independent entrance is at the northwest corner, interplays with commercial atmosphere along Minzu Road and extends to the west side, which lasts the commercial activities on Zhujin Road and provides convenience for the residents.

项目概况

地块总体规划以韩国太极旗格局为布置基础。以建筑单体形塑各个独立又相连的外部开放空间，中轴以人行步道及住宅底部架空南北连贯，中心布置大型开放空间，形成小区生活的中心。向东于风岭立交及朱槿路交汇处布置联络处，凸显建筑的重要与代表性。沿民族大道设立商业和办公，中轴东面为办公，西面为商业。在西北角设独立出入口营造民族大道沿街的商业氛围，并延伸至地块西侧，既延续了朱槿路的商业活动，更由于紧邻地块住宅小区主入口，为小区居民生活提供了便利。

Architectural Design

The commercial building is varied in heights on two sides. The one-storey part on the west side faces Malay Garden and there is a pedestrian street scene in appropriate scale between them. The two-storey part on the north side shapes an interesting commercial atmosphere via the concave-convex changes on the architectural form. The scale of 8-10 m unit ensures the commercial space is utilized flexibly.

建筑设计

建筑平面设计上，商业楼为一至二层。西侧一层商业和马来园间留设尺度合宜的步行街景色，北侧凸显二层商业楼建筑形体上的凹凸变化，营造丰富、有趣的商业氛围。商业以 8 m 至 10 m 为单元，保持商业使用空间的使用弹性。

住宅区出入口

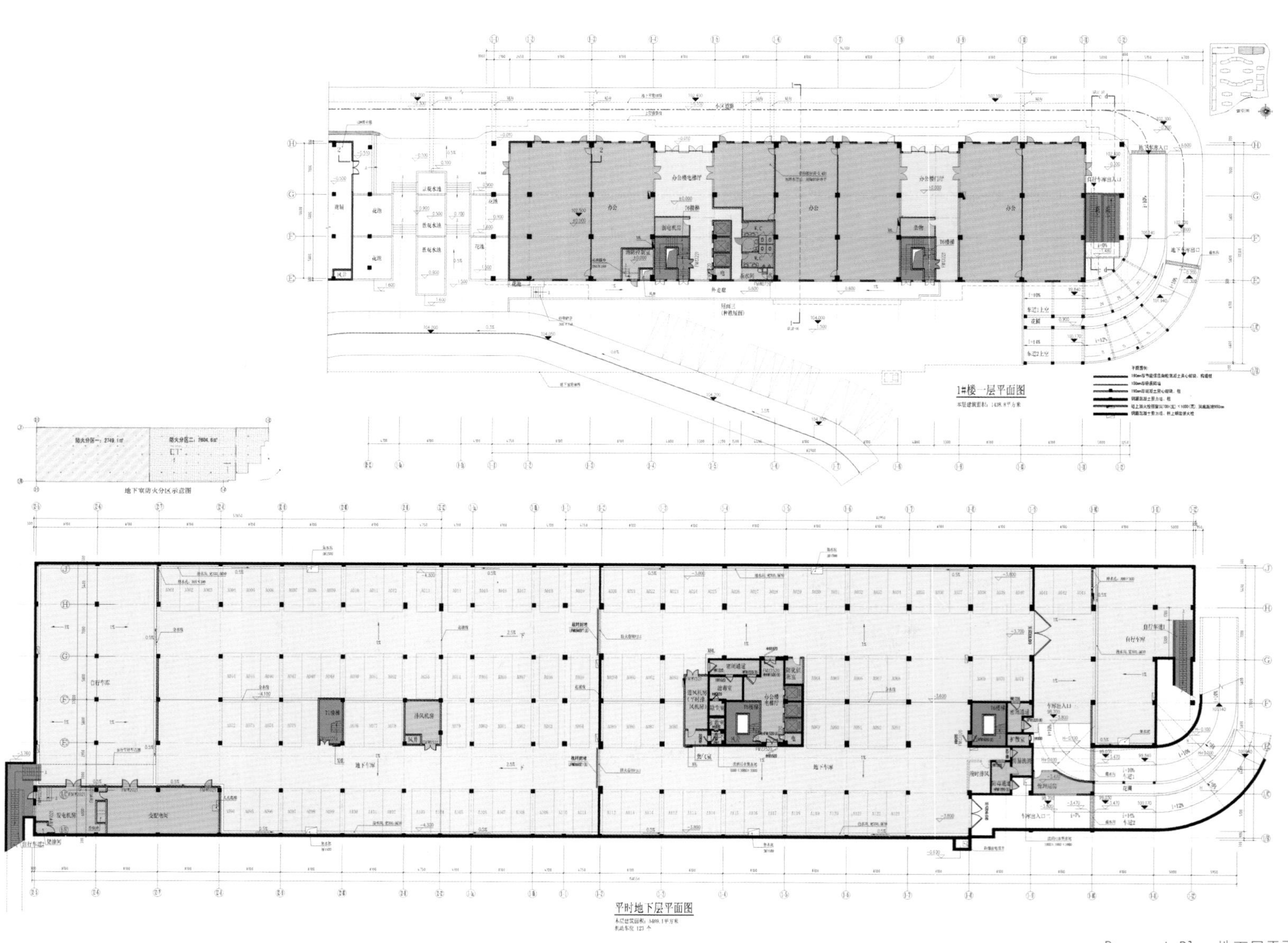

Basement Plan 地下层平面图

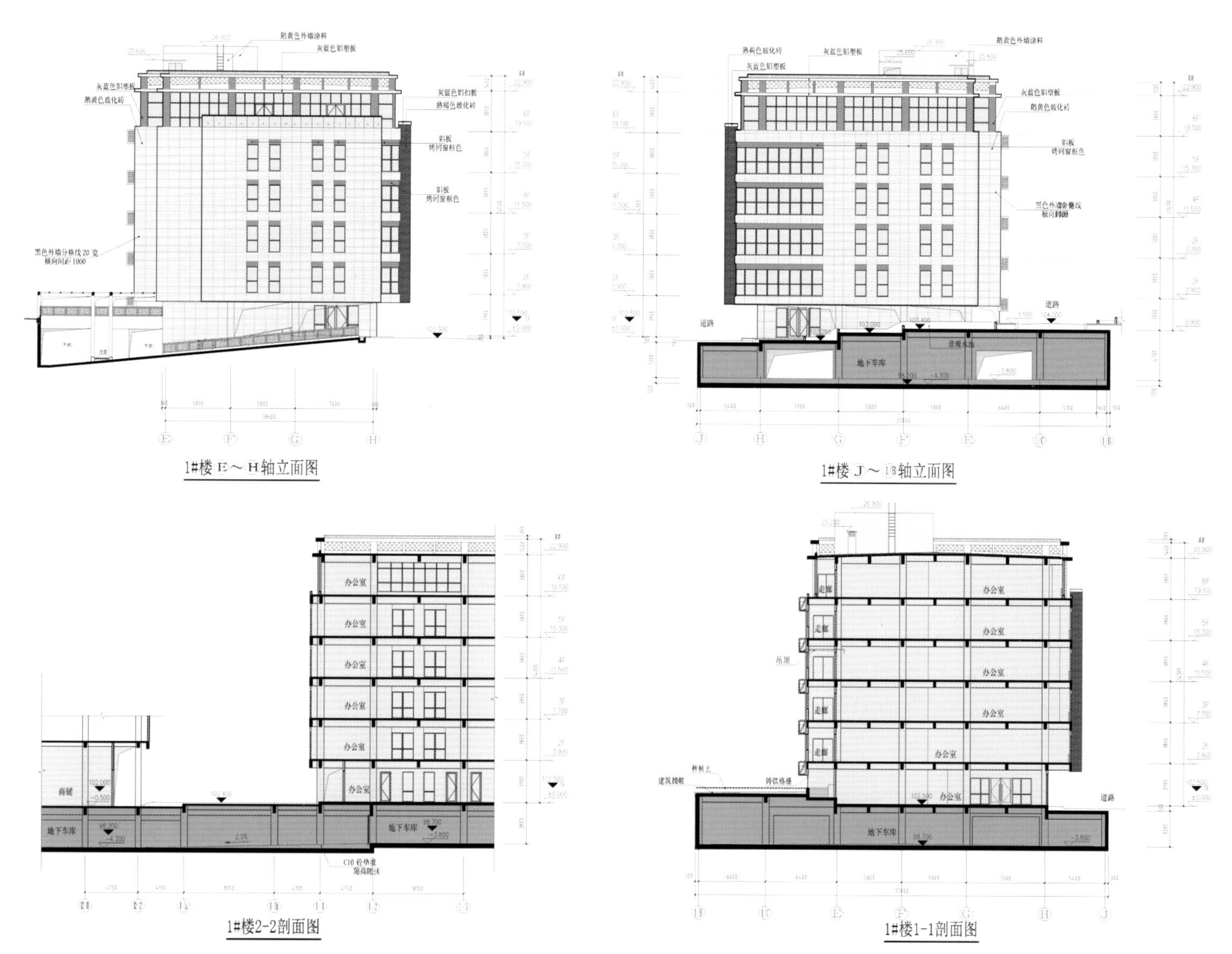

1#楼 E～H轴立面图

1#楼 J～1/B轴立面图

1#楼2-2剖面图

1#楼1-1剖面图

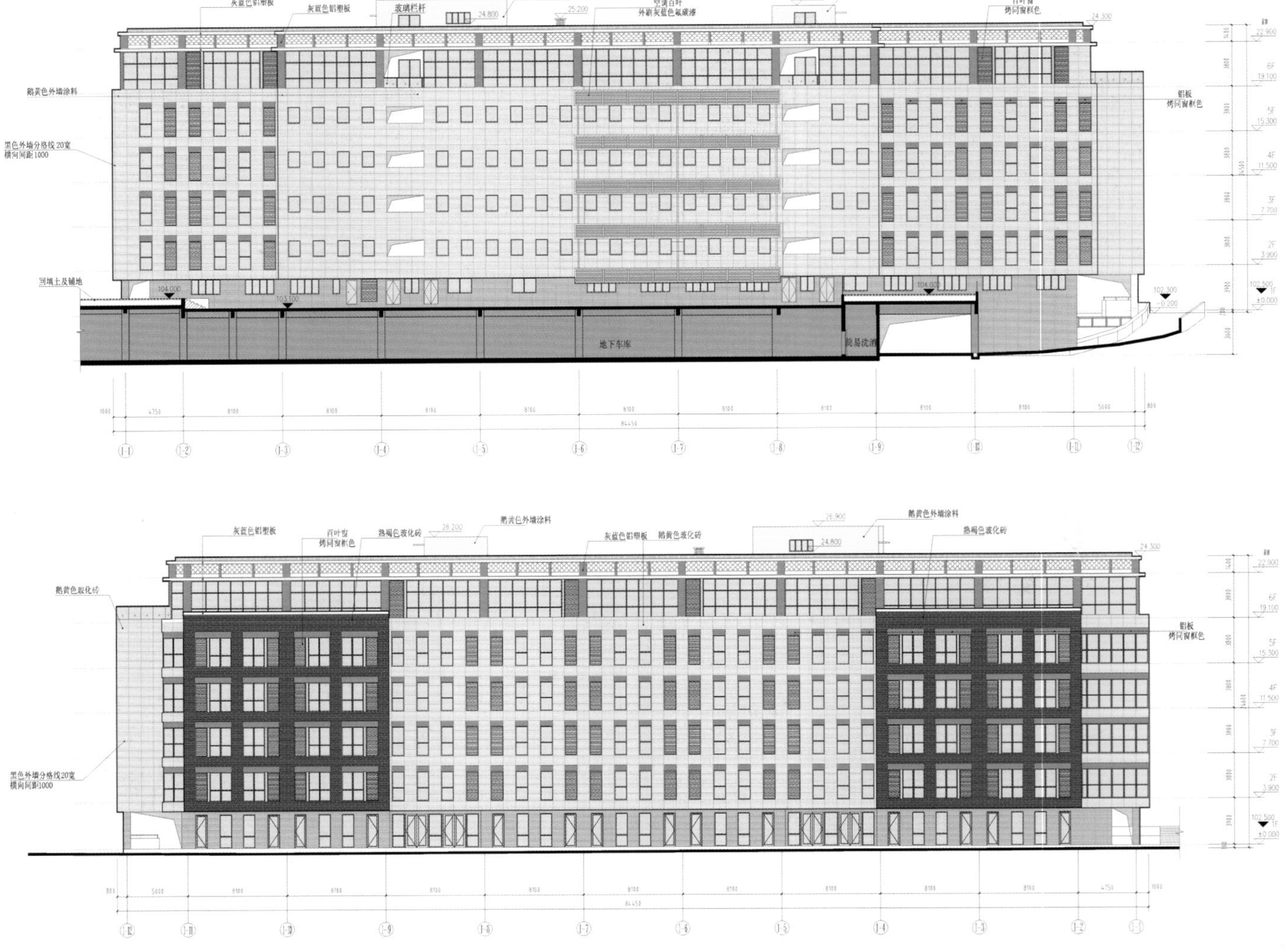

Axis Elevation 轴立面图

世洋御豪园
世洋
首尔街
世洋大厦
销售中心

KEY WORDS
关键词

Cultural Presentation 文化展示

Old Home Protection 旧居保护

Original Feature 原始风貌

Renovation of Plot B on Aomen West Road in Three Lanes & Seven Alleys, Fuzhou
福州三坊七巷澳门西路 B 地块保护更新设计

FEATURES
项目亮点

Cornice, a variety of roof lines and terraces, quadrangle courtyard in proper scale, warm-toned clincher boarding wall, adumbral louver and rain tent, all these materials reflect the keynote of a natural building, presenting vitality and charm.

挑檐和丰富的屋顶轮廓线、多样的露台，尺度宜人的四合院结合暖色调的“鱼鳞板”木墙面、具有遮阳功能的百叶和雨篷栅等，体现了亲近自然的建筑基调，赋予建筑生机与魅力。

Location: Fuzhou, Fujian, China
Architectural Design: An-design Architects
Total Land Area: 2,505 m^2
Total Floor Area: 5,360.6 m^2
Overground Floor Area: 3,505.6 m^2
Underground Floor Area: 1,855 m^2

项目地点：中国福建省福州市
设计单位：北京华清安地建筑设计事务所有限公司
总用地面积：2 505 m^2
总建筑面积：5 360.6 m^2
地上建筑面积：3 505.6 m^2
地下建筑面积：1 855 m^2

Overview

Conditioned on "protection, folk custom, integration, innovation", this building is 63 m long on the east-west direction and 40 m wide on the north-south direction. Located south of Antai River and playing the main role of presenting original feature, it shoulders the important task to present the architectural form along the main axis and the folk culture. Therefore, the characteristic local "CHAI LAN CUO" is arranged along the river and alleys on the east and west sides. In addition, designers did a lot of research and survey work for the replacement of Zhang Tianfu Residence in accordance with the principles of respecting history and coordinating protection and renovation, so as to preserve the original features and the profound cultural connotation.

项目概况

工程整体以“保护、民俗、整合、创新”为前提，建筑东西长约63 m，南北长约40 m。建筑北侧紧邻安泰河，为主要风貌控制面，担负着片区整个发展主轴的建筑形态及民俗文化展示的重任。故在沿河方向及东、西侧街巷以传统风貌建筑形式设计，即采用福州特有的临街建筑形式——“柴栏厝”，进行空间设计组合。融合张天福迁建旧居，张天福旧居为迁建保护项目，本着尊重历史、保护与更新相协调的原则，设计师对张天福旧址在搬迁前进行了详细的建筑测绘、记录与调研，并对迁建及修复方案进行深入、详细的设计，力求落成后的旧居保留原始的风貌特点及厚重的文化内涵。

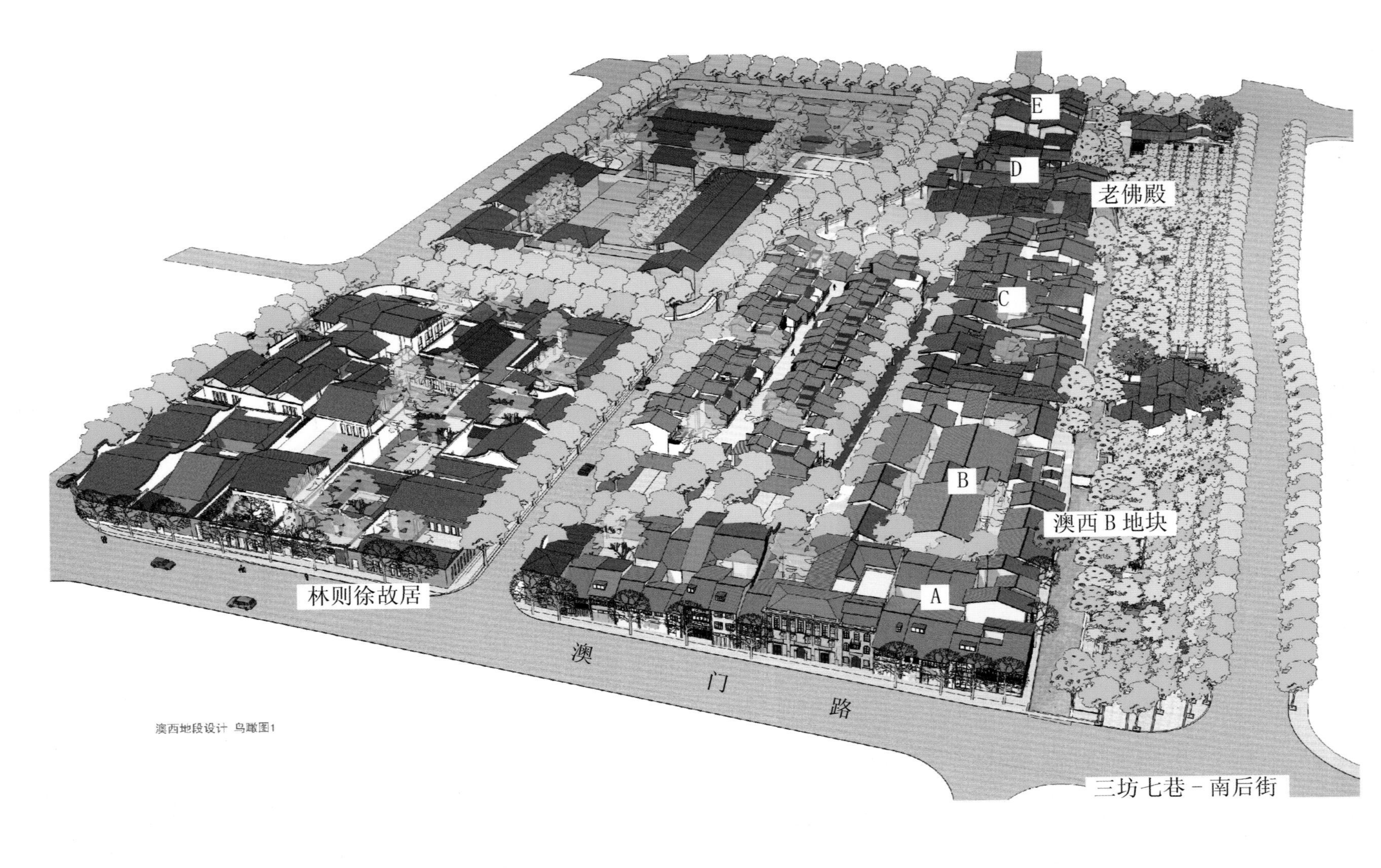

澳西地段设计 鸟瞰图1

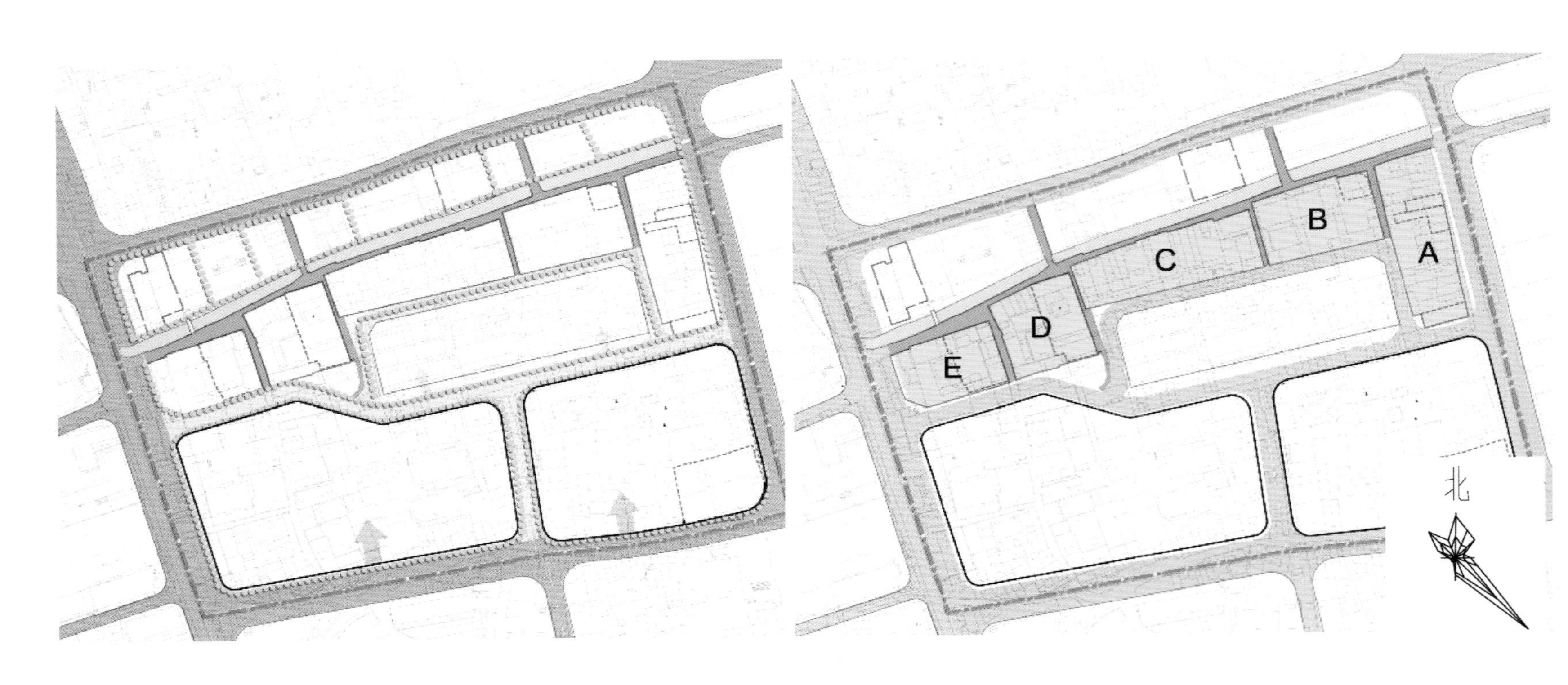

南立面实景图

Architectural Design

As renovation work in a historic district, it is more limited than the usual ones. Therefore, green courtyards are created as much as possible to guarantee a pleasant environment. Inner courtyards are well arranged respectively on the underground, first floor, second floor and the third floor to connect the interior and exterior and provide well-aligned space, which allows the owners taste traditional features better in a limited space. Cornice, a variety of roof lines and terraces, quadrangle courtyard in proper scale, warm-toned clincher boarding wall, adumbral louver and rain tent, all these materials reflect the keynote of a natural building, presenting vitality and charm.

建筑设计

作为历史街区更新类项目，受风貌协调限制，故在地块内尽可能多设置庭院绿化，以保证建筑环境优美、宜人。建筑空间运用连贯、韵律、组合的手法，分别在地下一层、地上一层、地上二层及三层设计了内庭院，作为联系内外空间、组织空间变化和保证各空间秩序的有效衔接，在有限的空间里，使业主更好地领悟传统空间特色。挑檐和丰富的屋顶轮廓线、多样的露台，尺度宜人的四合院结合暖色调的“鱼鳞板”木墙面、具有遮阳功能的百叶和雨篷栅等，体现了亲近自然的建筑基调，赋予建筑生机与魅力。

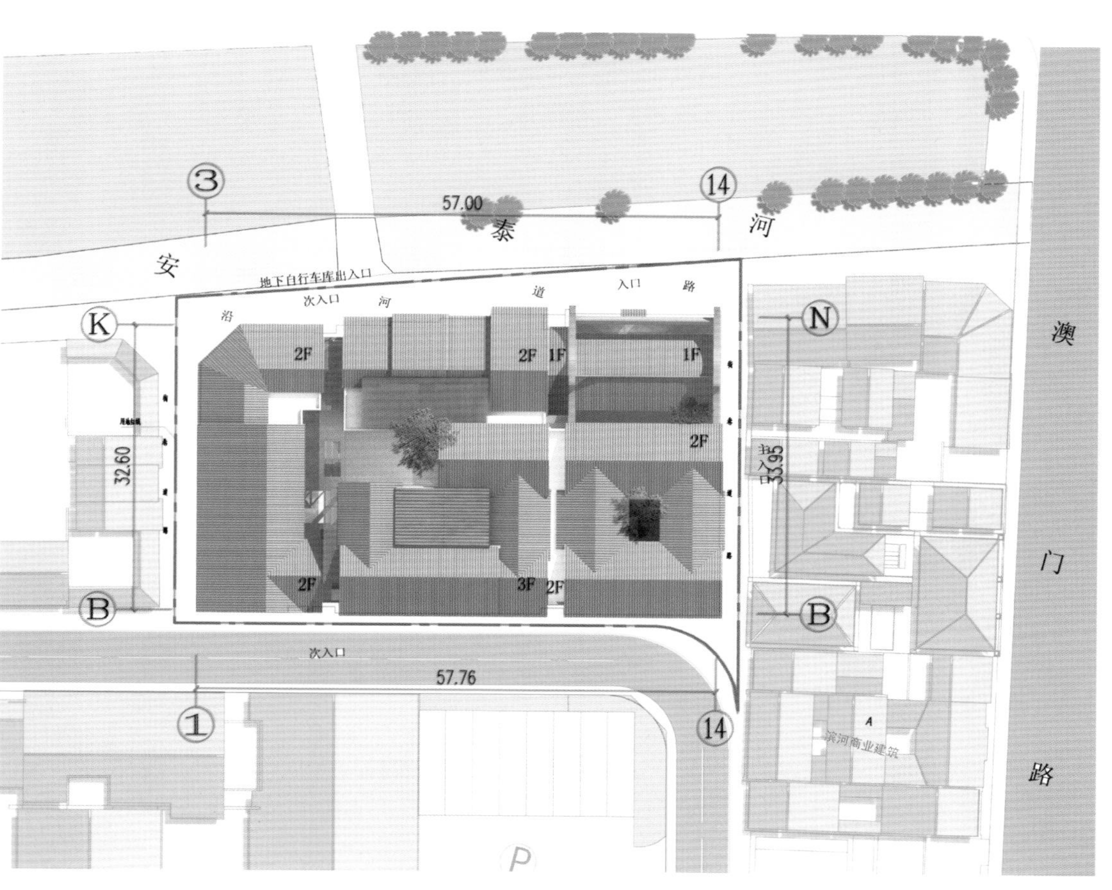

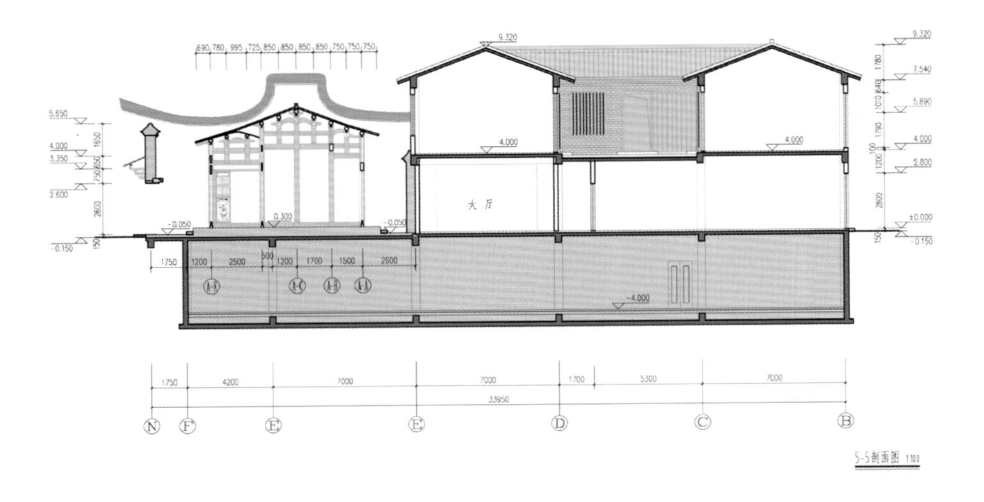

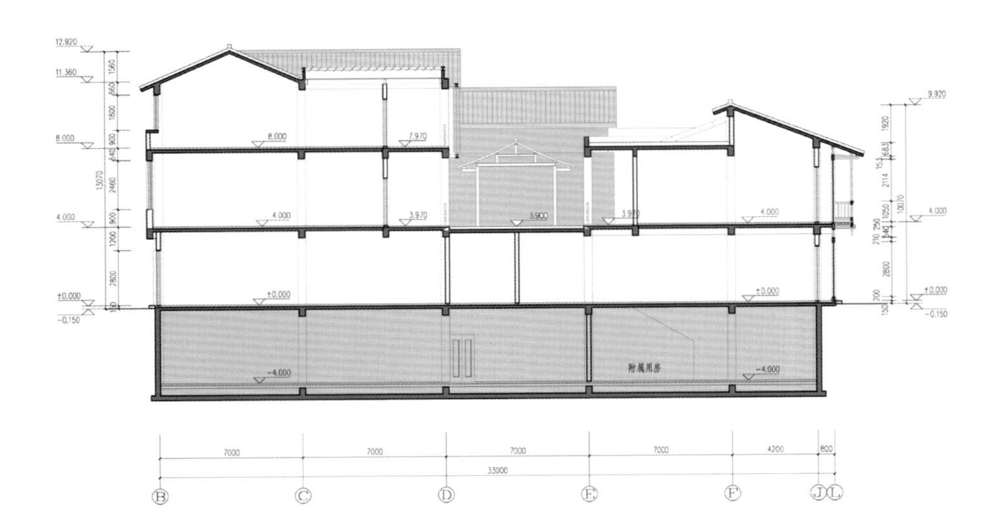

Sectional Drawing 剖面图

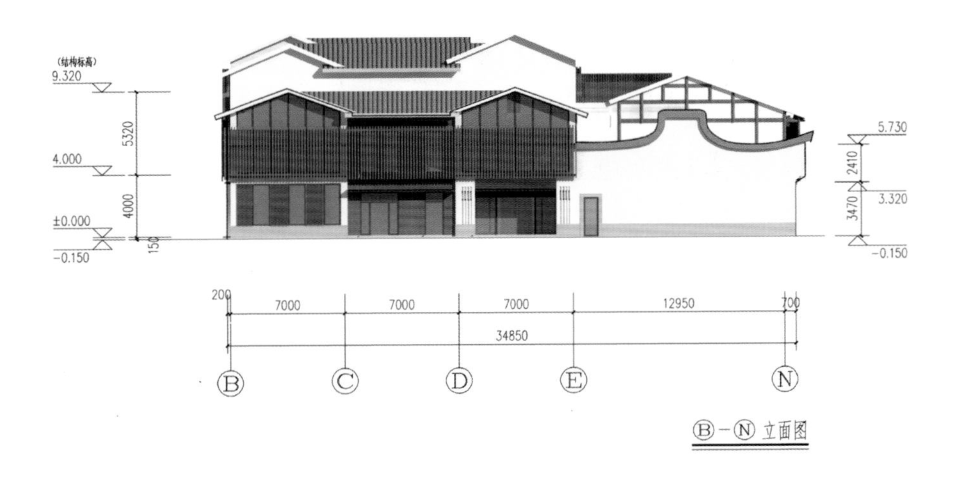

Ⓑ-Ⓝ 立面图

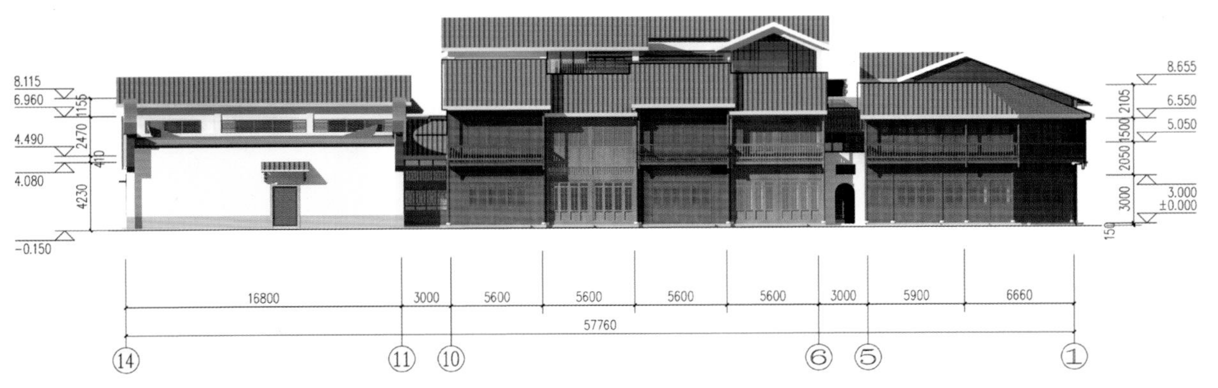

Sectional Drawing 剖面图

South Elevation 南立面图

6-8

西巷实景图 08

KEY WORDS
关键词

Quasi Symmetric Layout 准对称式布局

Tree-forest Modeling 群树造型

Period Flavor 时代气息

JSWB International Green Land Furniture Village Phase II
吉盛伟邦绿地国际家具村二期

FEATURES
项目亮点

The theme pavilion adopts a quasi symmetric layout that it uses the expressive force of tree-forest modeling to create another climax for the whole regional axis, forming virtual and real, high and low contrast with the landmark information center, Phase I.

主题展馆采用准对称式布局，以具有表现力的群树造型形成整个区域轴线上另一个高潮，与轴线一期的标志建筑信息中心形成虚与实、高与低的对话。

Location: Qingpu District, Shanghai, China
Architecture Design: China Architecture Design & Research Institute, Shanghai Johnson Architectural & Engineering Designing Consultants LTD
Total Land Area: 170,210 m^2
Total Floor Area: 196,418 m^2
Plot Ratio: 1.37

项目地点：中国上海市青浦区
设计单位：中国建筑设计研究院、上海中森建筑与工程设计顾问有限公司
总用地面积 :170 210 m^2
总建筑面积 :196 418 m^2
容积率：1.37

Overview

The project land is the Plot 1 to 3 of the Zhaoxiang super commercial gathering area in Qingpu District, Shanghai. The land is planed to be used for commercial finance, and mainly functioned as large furniture exhibition, small commercial exhibition and office. The commercial part is positioned for professional furniture & household products exhibition and catering business. The office is allocated in the standard of Class A, including a complete set of conference facilities.

项目概况

项目用地为上海青浦区赵巷超级商业商务聚集区 1–3 号地块。本项目规划用地性质为商业金融，主要功能为大型家具业展览、小型商业展览和办公。其中，商业定位为专业家具及家居用品展览和餐饮服务。办公为甲级配置的写字楼，含配套会议设施。

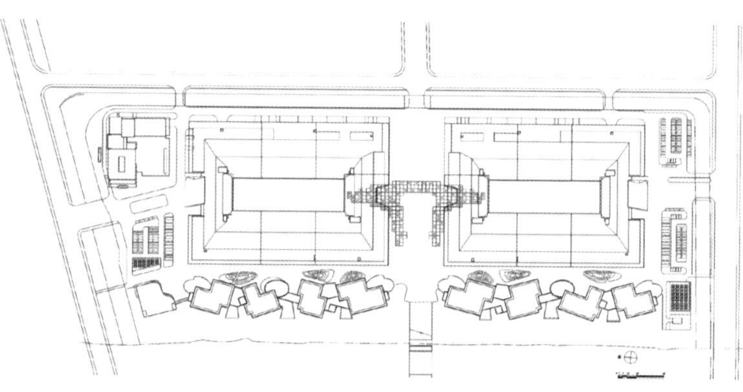

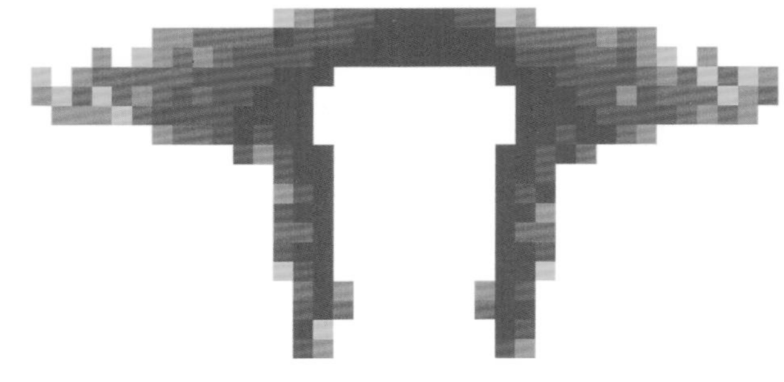

“红树林”色彩研究

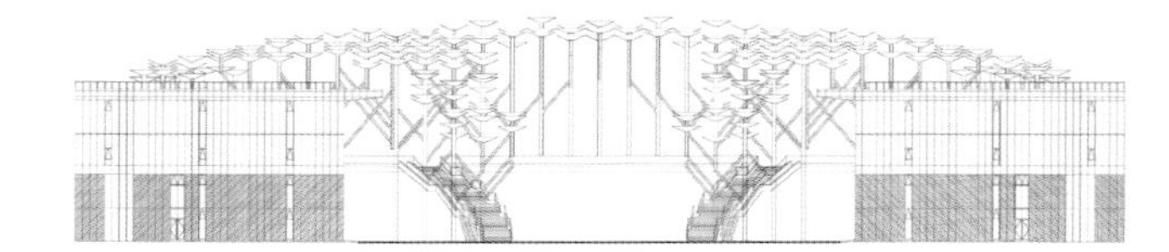

“红树林”立面图

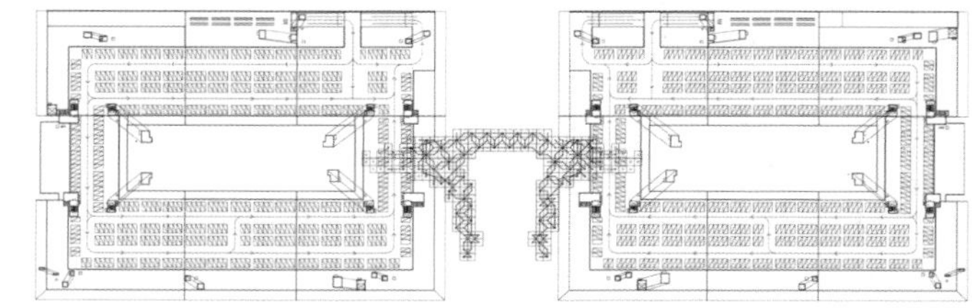

展览屋顶平面图

Design Concept

1. Tree. Taking tree as the theme of the furniture exhibition area and the main landscape axis, it will correspond to the origin of the industry and have a good implied meaning as will. Trees symbolize the flourish and endlessness, and also symbolize the full vitality in the enterprise of JSWB International Green Land Furniture Village.

2. City. The project is called city not only because of its large scale, but also because its form and structure adopt modern techniques to embody the city space, which means more possibilities provided for the merchants and customers. Comprehensive display, complete services, comfortable rest area, landscape inside or outside the city, and the mutual links all make it a real city.

设计理念

1. 树。以树作为家具展示区的主题，并作为轴线上的主要景观，既契合行业的本源，又有美好的寓意。它象征着枝繁叶茂、生生不息，也象征着吉盛伟帮绿地国际家具村的事业蒸蒸日上。

2. 城。不仅因大而为城。在这里，形态本身以现代的手法体现了城的空间，城更为商家和顾客提供更多可能。包罗万象的展示、齐全的服务、舒适的休息区、城外城内的景观，城上城下的联系等都使这里成为一个名副其实的城。

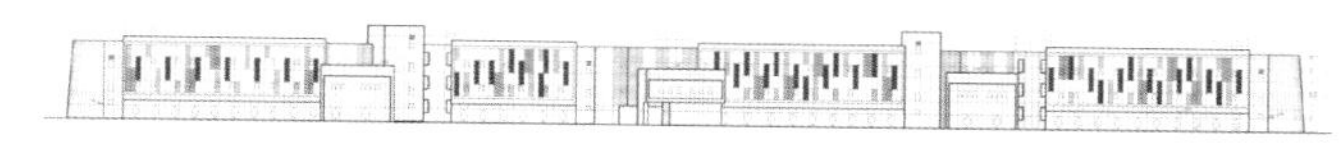

商业西立面图

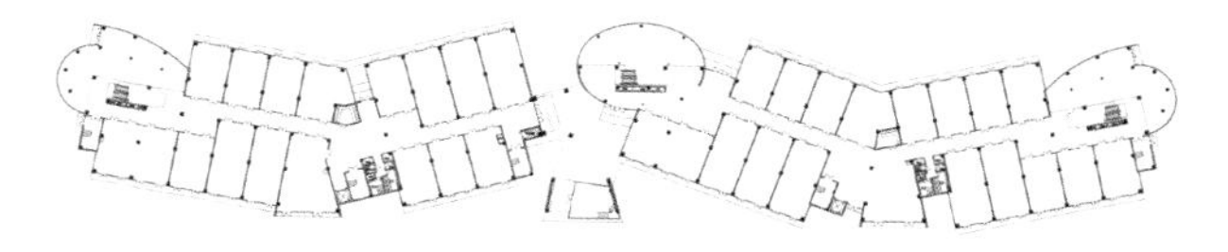

商业一层平面图

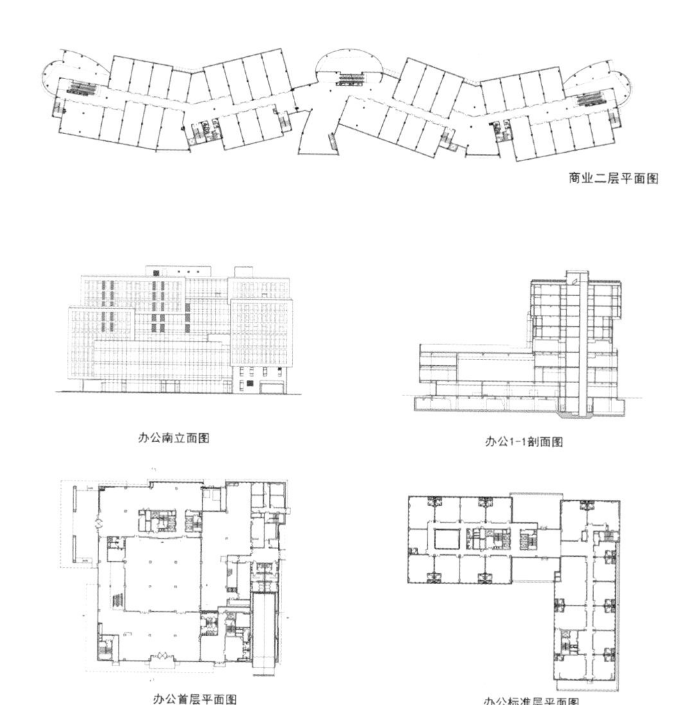

商业二层平面图

办公南立面图

办公1-1剖面图

办公首层平面图

办公标准层平面图

Design Methods

1. New Constructivism. Different from the previous architectural design, this project, based on according with the functional requirements, prefers to take methods of continuous texture, density distribution of color and body and block logic to form fresh new form and achieve the unity on the concept and expression with the overall modern design trend (including furniture design).

2. Space. The concise structure contains abundant space design. Tree space on the central axis, water space inside the city, transition space around the second floor, green space between the flagship store and rest space inside the building together form space full of changing rhythm, and bring new spatial experience as well.

Planning and Layout

The theme pavilion adopts a quasi symmetric layout which uses the expressive force of tree-forest modeling to create another climax for the whole regional axis, forming virtual and real, high and low contrast with the landmark information center, Phase I. Independent fine crystal hall and professional exhibition hall is fixed on the west side, echoing with the international exhibition hall Phase I, and its building materials and structure fully embody the modern fashion, forming pure contrast with the main exhibition hall; besides, it sets recreational space of coffee house and drinking bar on one side of the river bank. Office and conference functions appear in the northeast corner in the form of building blocks, together with the northern entrance of the exhibition hall to form landmark mass along the Yinggang Road.

设计手法

1. 新建构主义。和以往的建筑设计有所不同，在符合功能的基础上，这里更多地采用连续的肌理、色块及形体的密度分布、体块逻辑等方法，打造新颖的形态，与现代化总体设计思潮（包括家具设计）形成观念与表达上的统一。

2. 空间。简洁的形态包含丰富的空间设计。中轴线上的树空间、城内的水空间、二层外围的灰空间、旗舰店之间的绿空间、建筑内部的休息空间等形成空间节奏的变化，并带来全新的空间体验。

规划布局

主题展馆采用准对称式布局，以具有表现力的群树造型形成整个区域轴线上另一个高潮，与轴线一期的标志建筑信息中心形成虚与实、高与低的对话。独立式精晶展厅与专业展厅布置在用地西侧，与一期国际展厅相呼应，建筑材料与形式充分体现时尚、现代特色，与主展馆的完整、纯粹形成对比，同时在河岸一侧设置咖啡店、水吧店等休闲空间。办公及会议配套功能以积木搭建的形式出现在用地东北角，与展厅北入口共同构成项目沿盈港路的标志体量。

KEY WORDS
关键词

Curved Lines 流线元素
Convex and Concave 虚实对比
Convex and Concave 挺拔感

Shenzhen NEO Tower
深圳绿景纪元大厦

FEATURES
项目亮点

NEO Tower is built as a conversation between bold lines and sharp curves. A contrast between curved and flat shapes, horizontal and vertical lines, and dark and light color, forms this active and bold design.

建筑平面南北向以独特的凹凸曲面呈现相互环绕的视觉造型，立面设计注入了平面流线元素，使建筑充满动感又不失大方、稳重。

Location: Shenzhen, Guangdong, China
Architectural Design: CCDI
Land Area: 8,681 m²
Total Floor Area: 130,300 m²
Completion: 2011

项目地点：中国广东省深圳市
设计单位：中建国际（深圳）设计顾问有限公司
用地面积：8 681 m²
总建筑面积：130 300 m²
竣工时间：2011 年

Architectural Design

In the visual arts, contrast is what gives a shape definition and meaning. With this in mind, the NEO Tower is built as a conversation between bold lines and sharp curves. A contrast between curved and flat shapes, horizontal and vertical lines, and dark and light color, forms this active and bold design. With its sharp edges, a play of convex and concave lines distinguish each facade; while one side curves inward, the opposite side curves outward; dark glass and steel on one side is balanced by light materials on the other. The roof area is a complex play of angles which complement each other through a triangular structure, like dueling forces of light and dark. This aesthetic design presents different views from Shennan Road.

建筑设计

建筑平面南北向以独特的凹凸曲面呈现相互环绕的视觉造型，立面设计注入了平面流线元素；曲面与平面，竖向与横向，材质的深与浅，对比的元素使建筑充满动感又不失大方稳重。塔楼顶部以斜向曲面勾勒出流线形象，彻底打破四方形塔楼的呆板，增加动感和时尚，并借助视觉分析增强建筑的高度感。竖向的金属杆件，让塔楼更显高耸、挺拔、简洁、明快，同时，还让整组建筑从深南大道不同的位置呈现出微妙的虚实变化。

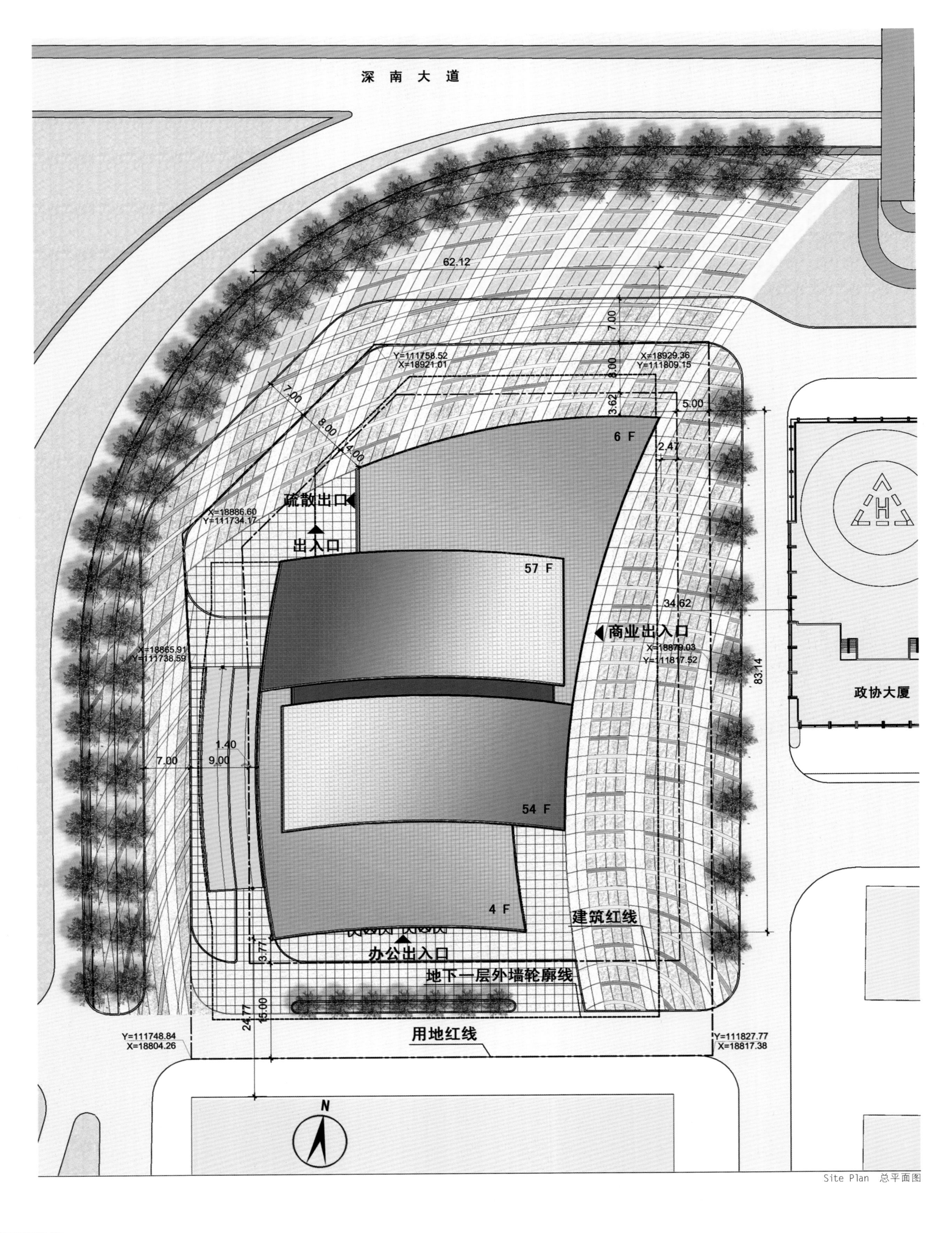

Site Plan 总平面图

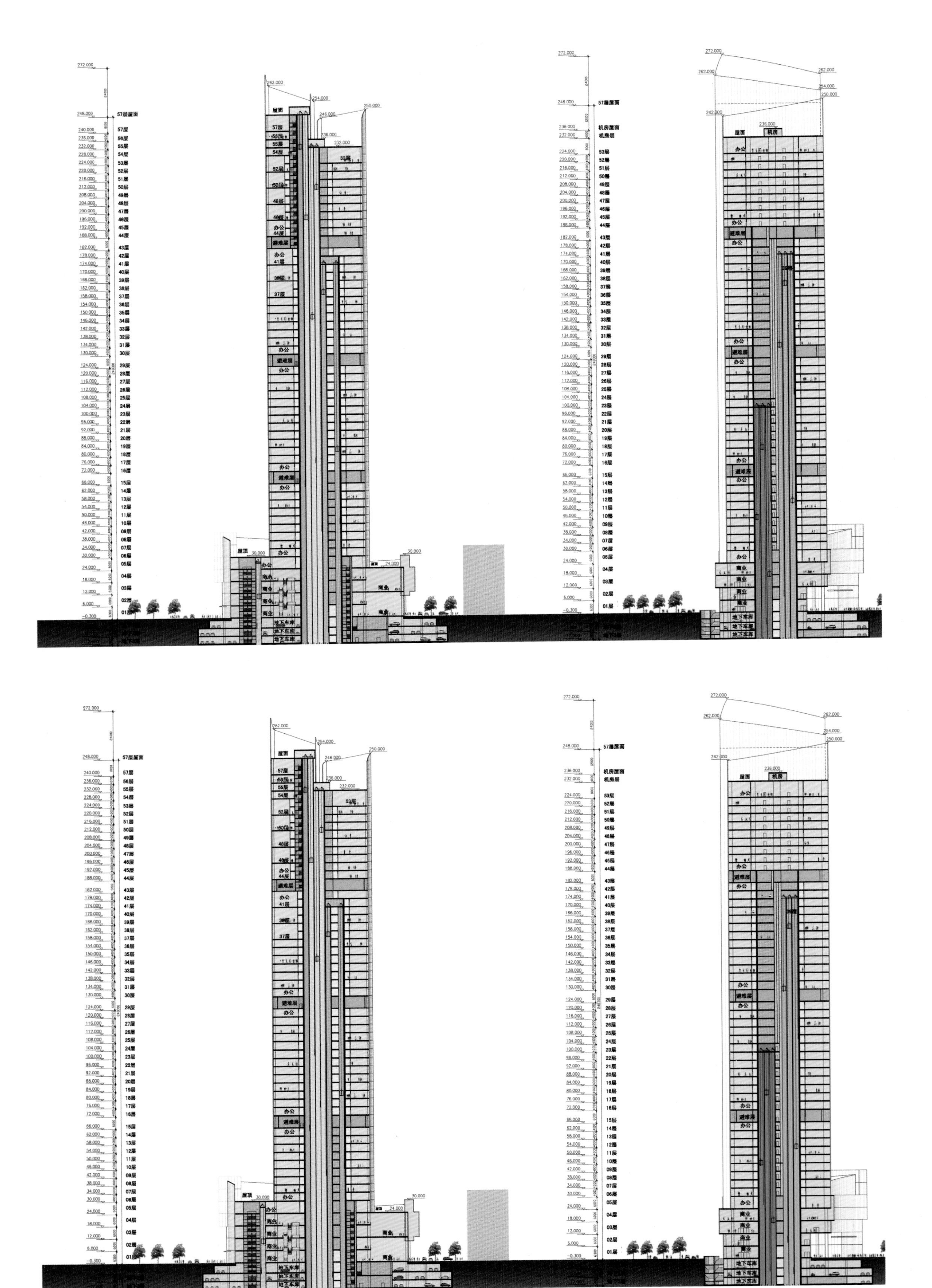

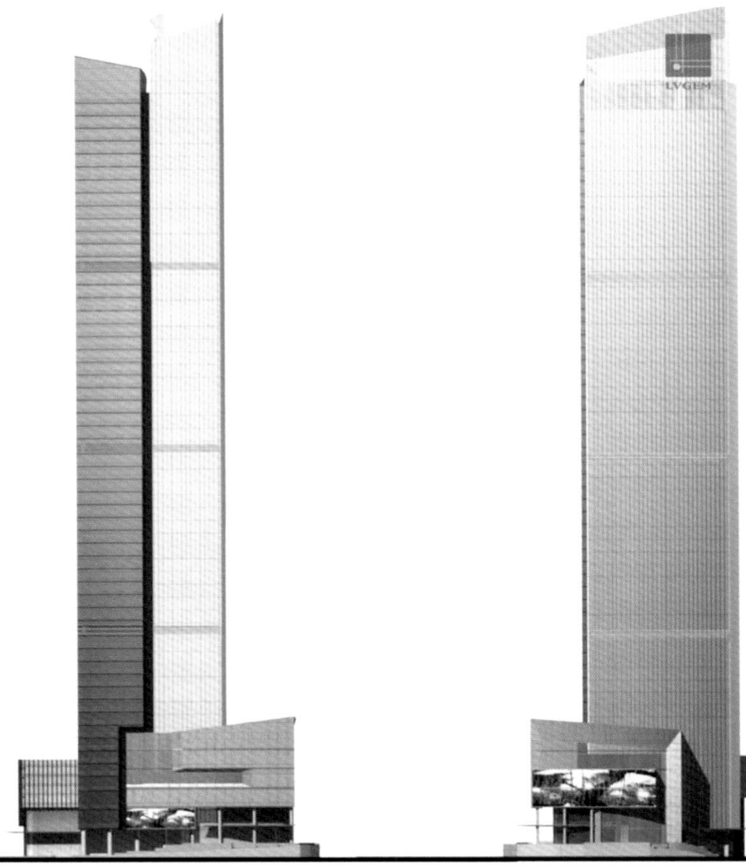

East Elevation 东立面图

North Elevation 北立面图

KEY WORDS
关键词

Double-skin Facade 双层幕墙

Green Roof 屋顶绿化

Low Carbon and Energy-saving 低碳节能

China Merchants Bank Building in Suzhou Industrial Park
苏州工业园区招商银行大厦

FEATURES
项目亮点

The west facade of the building adopts double-skin curtain wall system, between which is the greenhouse garden; the outer facade sets electric glass ventilation shutters which can effectively adjust the indoor micro climate, low-carbon and energy-saving.

建筑西立面采用双层“呼吸式”幕墙系统，双层幕墙之间是温室花园，外层幕墙设置电动玻璃通风百叶，有效地调节建筑室内的微气候，低碳节能。

Location: Suzhou, Jiangsu, China
Architectural Design: Suzhou Institute of Architectural Design Co., Ltd.
Land Area: 7,837 m^2
Total Floor Area: 44,850.57 m^2
Construction Scale: 44,850.57 m^2
Building Height: 98.2 m
Plot Rate: 3.96
Green Coverage Ratio: 12%
Completion: 2011

项目地点：中国江苏省苏州市
设计单位：苏州设计研究院股份有限公司
用地面积：7 837 m^2
总建筑面积：4 4850.57 m^2
建筑规模：44 850.57 m^2
建筑高度：98.2 m
容积率：3.96
绿化率：12%
竣工时间：2011 年

Overview

Floors 1 to 4 are the annexes of the building, and its function mainly includes two parts, i.e. business area and supporting auxiliary function area. Business area is on the 1st and 2nd floor of the annex building, and its entrances is respectively set in the west and north sides that is convenient for external business. There is door control between the office entrance and business area, which make the two functional areas enable to be apart or combined with independent management. Each functional area in the annex building, except the lobby, sets up monitoring center, conference room, dining room, small coffee shop, staff bathroom, home for the employees, management rooms and other auxiliary functional rooms. Above the annex is the administrative office area, mainly as large, open office space. The whole space is arranged flexibly around the core tube that enables to be apart or combined. Floors 19 to 20 are considered to be the governor's office.

项目概况

建筑 1 至 4 层为大厦裙房，其功能主要分为两部分：营业区和配套辅助功能区。营业区位于大楼裙房一、二层，入口分别设置在建筑的西面和北面，方便对外经营。办公入口到顶与营业区之间设置门禁，使两功能区可分可合，独立管理。裙房的各功能区除大堂外，设置监控中心、全行会议室、大楼食堂、小型咖啡厅、员工浴室、职工之家、管理用房等辅助功能用房。大楼裙房以上为行政办公区，以大开间办公为主，为开放式办公空间。整个空间围绕核心筒布置，并且各单元可分可合，布置灵活。同时，在主楼 19 至 20 层设置行长办公室。

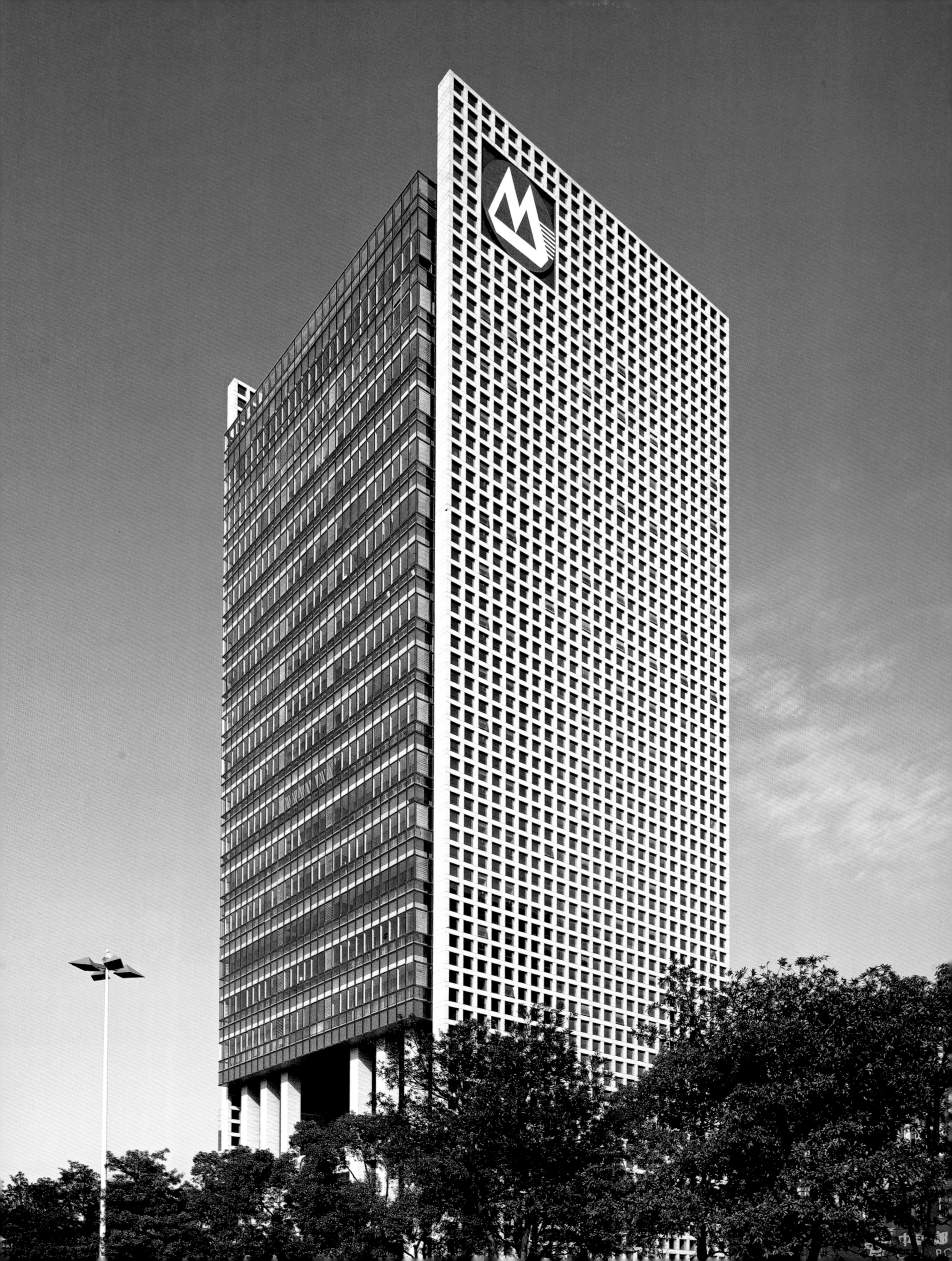

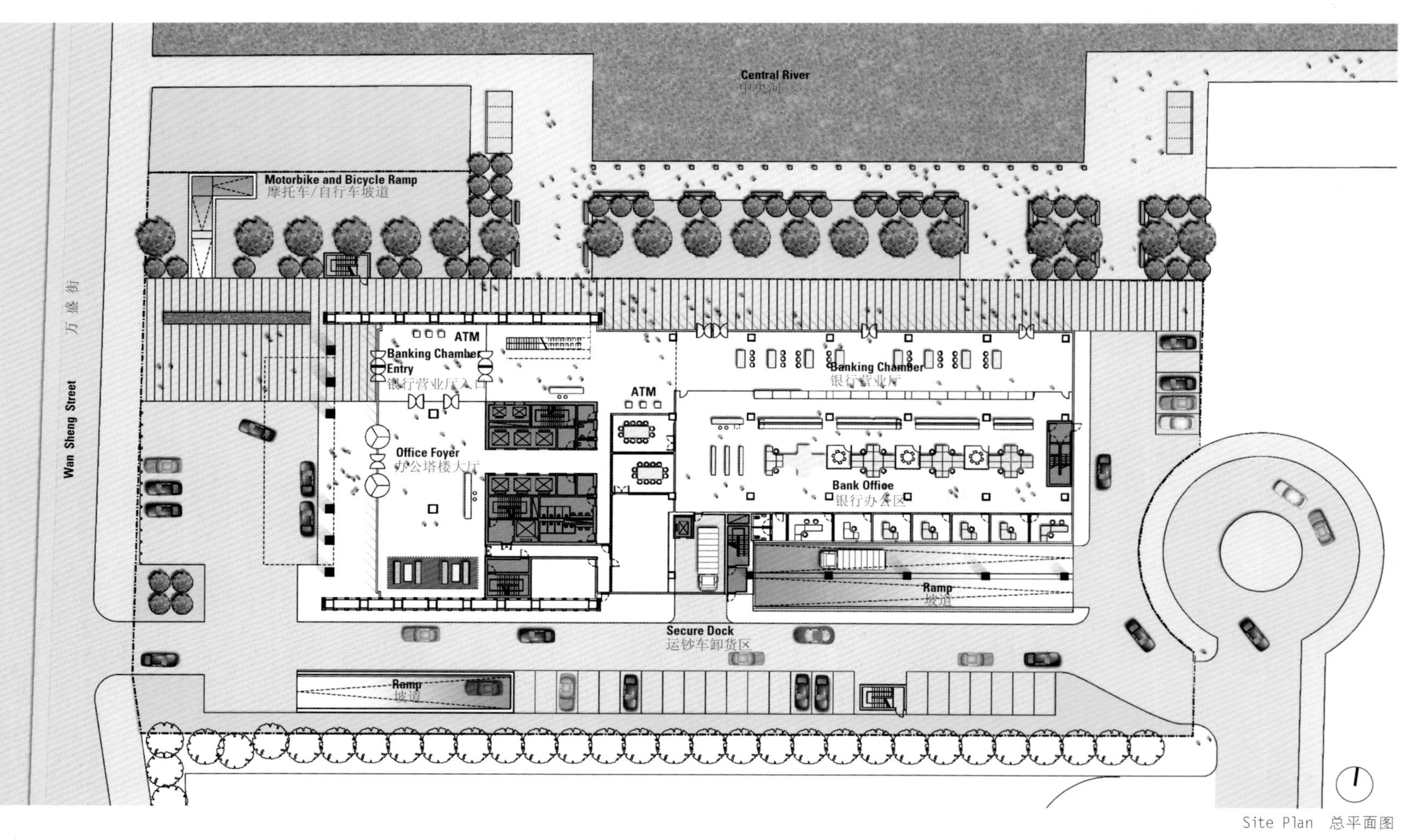

Site Plan 总平面图

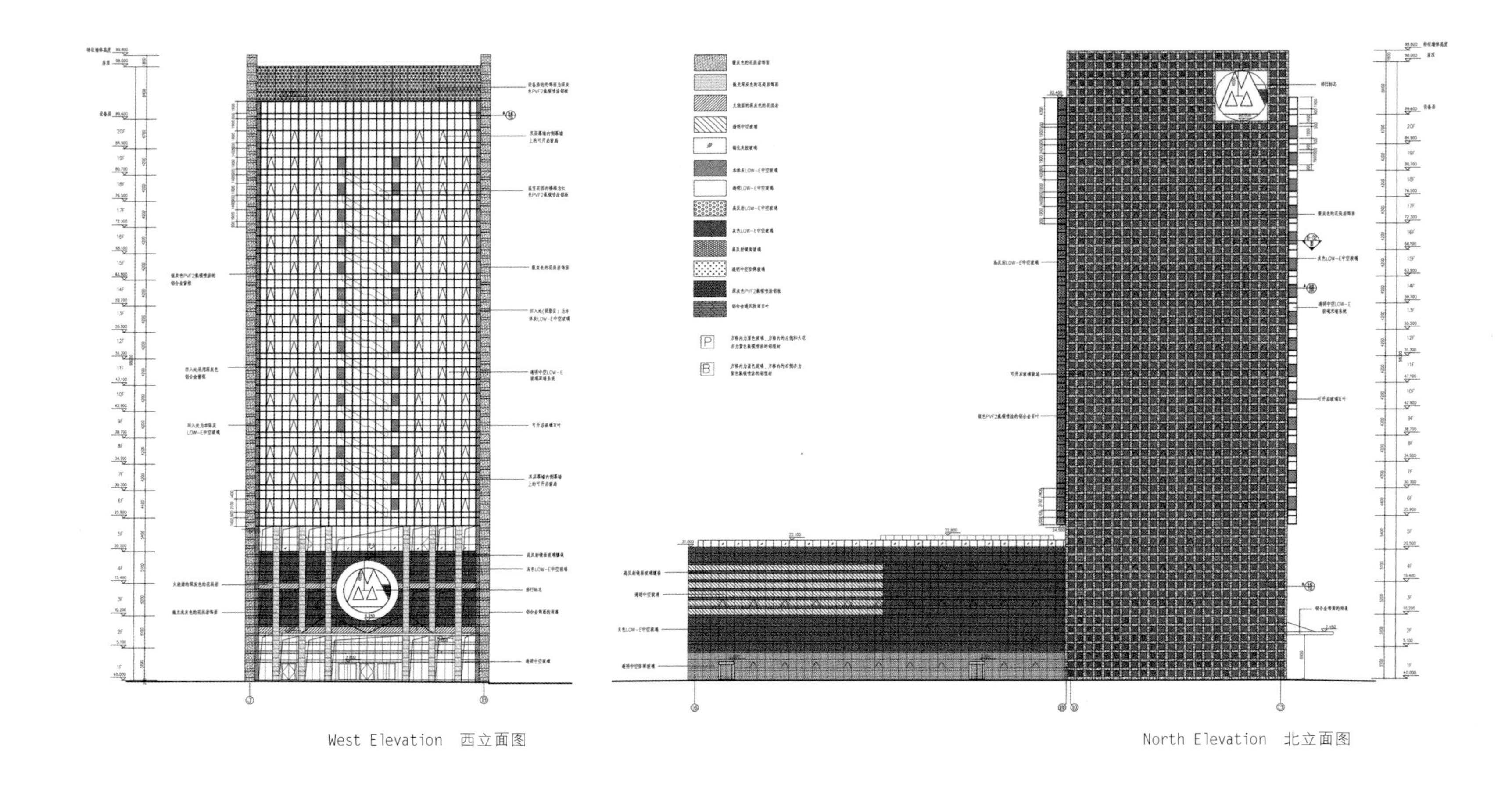

West Elevation 西立面图

North Elevation 北立面图

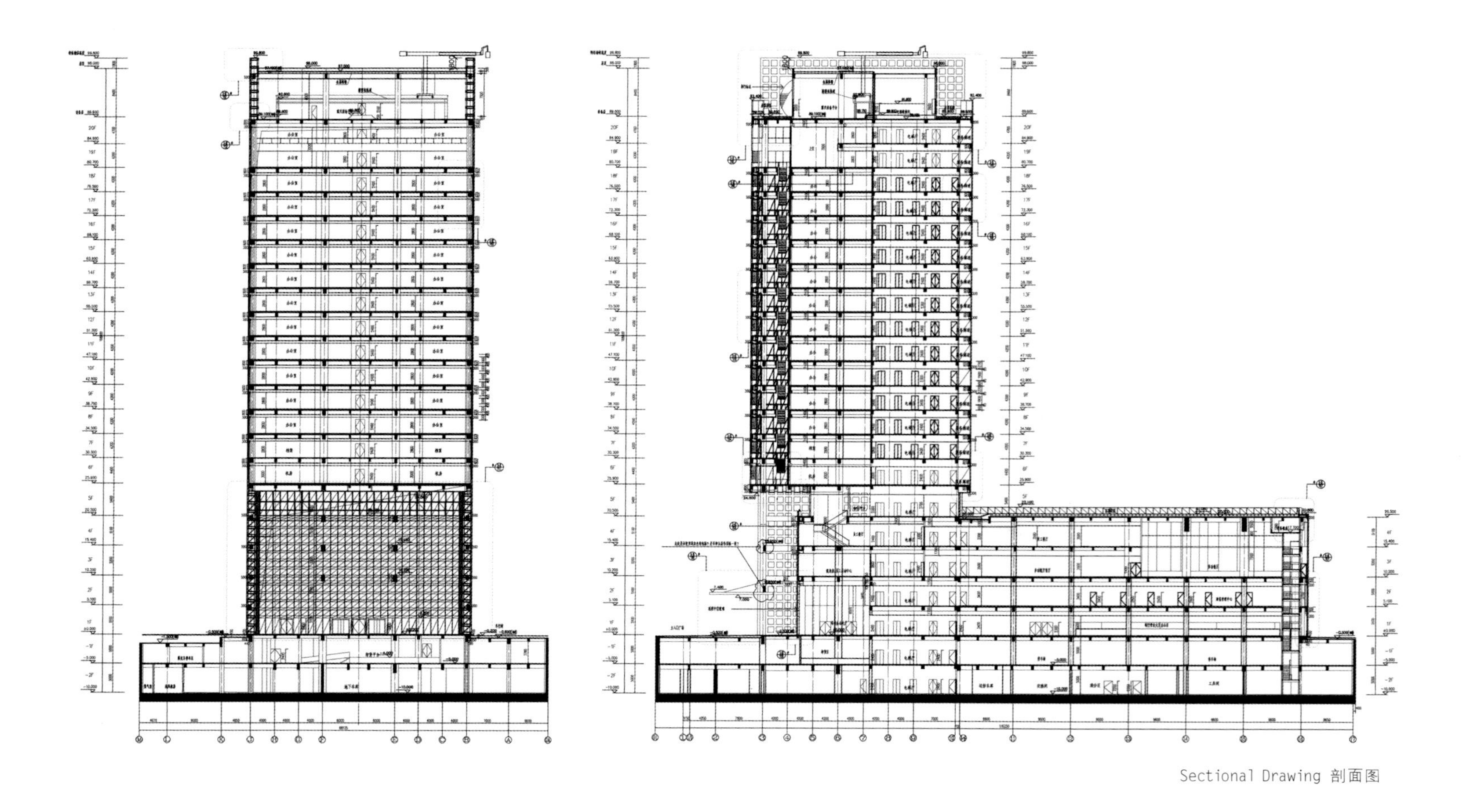

Sectional Drawing 剖面图

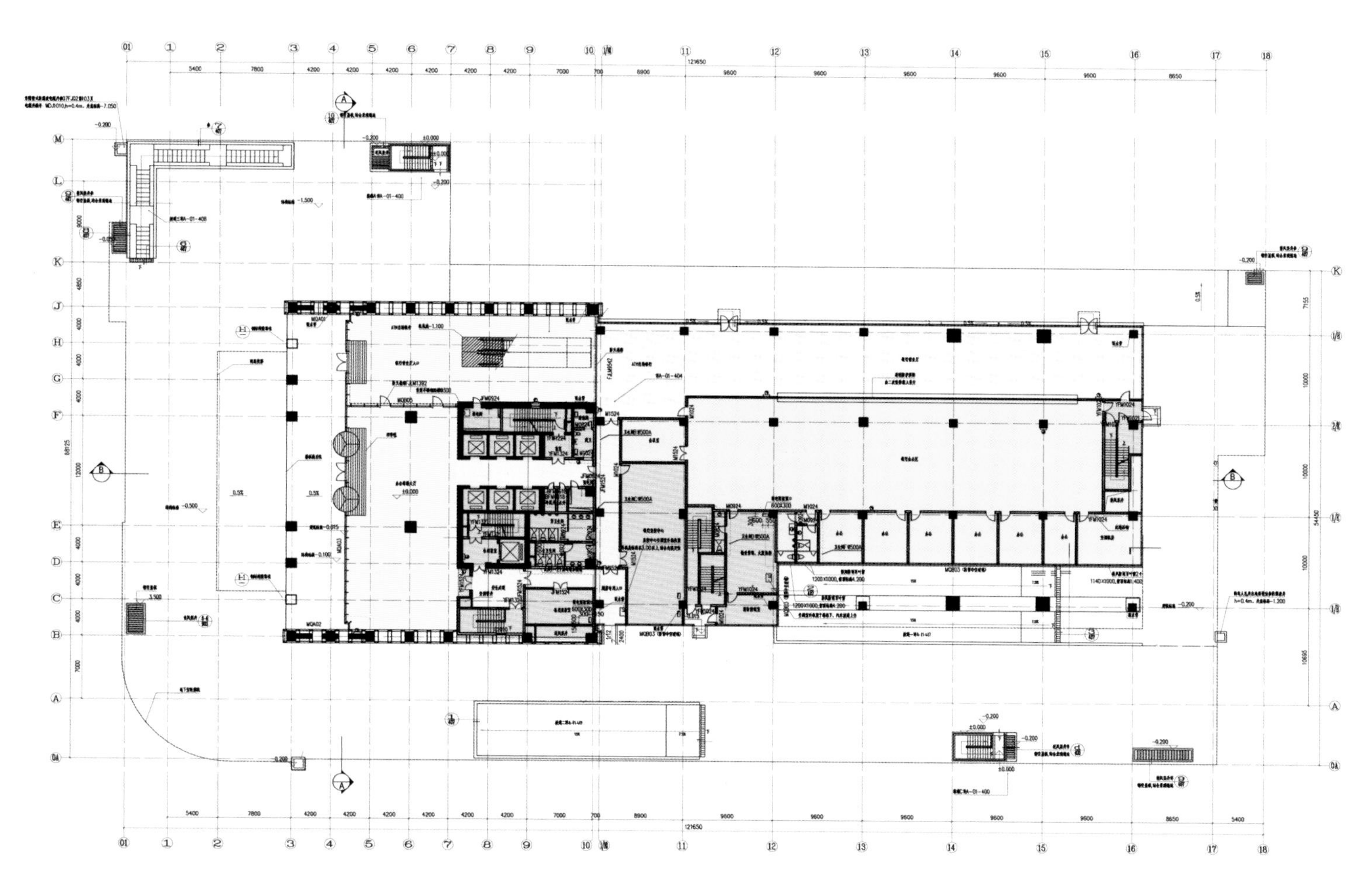

First Floor Plan　一层平面图

Architectural Design

The west facade of the building adopts double-skin curtain wall system, between which is the greenhouse garden; the outer facade sets electric glass ventilation shutters which can effectively adjust the indoor micro climate. The annex building is equipped with solar heat collection panels which provide hot water for the building. There is also planting and greening roof, and the air conditioning system uses air-cooled heat pump variable fresh air system of multi-connected unit of refrigerant flow. The multi-connected unit is allocated with reasonable system according to the requirements of using time and functions, so as to achieve the goal of energy saving and operating flexibility.

建筑设计

建筑西立面采用双层"呼吸式"幕墙系统，双层幕墙之间是温室花园，外层幕墙设置电动玻璃通风百叶，有效地调节建筑室内的微气候。建筑裙房上设太阳能集热板，供本建筑热水使用。屋顶采用种植绿化，空调系统采用风冷热泵型变冷媒流量多联机组加新风系统。多联机组根据使用时间和功能等要求合理配置系统，以达到低碳节能、灵活运行的目的。

KEY WORDS
关键词

Tang Architecture Style 唐式建筑风格

Culture Tradition 文化传统

Building Materials 建筑材料

Tuanbo Scenery Holiday Hotel and Tianfang Tuanbo Club House
团泊风景假日及天房团泊会馆

FEATURES
项目亮点

The project adopts Tang architecture style, and its well-proportioned structure that tier upon tier forms natural fusion with the surrounding environment; the composition technique of courtyard and space also brings out the best with Tang style in each other, highlighting the theme of "hermit" in Chinese culture.

建筑采用唐式风格，层层叠叠、错落有致的形体与环境形成自然的融合，建筑院落和空间的构成手法也与唐式风格相得益彰，凸现出“隐”文化这一中国文化的主题。

Location: Tianjin, China
Architectural Design: Beijing Victory Star Architectural & Civil Engineer Design Co., Ltd.
Total Land Area: 20,181.4 m^2
Total Floor Area: 5,995.88 m^2
Building Density: 16%
Plot Ratio: 0.229
Green Coverage Ratio: 40%
Completion: 2012

项目地点：中国天津市
设计单位：北京维拓时代建筑设计有限公司
总用地面积：20 181.4 m^2
总建筑面积：5 995.88 m^2
建筑密度：16%
容积率：0.229
绿化率：40%
竣工时间：2012 年

Overview

The project is a private-use club of Tianfang Group; according to the geographical environment, the island courtyard and concealed springs combine the building and surrounding environment perfectly; the building style follows the example of Chinese traditional hot springs health culture; it also adopts Tang architecture style, and its well-proportioned structure that tier upon tier forms natural fusion with the surrounding environment; the composition technique of courtyard and space also brings out the best with Tang style in each other, highlighting the theme of "hermit" in Chinese culture. This building adopts stone materials and the collocation of metal decorative components with pure and natural tonal, which reflect the exquisite feeling and lightness of modern materials from its traditional charm, and achieve the effect of reflecting the essence of traditional culture by using modern building materials.

项目概况

本项目为天方集团自用会所，根据地理环境，通过“岛院隐泉”的脉络使建筑与环境有了良好的结合，建筑风格取法中国温泉养生的传统文化，采用唐式风格，其层层叠叠、错落有致的形体与环境形成自然的融合，其建筑院落和空间的构成手法也与唐式风格相得益彰，凸现“隐”文化这一中国文化的主题。建筑采用石材、金属装饰构件组合的搭配形式，色调清新、自然，于传统韵味中体现现代材料的精致感和轻盈感，达到用现代建筑材料体现传统文化精神的效果。

Site Plan 总平面图

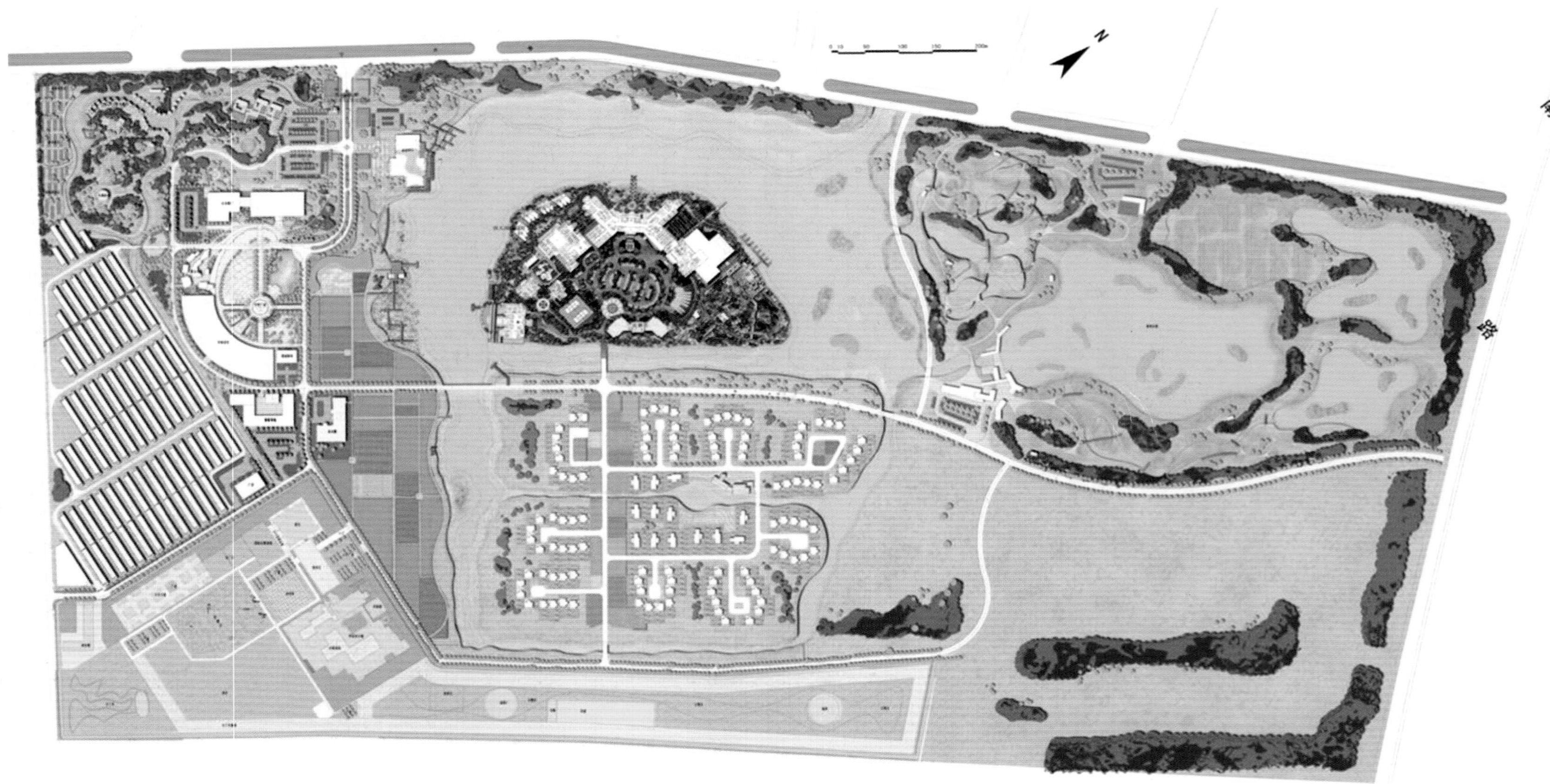

Site Plan　总平面图

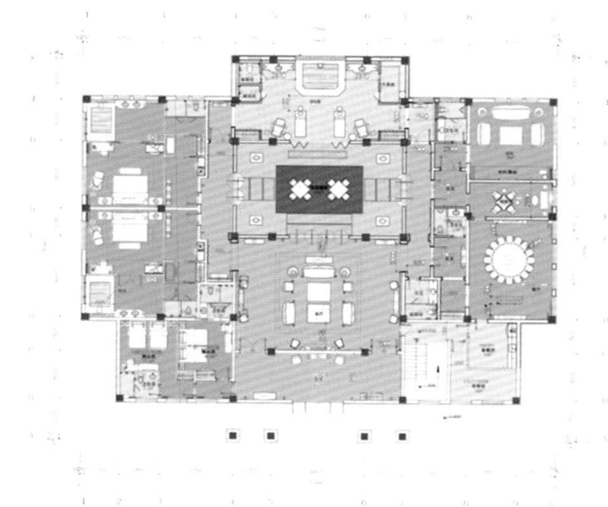

First Floor Plan　一层平面图

Second Floor Plan　二层平面图

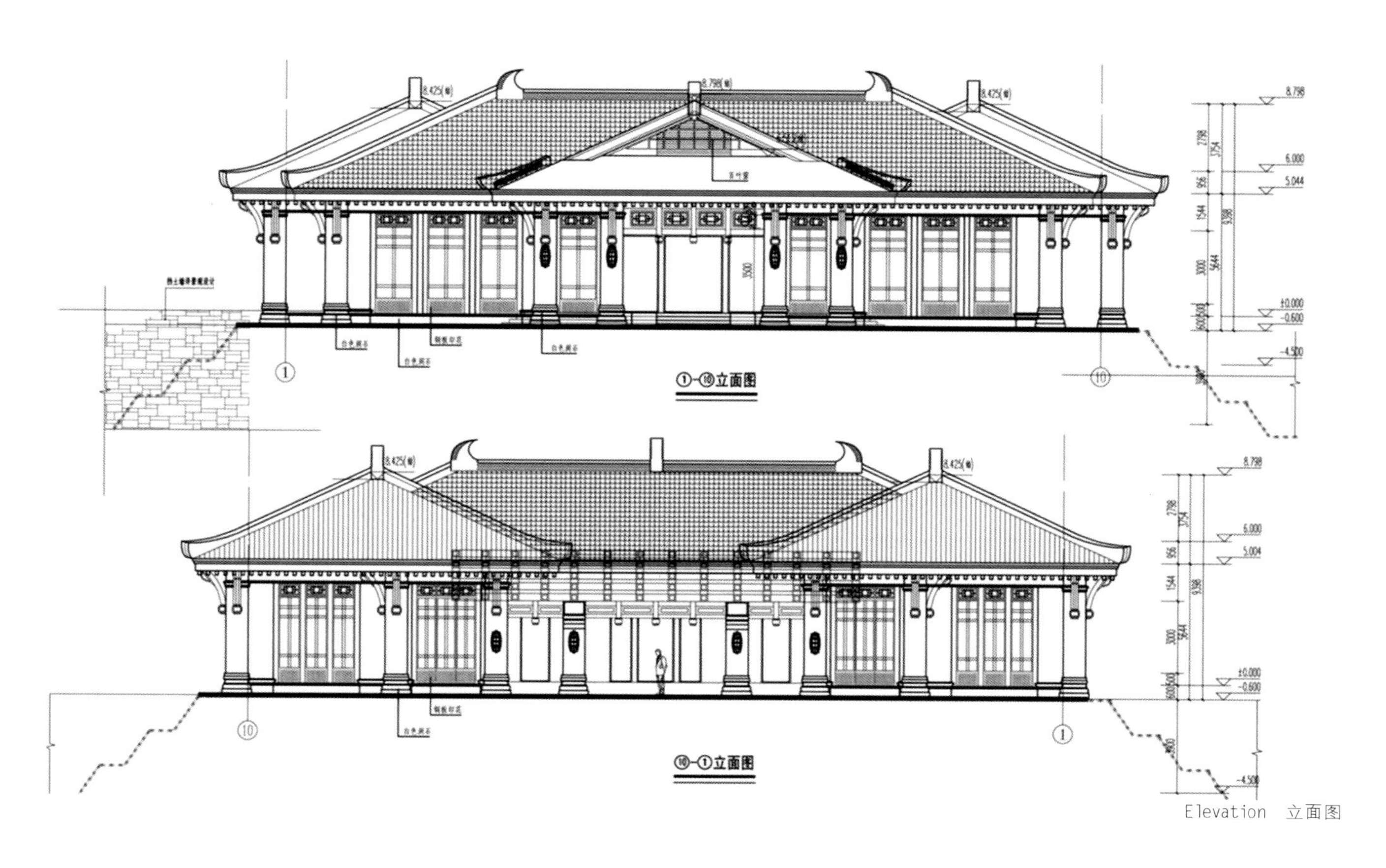

Elevation　立面图

KEY WORDS
关键词

Glass Curtain Wall 玻璃幕墙

Structural Design 结构设计

Mixed Use 功能综合

China Minmetals Tower
中国五矿商务大厦

FEATURES
项目亮点

The facade is designed with single-layer glass curtain wall, on which horizontal lines and vertical lines interweave to form a lattice pattern. This kind of facade makes the building look transparent and graceful, and presents a great visual effect.

立面上设置单层网玻璃幕墙，由水平向索和竖向索正交布置，构成索网格结构，营造出一种轻盈通透和内外交融的高透视觉效果。

Location: Tianjin, China
Architectural Design: Tianjin University Research Institute of Architectural Design
Total Land Area: 20,800 m^2
Total Floor Area: 183,267 m^2
Completion: 2011

项目地点：中国天津市
设计单位：天津大学建筑设计研究院
总占地面积：20 800 m^2
总建筑面积：183 267 m^2
竣工时间：2011 年

Overview

The site is situated in the center of Xiangluowan CBD, Tanggu District of Tianjin City, with Yingbin Avenue on the east, Tuochangnan Road on the south, the planning road on the west and Tuochang Road on the north. The Minmetals Tower includes the programs like commerce, serviced apartment, SOHO, convention, banquet, recreation and entertainment.

项目概况

中国五矿商务大厦位于天津市塘沽区响螺湾商务区核心区域，东至迎宾大道，南至坨场南道，西至规划路，北至坨场道。五矿商务大厦主要功能包括商业、酒店式公寓、SOHO 办公、会议、宴会、休闲娱乐等。

Planning

The complex consists of the office building A and serviced apartment building B, as well as two high-rise buildings for SOHO. Building A has a floor area of 88,685 m^2, building B has 94,582 m^2. The podium houses three floors for commercial use. Tow high-rise buildings are connected by sky corridor on the third floor.

平面布局

大厦包括 A 座的商务办公楼和 B 座的酒店型公寓及 SOHO 办公楼两栋高层建筑，其中 A 座为 88 685 m^2，B 座为 94 582 m^2。裙房主要为 3 层商业，局部 4 层，两座塔楼之间在 3 层以通廊连接，整体格局一气呵成。

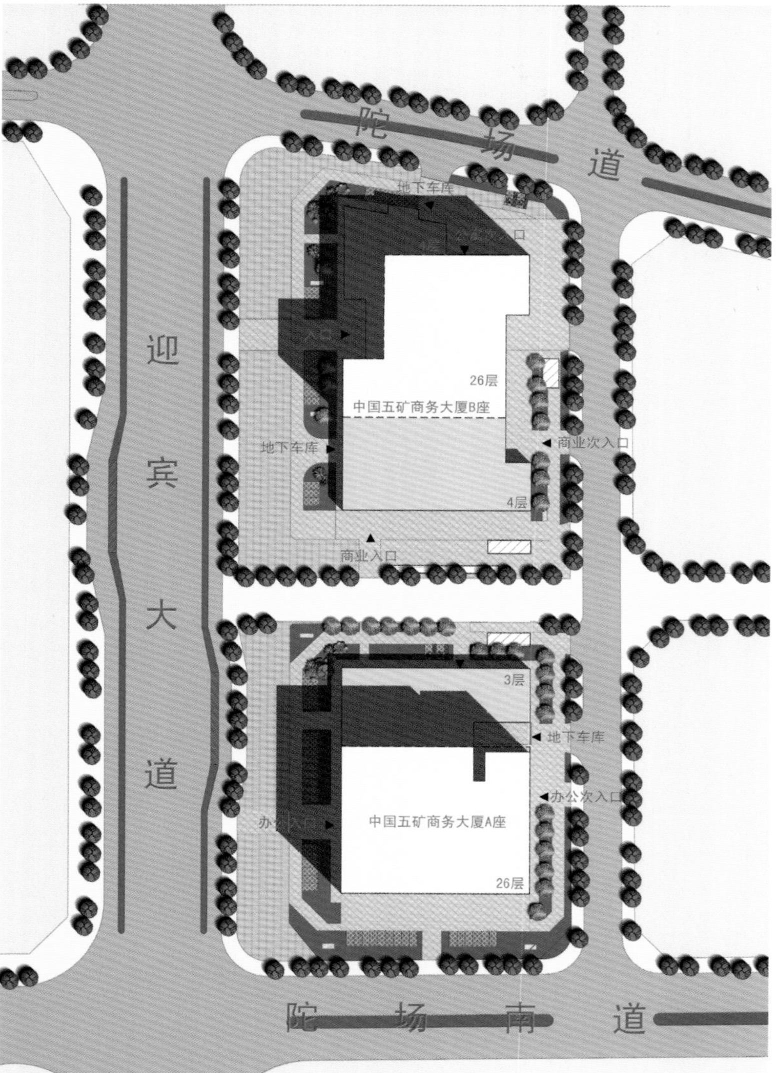

Site Plan 总平面图

Architectural Design

The complex is in rectangular shape. To create an enclosed but bright atrium inside the building, the facade is designed with single-layer glass curtain wall. Horizontal lines and vertical lines interweave to form a lattice structure. This kind of facade makes the building look transparent and graceful, presenting a great visual effect.

建筑设计

建筑平面呈矩形，为了使建筑物内部形成封闭、明亮的中庭，在立面上设置单层网玻璃幕墙，由水平向索和竖向索正交布置，构成索网格结构。单层索网幕墙结构的运用营造出一种轻盈通透和内外交融的高透视觉效果。

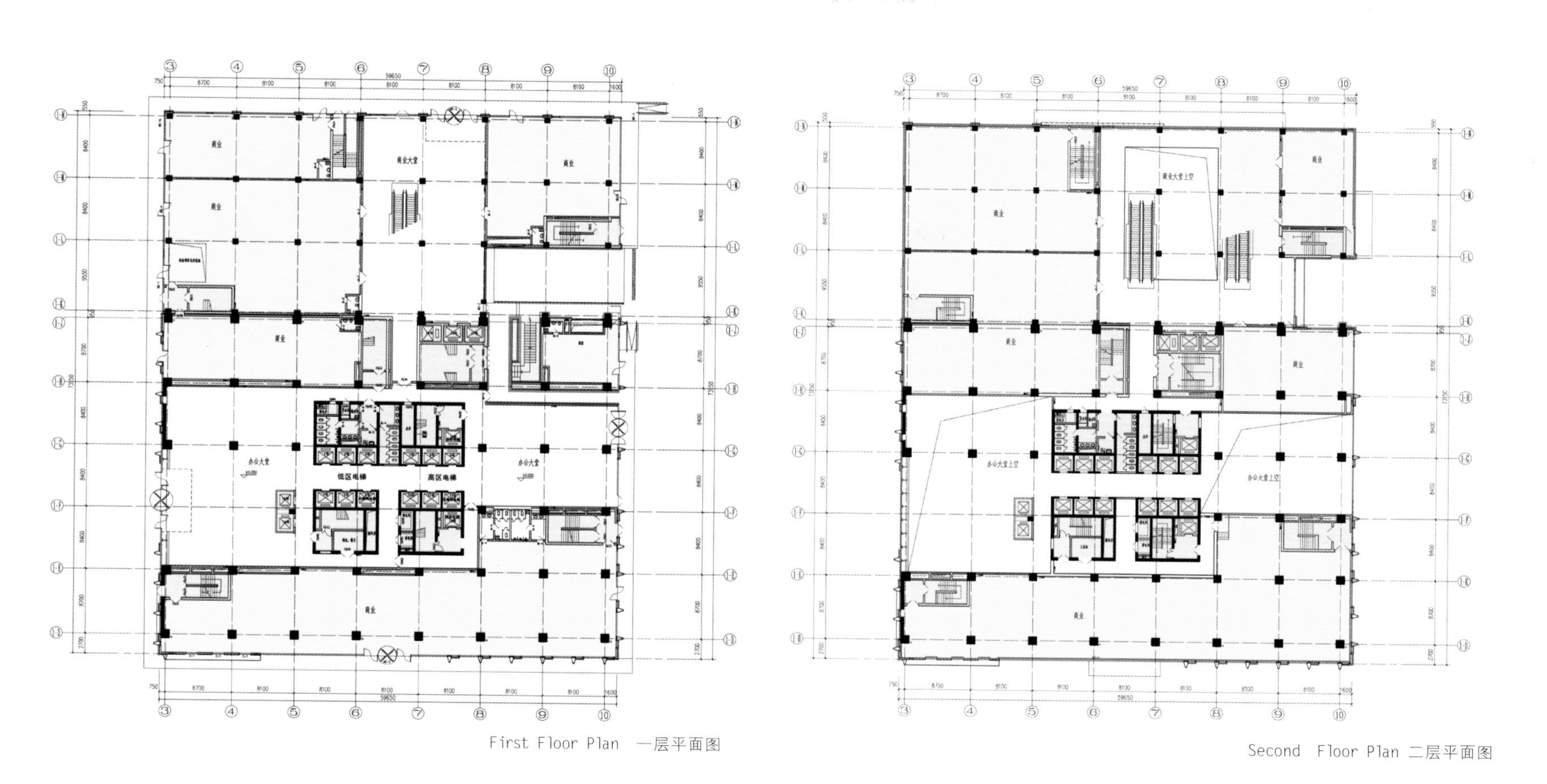

First Floor Plan 一层平面图

Second Floor Plan 二层平面图

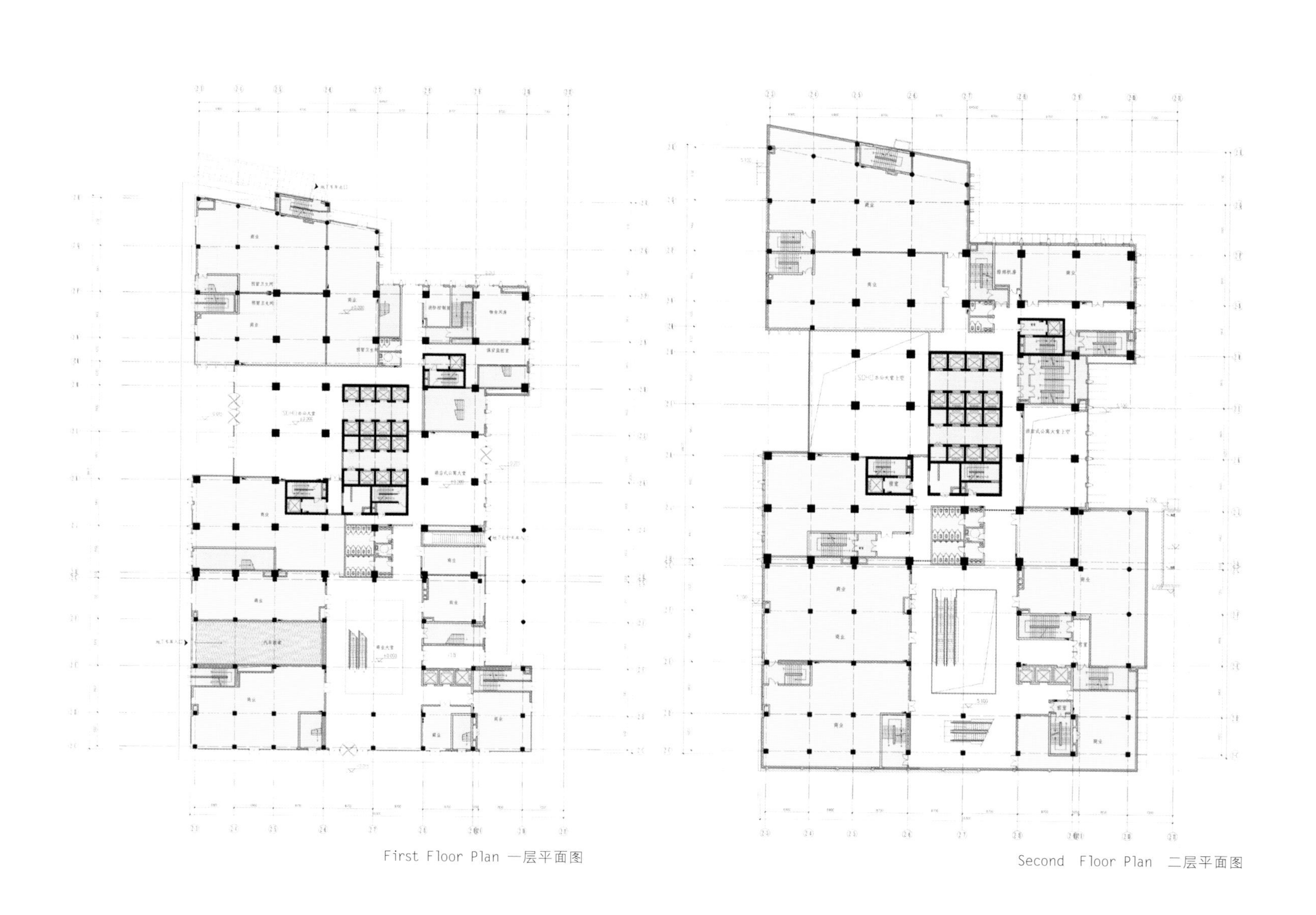

First Floor Plan 一层平面图

Second Floor Plan 二层平面图

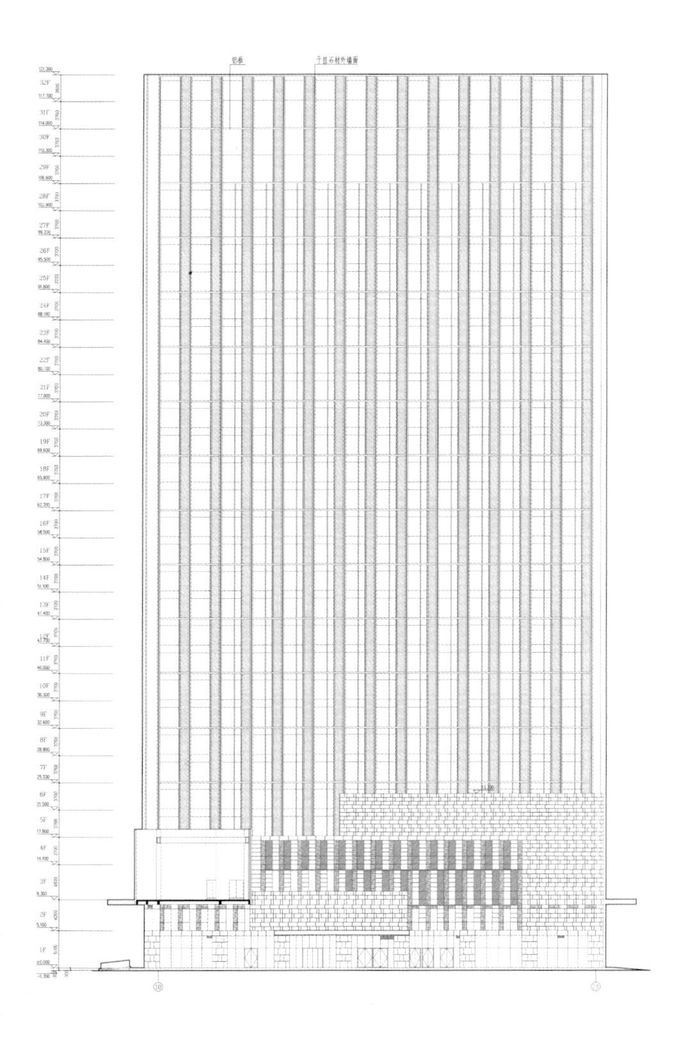

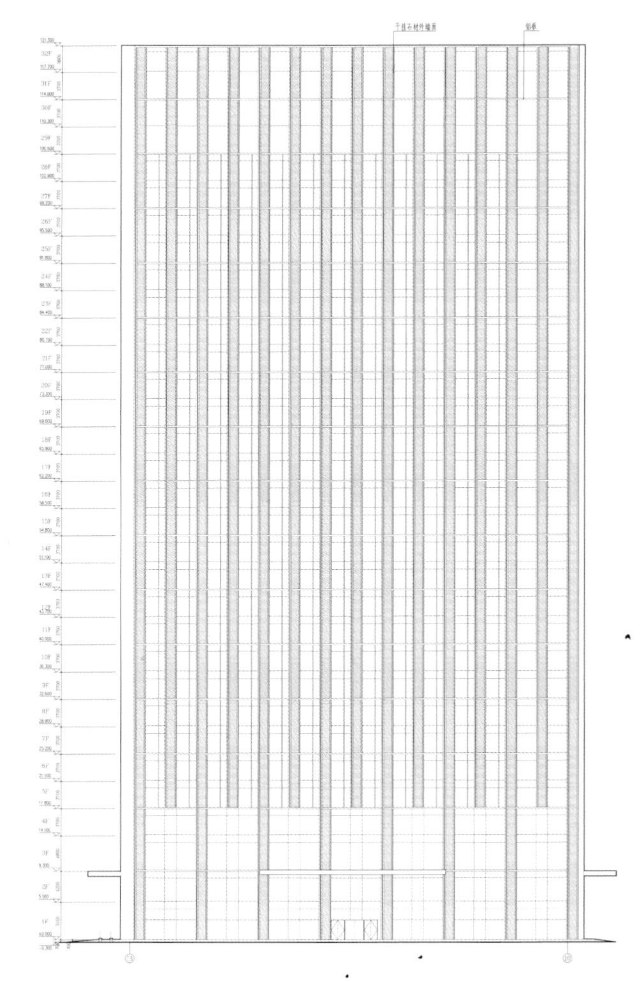

Elevation　立面图

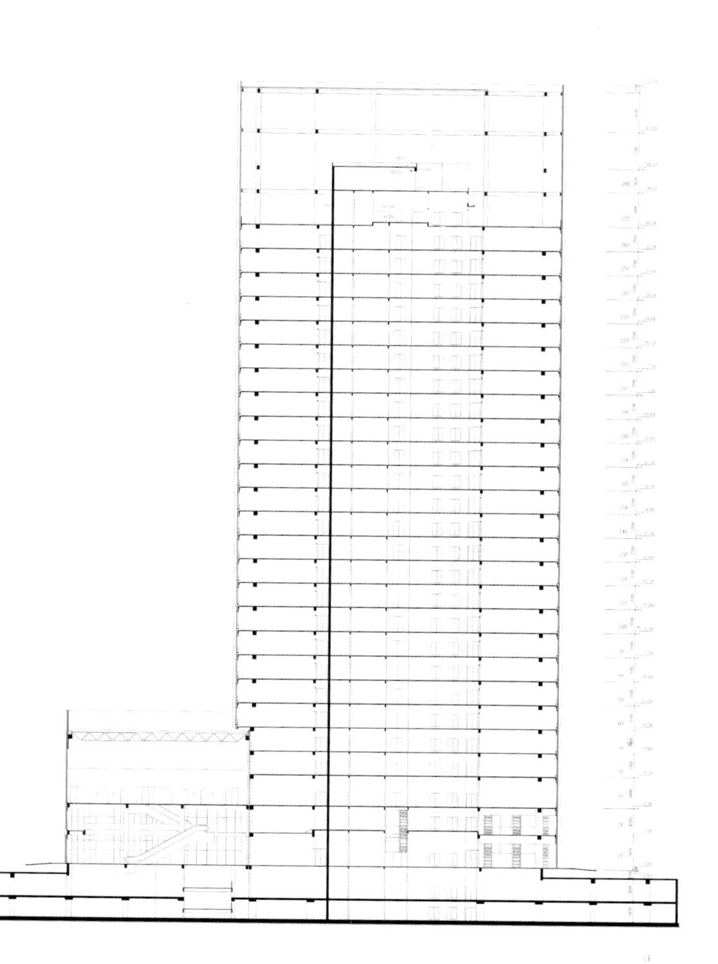

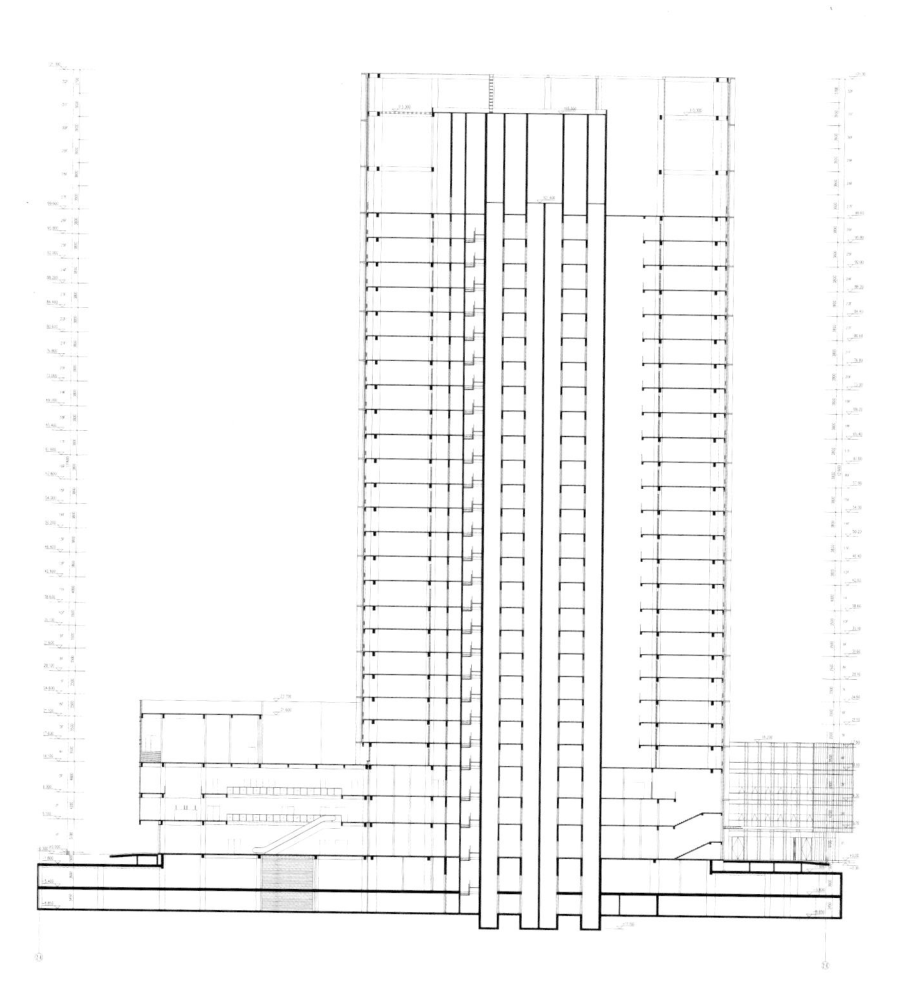

Elevation 立面图

KEY WORDS
关键词

Glass Facade 玻璃立面
Abstract Pattern 抽象图案
Sense Of Science And Technology 科技感

Chongqing Xianwai SOHO and Club
重庆线外 SOHO 及会所

FEATURES
项目亮点

The architects design a pattern that abstracted from the circuit board, and make it throughout the indications of the building and community with obvious theme and visual effect.

建筑师设计了由电路板抽象而来的图案，并将其贯穿于整个建筑及小区的标识中，主题及视觉效果突出。

Location: Chongqing, China
Architecture Design: China Architecture Design & Research Group
Completion: 2010

项目地点：中国重庆市
设计单位：中国建筑设计研究院
竣工时间：2010 年

Overview

Three SOHO buildings are all 32-layer high-rise towers with a height of 99.8 m; each tower building has a typical floor area of 495 m² and a ground floor area of 16,000 m²; the total ground floor area of the three towers is 48,000 m², and the underground construction area is 5,960 m². The club house has two layers with a total height of 10.2 m and a total construction area of 2,480 m².

项目概况

三座 SOHO 均为地上 32 层的高层塔楼建筑，建筑高度 99.8 m，每座塔楼标准层面积均为 495 m²，每座塔楼地上建筑面积均为 16 000 m²；三座塔楼地上总建筑面积为 48 000 m²，地下建筑面积共 5 960 m²。会所共两层，南低北高，总高度 10.2 m，总建筑面积 2 480 m²。

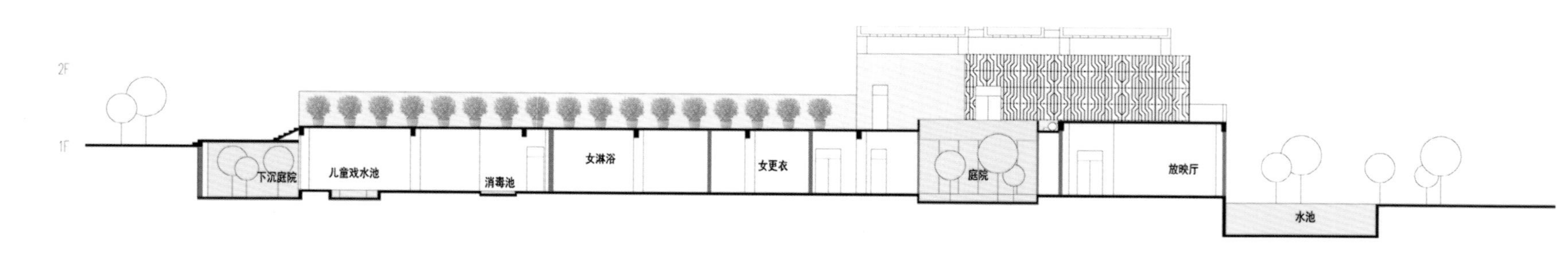

Sectional Drawing 剖面图

Architectural Design

The building looks like exquisitely carved crystal glass lightly floating on the surface of the water. It is located in a land with relief of high in the south and low in the north, and people can reach the second floor of the building in the south side through the roof garden in the north side to overlook the whole southern square. Three towers adopt the same facade style. The architectural pattern designed by the architects spreads throughout the building interior space, indications and LOGO of the community; therefore, it gets its name "Xianwai".

SOHO apartment buildings are vertical continuous reversers, which represent the interconnection and communication of the information age, and form a cross-vertical comparison with the horizontal reverse of the office building in the west side (uncompleted). The external wall selects shutter, grey brick, metal and other materials, which reflects the fusion of tradition and modern with strong sense of science and technology.

The club is located in the center of the community, early used as sales office, which reflects the theme and visual effect of the whole community. With a few of concise masses in different sizes, and pattern that abstracted from the circuit board, the club organizes pool in front, and the lights flown from the gap of the facade design pattern stretch all the way to the bottom of the pool, which forms strong dynamic lighting effect.

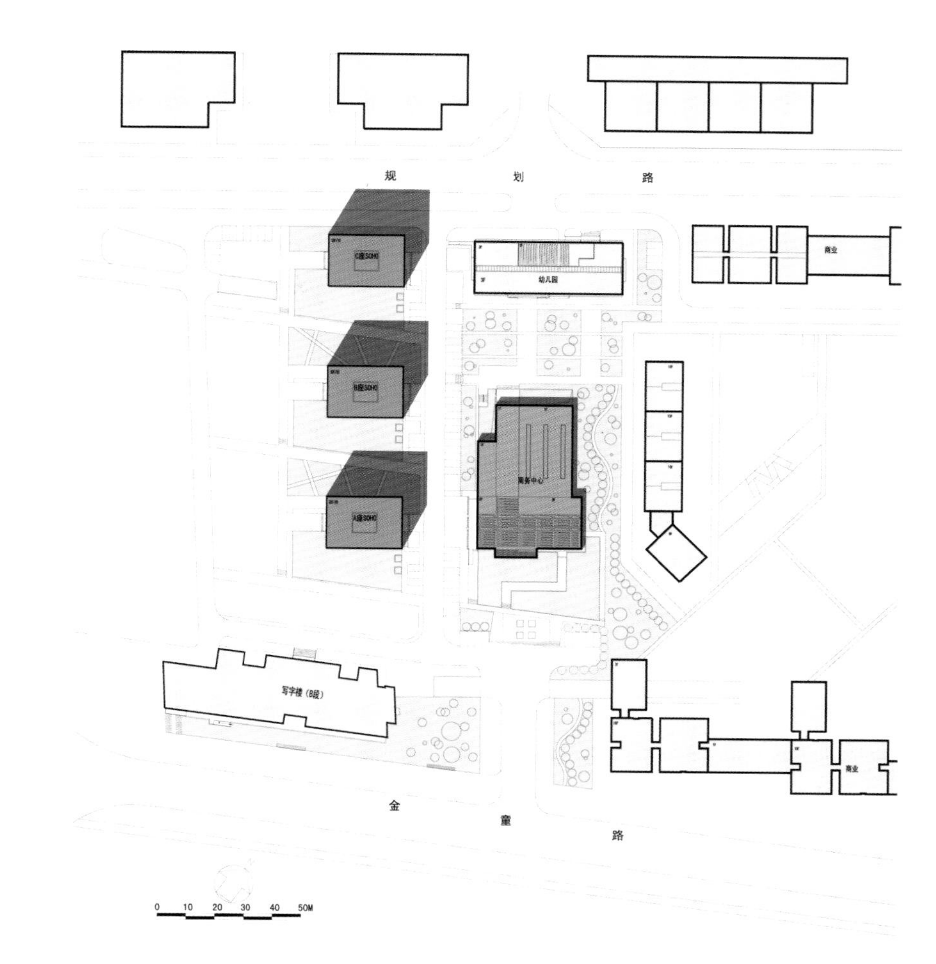

Site Plan 总平面图

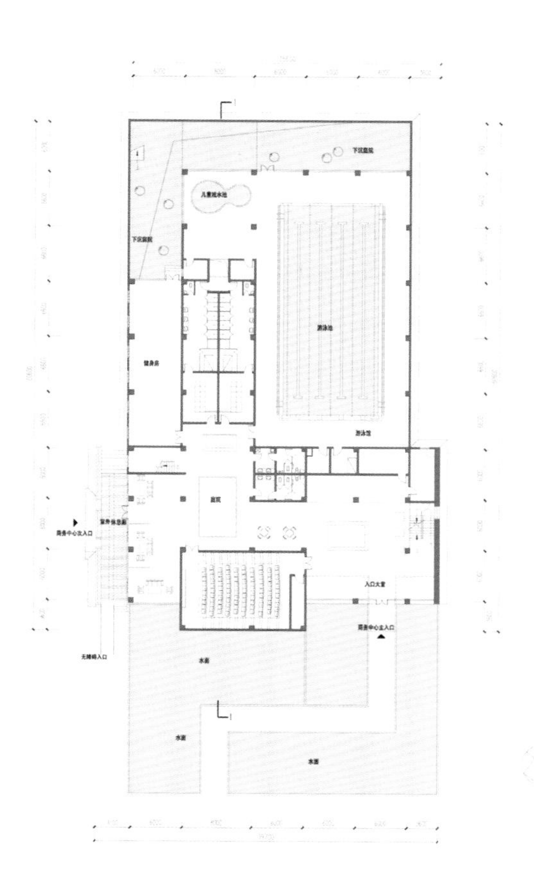

Club First Floor Plan 会所一层平面图

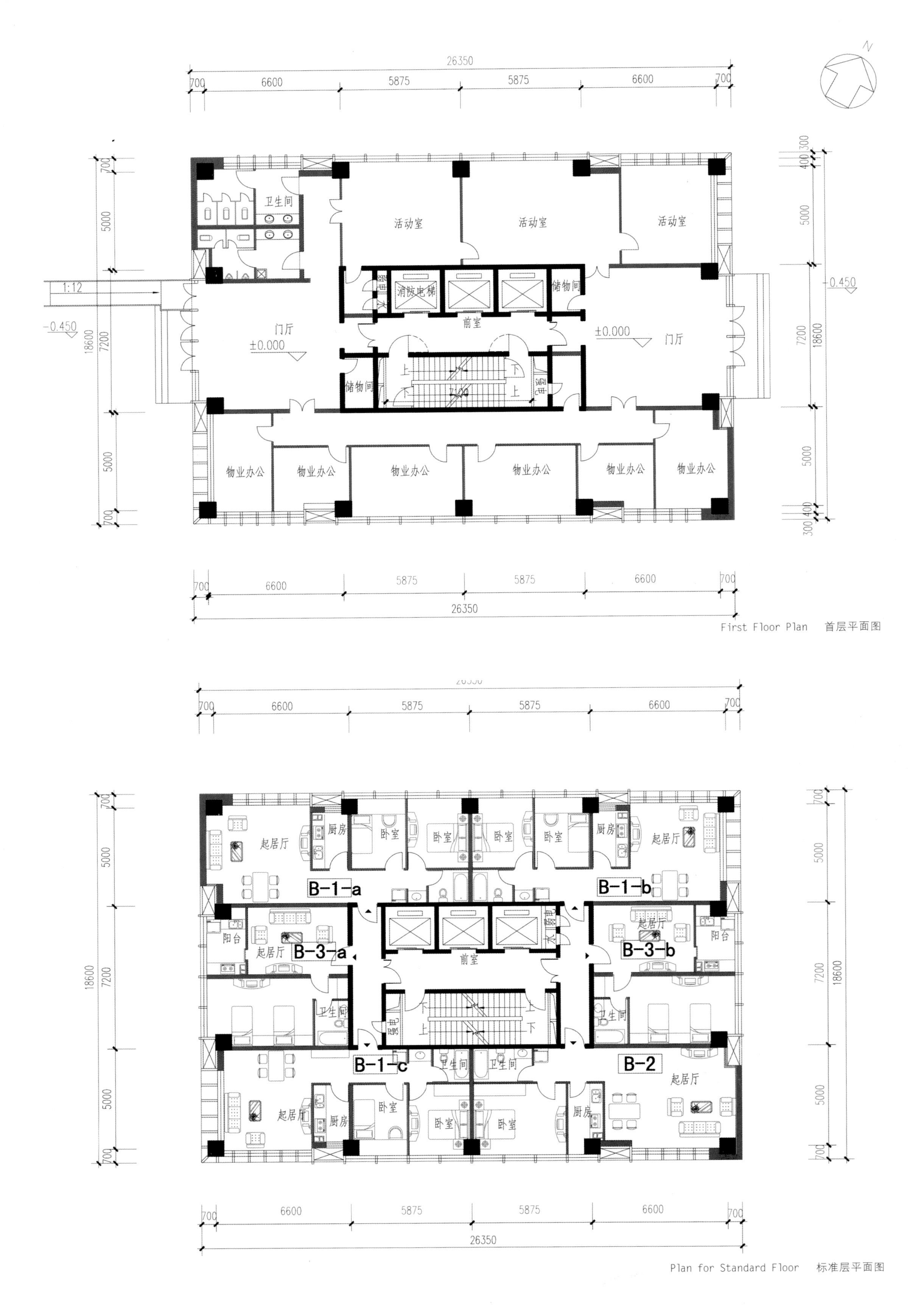

First Floor Plan　首层平面图

Plan for Standard Floor　标准层平面图

建筑设计

建筑犹如玲珑剔透的玻璃体，轻盈地浮在水面上。建筑坐落在南低北高的用地上，人们由北侧穿过屋顶花园可到达建筑南部的二层，借以远眺整个南区广场。三座塔楼采用相同的外立面风格。由建筑师设计的建筑图案贯穿于整个建筑的室内以及小区的标识与 LOGO 中，并由此定下案名“线外”。

SOHO 公寓楼为垂直连续扭转体，表现信息时代的互联与沟通，与西边写字楼（尚未建成）的水平扭转形成一横一竖的对比。外墙面选用百叶、灰砖、金属等材料，体现传统与现代的融合，具有强烈的科技感。

会所位于小区的中心，前期用作售楼处，体现整个小区的主题及视觉效果。设计以几个不同大小的简洁的虚实体块组成，并辅以同样的由电路板抽象而来的图案，馆前为水池。夜晚，由立面图案缝隙中流淌下来的灯光一直延伸到池底，有极强的动感光影效果。

KEY WORDS
关键词

Drum Suspension 筒式悬挂结构

Unique Structure Modeling 造型独特

Eco-friendly 生态环保

Guiyang International Conference and Exhibition Center 201 Tower
贵阳国际会议展览中心 201 大厦

FEATURES
项目亮点

The 201 tower is a modern building designed in the prototype of Guizhou minority Lusheng, when completion, it will shapes like a lush green tree. The building uses 12 CFT columns to support and is the world's tallest drum suspension structure building which technology currently is the first in the country.

201 大厦采用了贵州少数民族芦笙为原型的现代楼宇设计，建成后的造型像一棵绿色繁茂的大树，建筑采用 12 根钢管混凝土柱支撑，是全球最高的筒式悬挂结构建筑，目前在国内建筑中尚属首创。

Location: Guiyang City, Guizhou, China
Architectural Design: Shenzhen O'Brien Engineering Consultants Ltd.
Floor Area: 51 000 m^2

项目地点：中国贵州省贵阳市
设计单位：深圳市欧博工程设计顾问有限公司
建筑面积：51 000 m^2

Overview

By meeting international authority LEED green ecological assessment system and the national green building standards on energy saving, environmental protection and other ecological indicators, Guiyang International Conference and Exhibition Center is one of the most functionally complete, the most advanced multi-purpose facility Convention and Exhibition Centre in the country and an important window for Guiyang City's foreign trade activities.

项目概况

贵阳国际会议展览中心在建筑节能、生态环保等生态指标方面达到国际权威绿色生态评估体系 LEED 和国家绿色生态标准，是国内功能配套最完备、设施最先进的多功能会议展览中心之一，贵阳市对外经贸活动的重要窗口。

类博览会

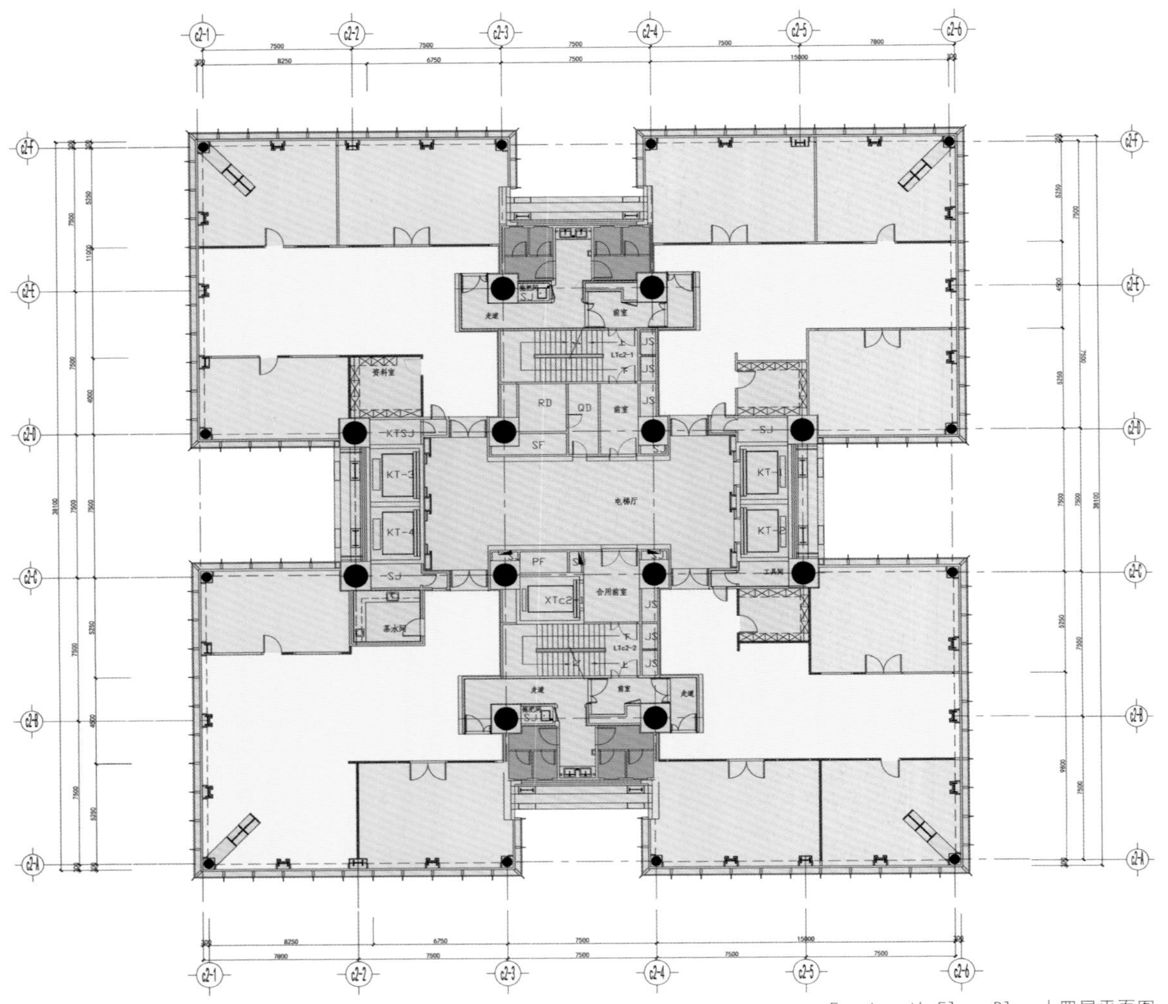

Fourteenth Floor Plan 十四层平面图

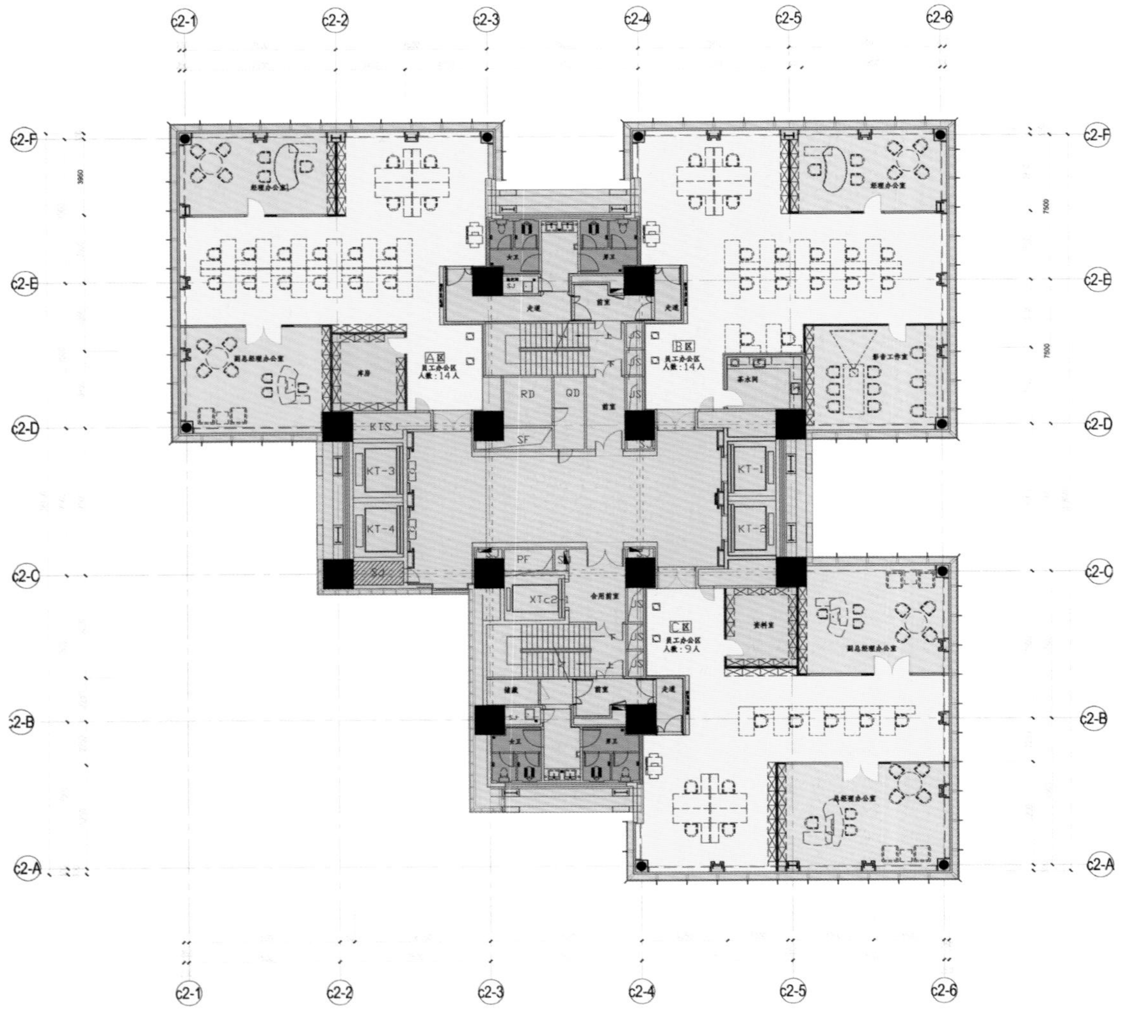

Twentieth Floor Plan 十五层平面图

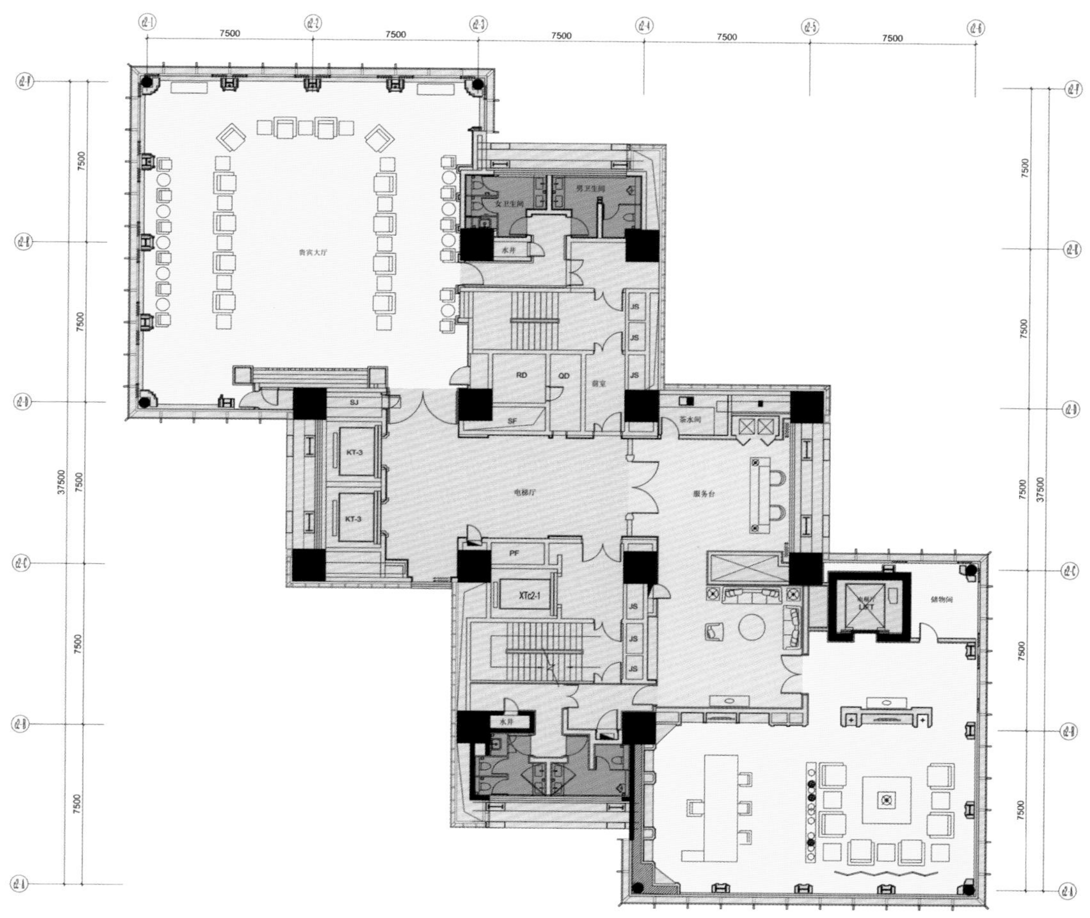

Thirtieth Floor Plan 三十层平面图

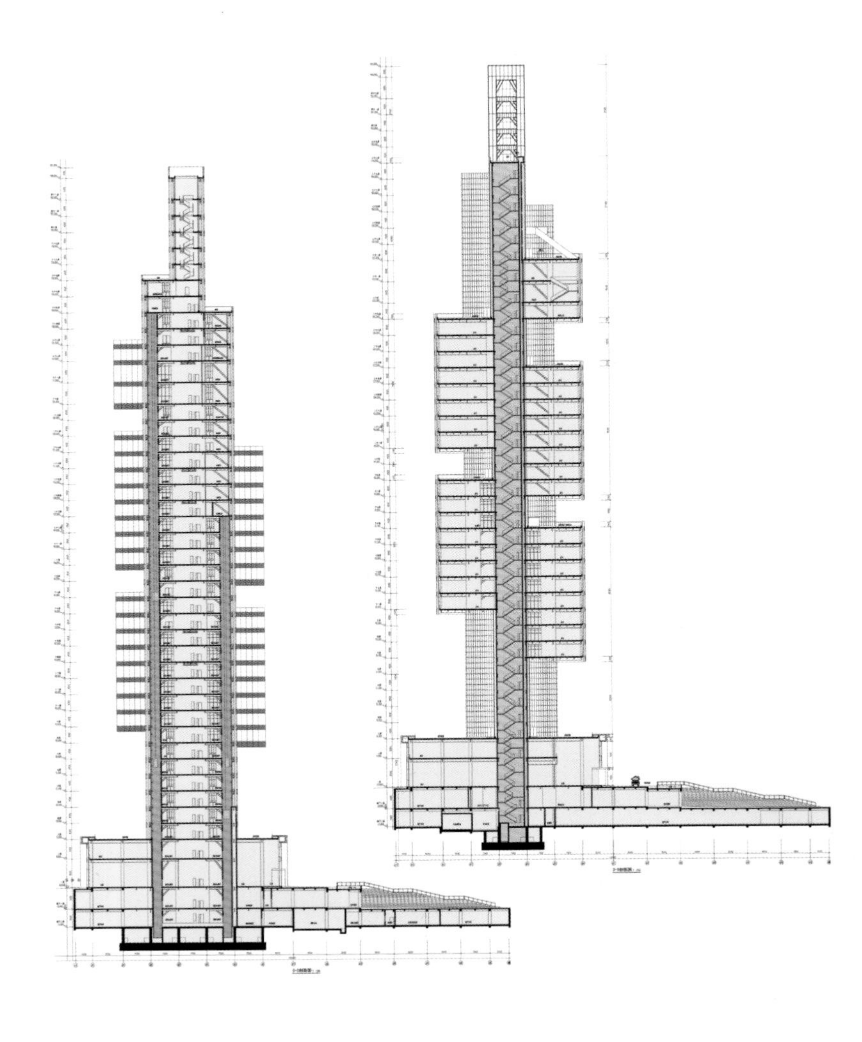

Architectural Design

With a full-height of 201 m, the Guiyang International Conference and Exhibition Center 201 Tower is a modern building designed in the prototype of Guizhou minority Lusheng, when completion, it will shapes like a lush green tree and become Jinyang even Guiyang's landmark. The building uses 12 CFT columns to support, which following the German car company BMW headquarters building and Singapore's the Treasury building, and is the world's third drum suspension structure building and the world's tallest drum suspension structure building, which technology currently is the first of its kind in the country. With unique new shape, the building will be put into use independent, high levels of 5A class office building in the future. Accompanying international ecological conference centers will use steel and concrete structure of the large span of 54 m steel truss system which is rare in architecture world. Not only that the 201 building were full use of green technology and to achieve ecological, environmentally friendly philosophy from the design, construction, and materials to the latter part of the operation.

建筑设计

贵阳国际会议展览中心 201 大厦全高 201 m，采用了贵州少数民族芦笙为原型的现代楼宇设计，建成后的造型像一棵绿色繁茂的大树，届时 201 大厦将成为金阳乃至贵阳的标志性建筑。建筑采用 12 根钢管混凝土柱支撑，是继德国宝马汽车公司总部大楼与新加坡国库大楼之后，全球第三个采用筒式悬挂结构的建筑，是全球最高的筒式悬挂结构建筑，目前在国内建筑中尚属首创。建筑造型独特、新颖，未来将投入运用为独立式、高水准的 5A 级办公写字楼。与之配套的国际生态会议中心则采用型钢混凝土结构和跨度达 54 m 的大型钢桁架结构体系，在世界建筑结构体系中属罕见。不仅如此，201 大厦从设计、施工、材料到后期的运行，均充分运用了绿色科技并实现了生态、环保的理念。

Hotel Building

酒店建筑

Difference Styles
风格各异

Modern Materials
现代材料

Local Characteristics
地域特色

Beautiful Environment
环境优美

KEY WORDS
关键词

Characteristic Watchtower 特色碉楼

Modern Material 现代材料

Rich-colored 色彩丰富

Huanglong Seercuo International Hotel
黄龙瑟尔嵯国际大酒店

FEATURES
项目亮点

Three grouped watchtowers and the single watchtower are echoing with each other, creating a sense of Tibetan and Qiang villages. And modern material, glass, is used in the towers of traditional sense, exuding the tinges of modernity.

三栋成组的碉楼与一栋独立的碉楼彼此呼应，母题的反复出现营造出藏羌山寨的空间意境。但这种地域传统“意境”因采用了现代材料——玻璃来表达，而散发出一股时代气息。

Location: Sichuan Aba Tibetan and Qiang Autonomous Prefecture, China
Architectural Design: School of Architecture, Tsinghua University / An-design Architects
Total Land Area: 30,800 m^2
Total Floor Area: 22,185 m^2

项目地点：中国四川省阿坝藏族羌族自治州
设计单位：清华大学建筑学院、北京华清安地建筑设计事务所有限公司
总占地面积：30 800 m^2
总建筑面积：22 185 m^2

Overview

The hotel accommodates 211 rooms and supporting amenities & services such as dining, meeting, entertainment and fitness. The building is in a low-to-high volume. The 2nd floor is at the lowest point and the 8th floor occupies the highest point.

项目概况

酒店包含211间客房，以及餐饮、会议、娱乐、康体等服务配套设施。建筑从低向高，层层叠落，最低处为2层，最高处为8层。

Architectural Design

In terms of architectural features, the project selected local unique Tibetan and Qiang residential towers as styling motif. Three grouped watchtowers and the single watchtower are echoing with each other, creating a sense of Tibetan and Qiang villages. And modern material, glass, is used in the towers of traditional sense, exuding the tinges of modernity. In the main building, designers took the form of Tibetan multi-storey Buddhist temple as a reference; beige wall derived from the color of "golden sand field" which is a famous spot in Huanglong Scenic Area; saffron on the moulding on the top of the wall is the usual color used for Tibetan regional building; part of copper sheet decorated wall imitated the Tibetan religious building; and the building bottom borrowed the conventional bricklaying from local residences.

建筑设计

在建筑特征上，本方案选用了当地藏羌民居独具特色的碉楼作为造型母题。三栋成组的碉楼与一栋独立的碉楼彼此呼应，母题的反复出现营造出藏羌山寨的空间意境。但这种地域传统“意境”因采用了现代材料——玻璃来表达，而散发出一股时代气息。在建筑主体上，本方案参照了藏族多层佛阁的形式特色；米黄色的墙面采自黄龙风景区内著名景点黄龙梯湖——“金沙铺地”的色彩；墙顶压条的藏红色来自藏族地域建筑惯用的色彩；部分采用铜皮装饰的墙面，是借鉴了藏族宗教建筑经常用金属皮做装饰的传统；建筑底层采用深色毛石墙面，则是参照了当地民居底层墙面惯用的砌法。

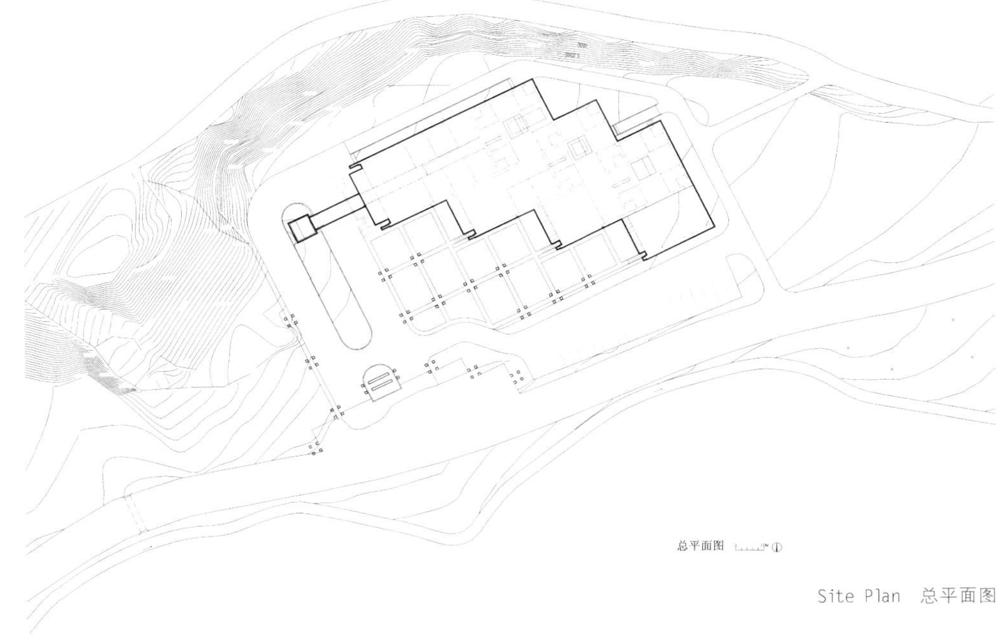

Site Plan 总平面图

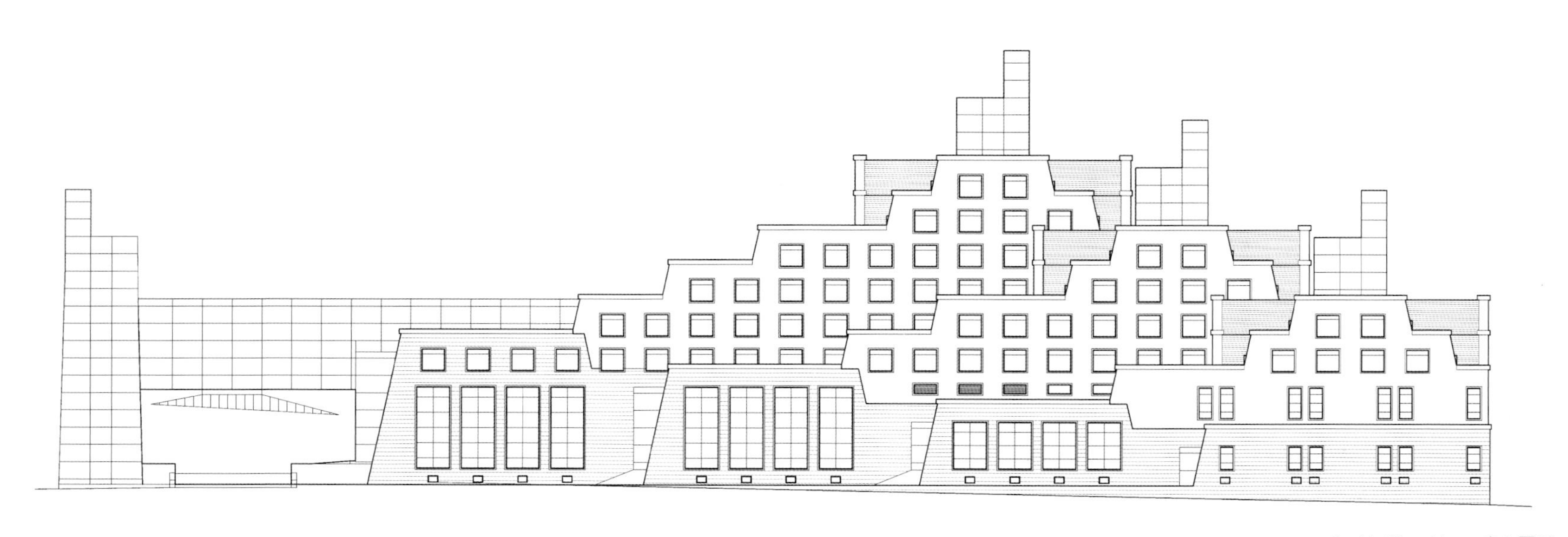

South Elevation 南立面图

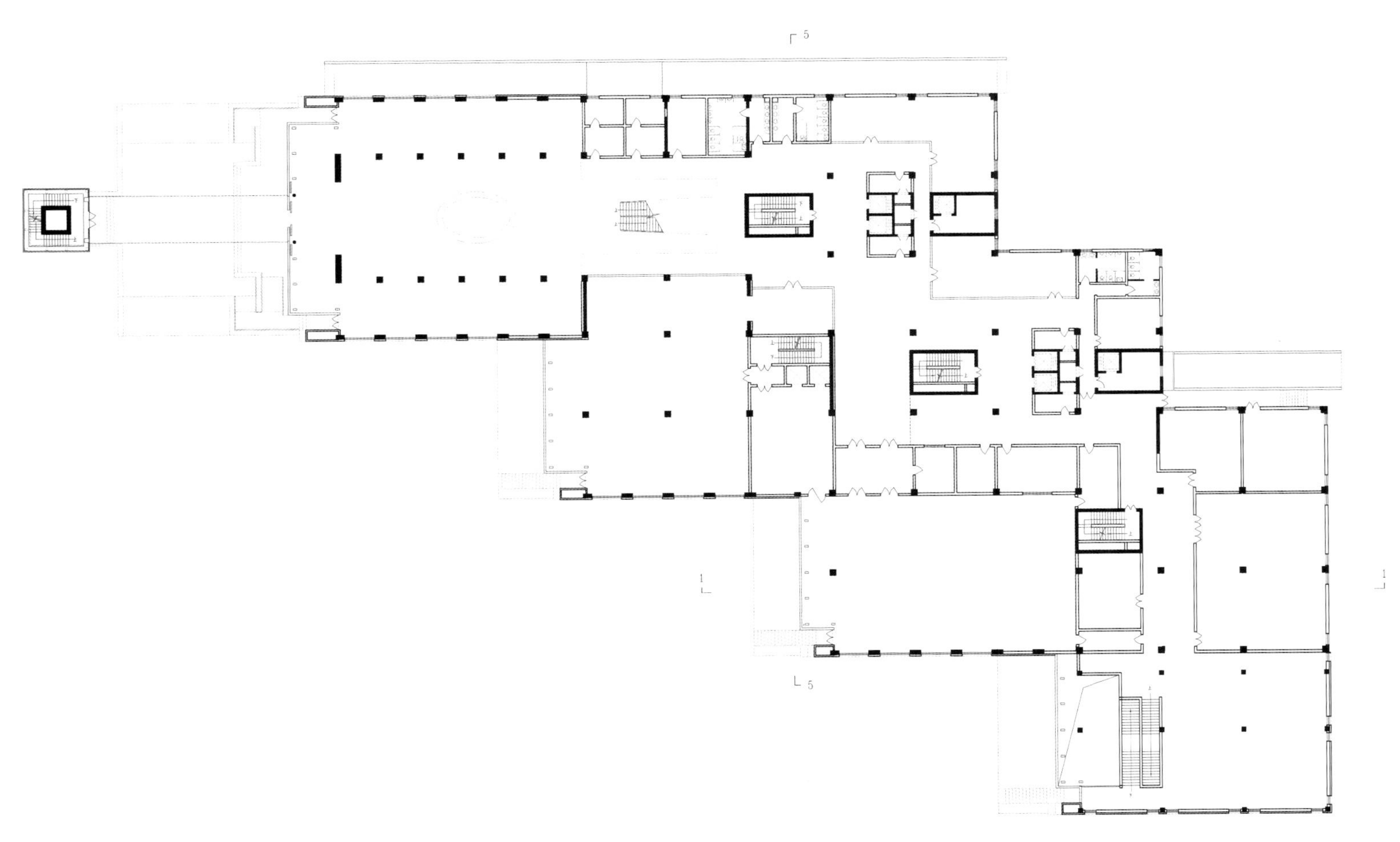

First Floor Plan 一层平面图

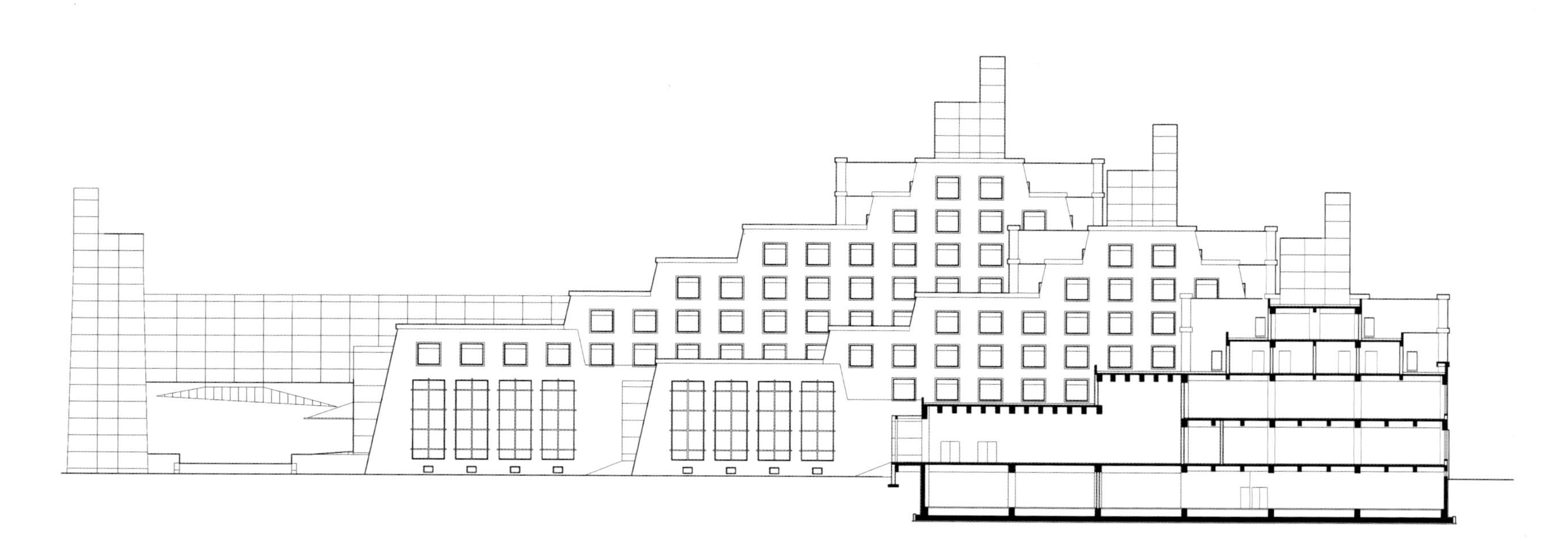

Elevation 立面图

KEY WORDS
关键词

Curve Shape 曲线造型
Glass Curtain Wall 玻璃幕墙
Sense of Rhythm 韵律感

Renaissance Tianjin Lakeview Hotel
万丽天津宾馆

FEATURES
项目亮点

The project design technique is lively and contracted with rich connotation. The overall modeling looks sedate and coherent, and it also uses stone architraves that bear rhythmical changes to make the outside eaves of the building feel rich and unified.

设计手法明快、简约，内涵丰富；整体造型稳重、连贯，利用石材线脚凹凸有韵律的变化，使建筑外檐既丰富又统一。

Location: Hexi District, Tianjin, China
Architecture Design: Tianjin Architecture Design Institute
Completion: 2010

项目地点：中国天津市河西区
设计单位：天津市建筑设计院
竣工时间：2010 年

Overview

The project has a construction height of 53.25 m, land area of 77,680 m^2, a total floor area of 95,776.8 m^2 and plot ratio of 1.44. It equips 368 guestrooms and 96 apartments.

项目概况

本项目建筑高度 53.25 m，用地面积 77 680 m^2，总建筑面积 95 766.8 m^2，容积率 1.44。拥有 368 套客房、96 套公寓。

Architectural Design

The project design technique is lively and contracted with rich connotation. The overall modeling looks sedate and coherent, and it also uses stone architraves that bear rhythmical changes to make the outside eaves of the building feel rich and unified. Central part in the north side uses curve modeling to fully combine with the landscape; and the south side applies complete straight-line modeling, which gives prominence to the concise and generous main entrance. Nine-floor high glass curtain wall at the main entrance forms strong contrast with the large-scale rain cover and solid walls on both sides, strengthening the visual effect of the entrance and highlighting the building properties and the identification of the main entrance.

建筑设计

设计手法明快、简约，内涵丰富。整体造型稳重、连贯。利用石材线脚凹凸有韵律的变化，使建筑外檐既丰富又统一。北侧中部采用曲线造型，与景观充分结合，南侧采用直线完整造型，突出主入口的简洁、大方。主入口处九层高的玻璃幕墙与大尺度的雨篷及两侧实墙形成强烈的对比，加强了入口的视觉效果，突出了酒店建筑的属性和主入口的标识性。

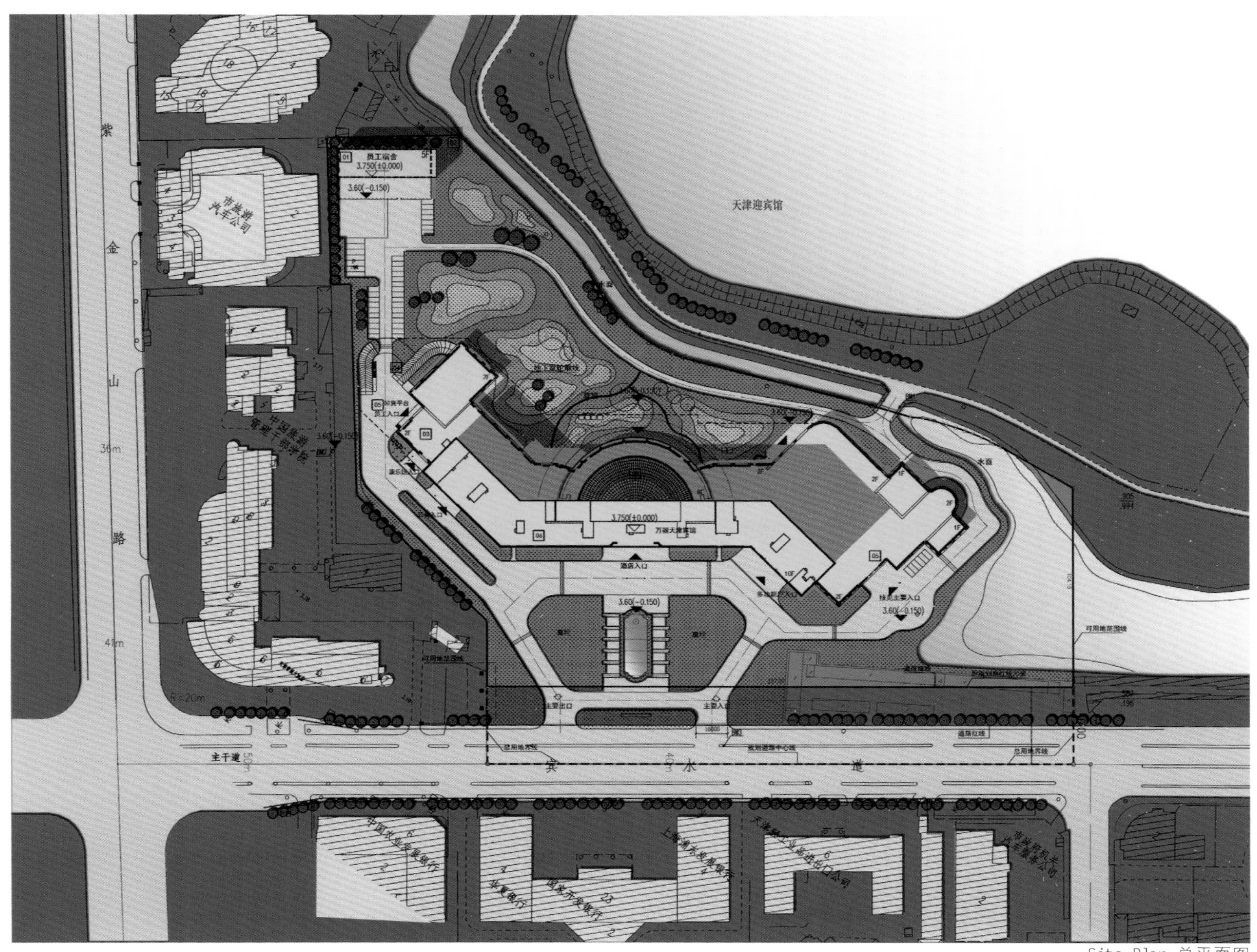

Site Plan 总平面图

Elevation Extending 立面展开图

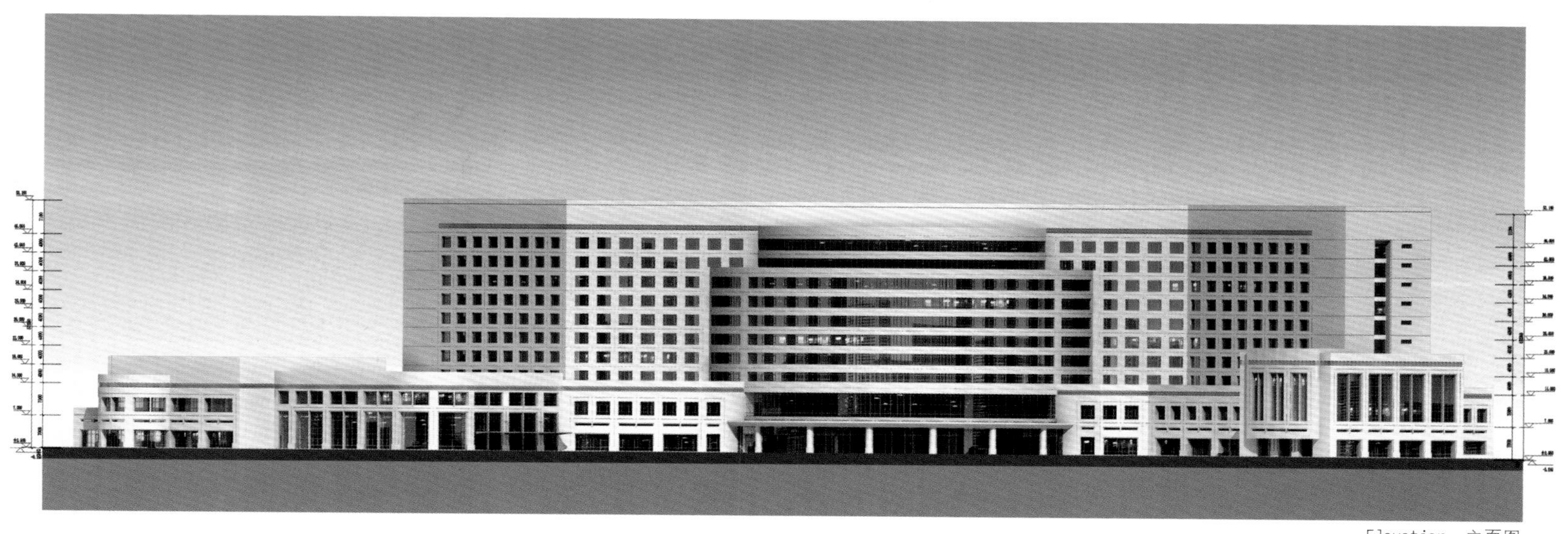

Elevation 立面图

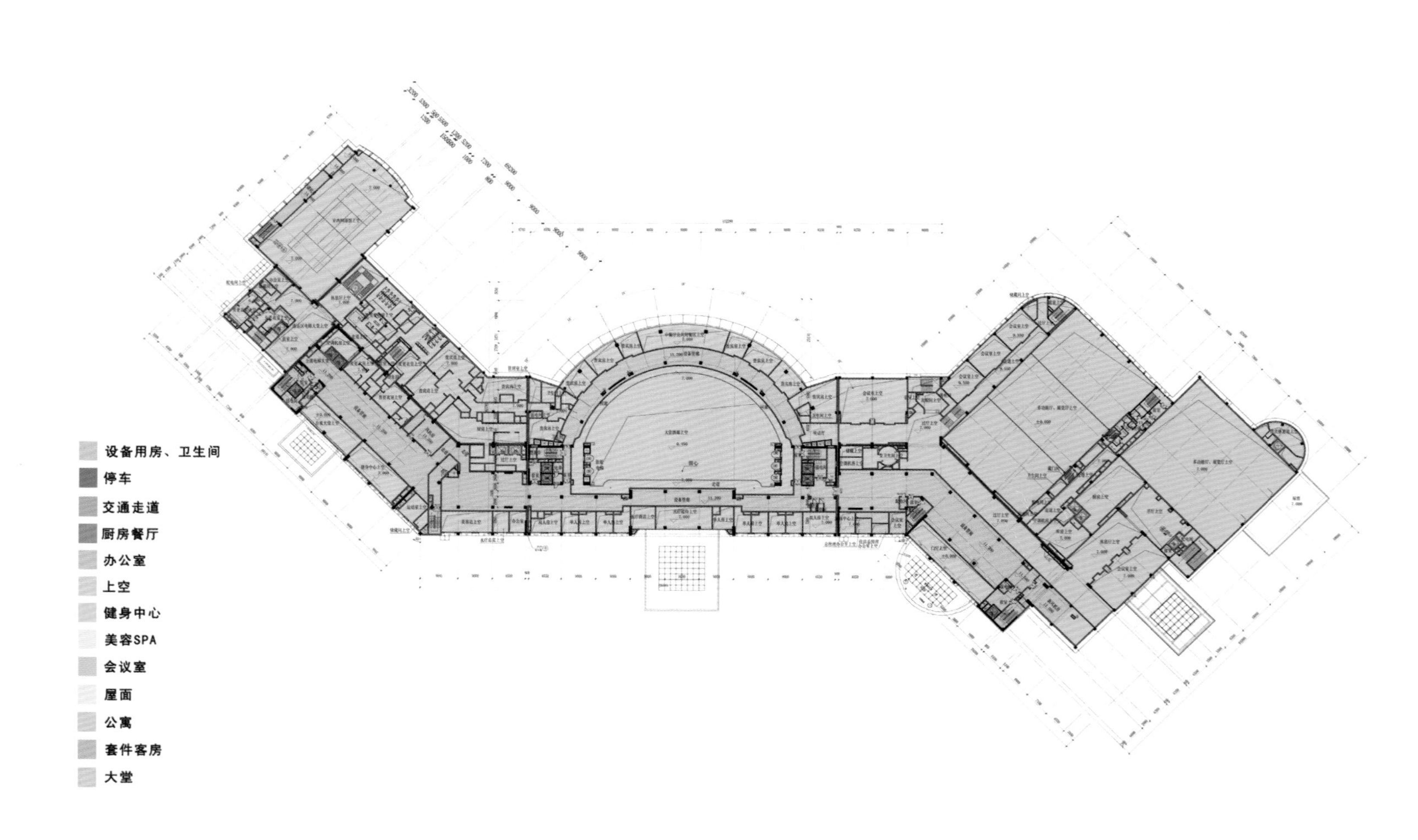
设备用房、卫生间
停车
交通走道
厨房餐厅
办公室
上空
健身中心
美容SPA
会议室
屋面
公寓
套件客房
大堂

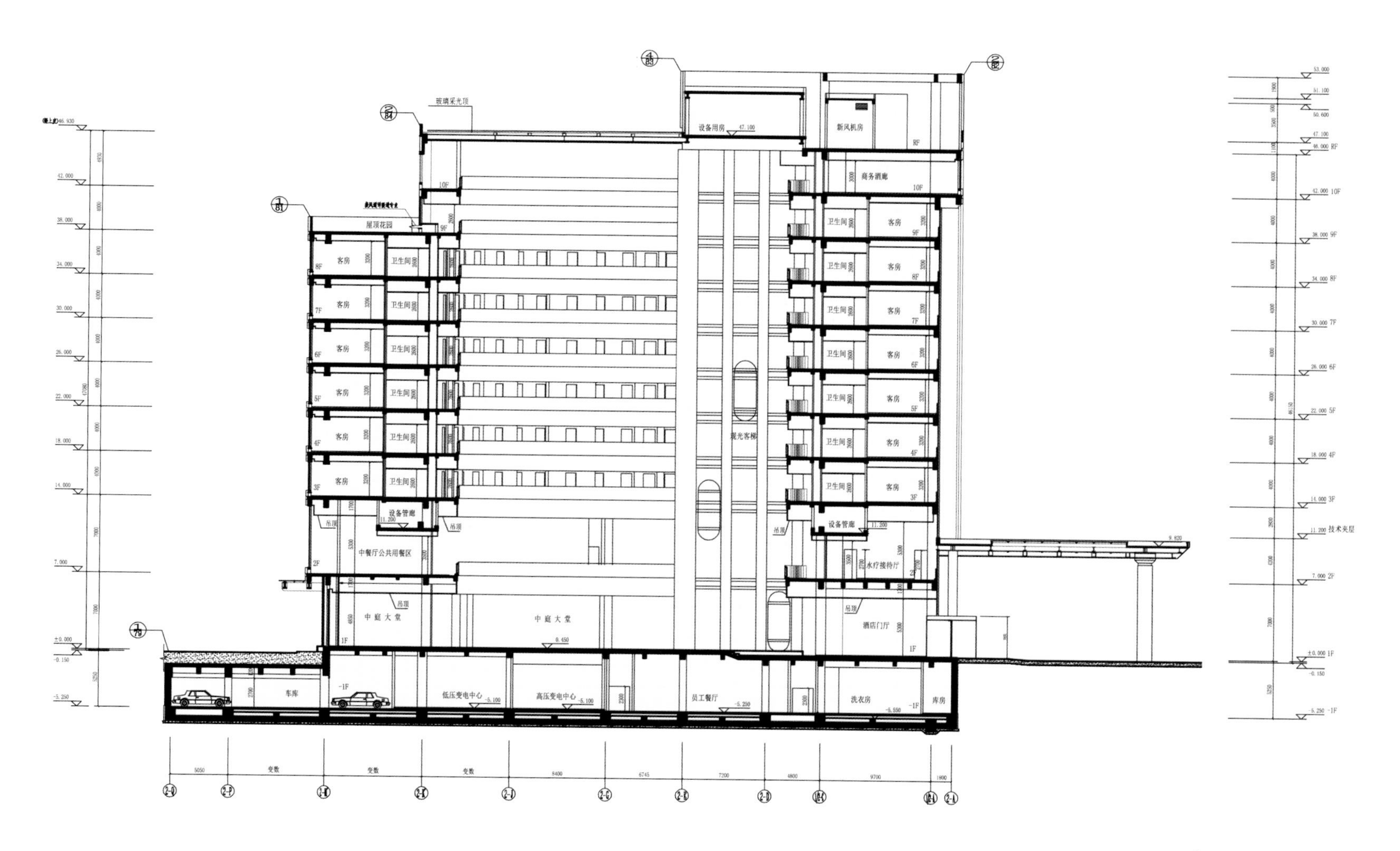

Sectional Drawing 剖面图

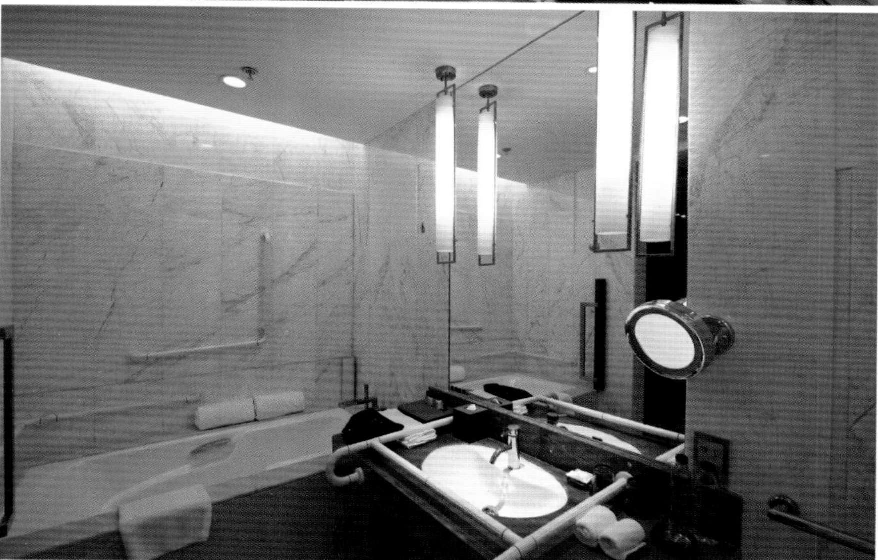

KEY WORDS
关键词

Into the Environment 融入环境

Stone Skin 石材表皮

Modeling Language 造型语言

Beijing Guquan Convention Center (CITIC Jingling Hotel)

北京谷泉会议中心（中信金陵酒店）

FEATURES
项目亮点

In consideration of favorable natural condition, designers imitated stone and stream and situated the building into the surrounding environment to realize a harmonious combination.

根据基地依山傍水、层峦叠落的自然条件，将石头和溪流作为酒店设计概念，使之与周边环境相呼应，将建筑融入环境之中。

Location: Beijing, China
Architectural Design: China Architecture Design & Research Group

项目地点：中国北京市
设计单位：中国建筑设计研究院

Overview

Located in Northern Dahuashan, Pinggu District where the fruit-bearing forest falling terrace by terrace, this convention center is about 90 km away from Beijing downtown. Situated at the north end of the ridge line, it overlooks West Valley Reservoir. In general, the entire site is surrounded by continuous mountains and lush vegetation.

项目概况

北京谷泉会议中位于北京市平谷区大华山北麓，距北京城区约 90 km。所在场地处于大华山面向西峪水库山脊线北端的山坳中，为台阶状层层跌落的果林，北侧正对西峪水库。整个场地周边山势连绵，植被茂密。

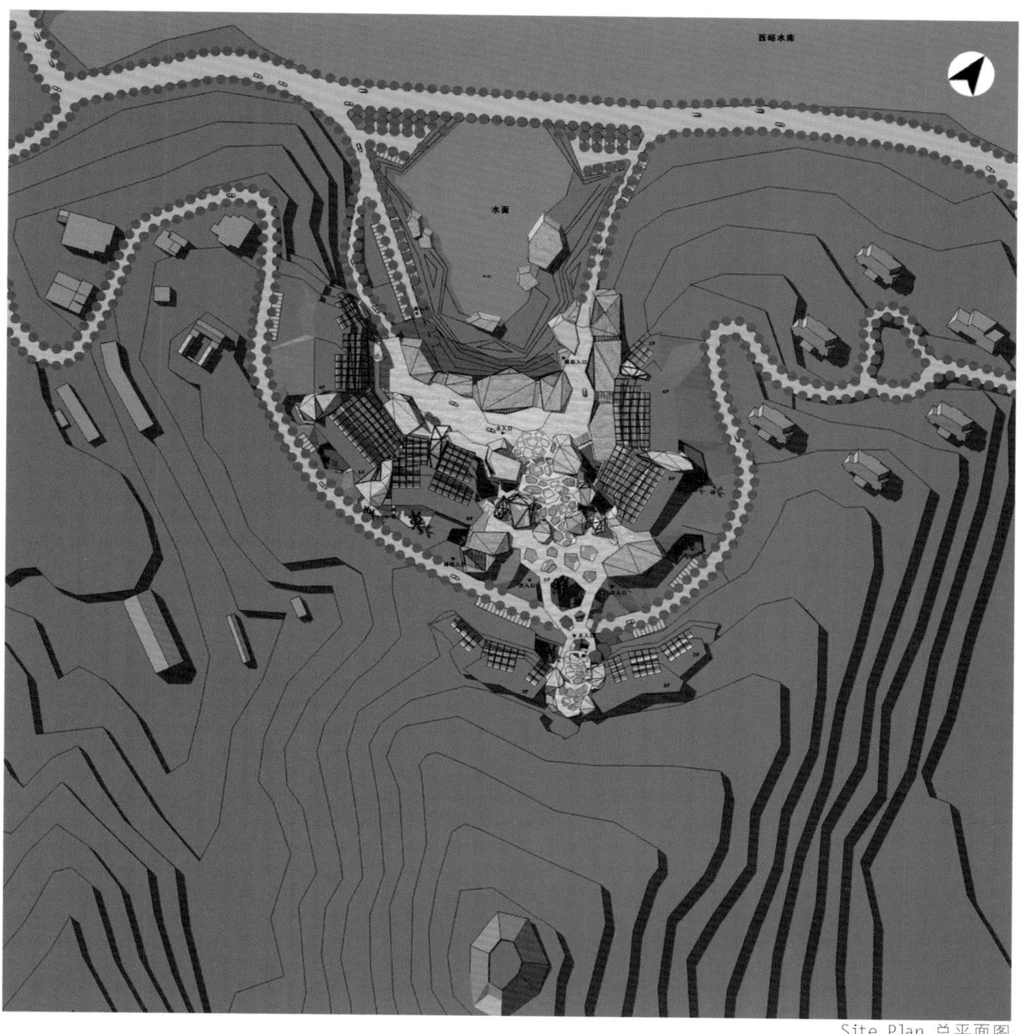

Site Plan 总平面图

Architectural Design

In consideration of favorable natural condition, designers imitated stone and stream and situated the building into the surrounding environment to realize a harmonious combination. Guest rooms are classified on the terrace falling level by level according to the gradient of the slope. Irregular volumes look like a pile of stones that salute to those magnificent stone buildings in mountain. Precast concrete dry-hang panel, which is viewed as real stone, is used for the building and being embedded with natural stone power to simulate natural color, thus to achieve the best combination. In addition, the dry-hang panel makes the building more wild and generous.

日景 .tif

建筑设计

根据基地依山傍水、层峦叠落的自然条件，将石头和溪流作为酒店设计概念，使之与周边环境相呼应，将建筑融入环境之中。根据山地坡度，将客房设计成层层跌落的退台形式，更好地利用现有地形条件。建筑意象来源于山间壮美的石砌建筑，通过不规则形体的组合呈现出巨石堆砌的意象。建筑材料选用预制混凝土干挂板，其表面肌理拓自山间真实山石，并加入天然石粉以模拟天然色彩，使建筑最大限度地融入环境之中。预制混凝土干挂板的运用给人以粗犷、大气的感受。

KEY WORDS
关键词

Stretch 体型舒展

Patchwork 错落有致

Into the Environment 融入环境

State Guesthouse of Hainan Branch of the People's Liberation Army General Hospital

解放军总医院海南分院国宾馆

FEATURES
项目亮点

The hotel mass is stretch, rhythmic patchwork roofs and echoes the surrounding mountains. The beige building facade looks close to the coastal color, the cold gray color roof coordinates with the mountain and sea hues, the overall low-key colors make the building blending with the environment.

酒店建筑采用舒展、水平的体型，富有节奏感的屋顶群体组合错落有致，与周边的山峦轮廓相呼应。建筑立面采用接近沙滩的米黄色调与海岸相融，屋顶采用冷灰色调与山海协调，整体低调的色彩使建筑与环境融为一体。

Location: Sanya City, Hainan, China
Architectural Design: Architectural Design Institute, South China University of Technology
Total Floor Area: 87,000 m²

项目地点：中国海南省三亚市
设计单位：华南理工大学建筑设计院
总建筑面积：87 000 m²

Overview

State Guesthouse of Hainan Branch of the People's Liberation Army General Hospital is a strategic project to promote international tourism in Hainan Island; the hotel is the center for the head of the Central Military Commission leadership, Hainan province and city leaders as a Health & Wellness Centre and the ambassadors' reception center. Located in Sanya Haitang Bay in the southern tip of Hainan, which shared the name "Oriental Hawaii", the hotel has 449 rooms and a total building height of 35.8 m². It includes a looby, rooms, restaurant, conference, sports, entertainment, spa room and other functions.

项目概况

解放军总医院海南分院国宾馆是推进海南国际旅游岛建设的一项战略工程，是中央军委首长、总部领导以及海南省、市领导的保健疗养中心和国宾接待中心。项目地处“东方夏威夷”的海南三亚市海棠湾南端，客房数为 449 间，总建筑高度为 35.8 m，酒店内设有大堂、客房、餐厅、会议、康体、娱乐以及水疗 SPA 等功能用房。

Architectural Design

Located in Sanya Haitang bay, the hotel with mountain, sea, sky, coconut, Silver, Green formed the base of the most unique natural landscape. The hotel mass is stretch, rhythmic patchwork roofs and echoes the surrounding mountains. The beige building facade looks close to the coastal color, the cold gray color roof coordinates with the mountain and sea hues, the overall low-key colors make the building blending with the environment.

建筑设计

项目位于三亚海棠弯，山、海、天与椰林、银滩、绿岛组成了基地最具特色的自然景观。酒店建筑采用舒展、水平的体型，富有节奏感的屋顶群体组合错落有致，与周边的山峦轮廓相呼应。建筑立面采用接近沙滩的米黄色调与海岸相融，屋顶采用冷灰色调与山海协调，整体低调的色彩使建筑与环境融为一体。

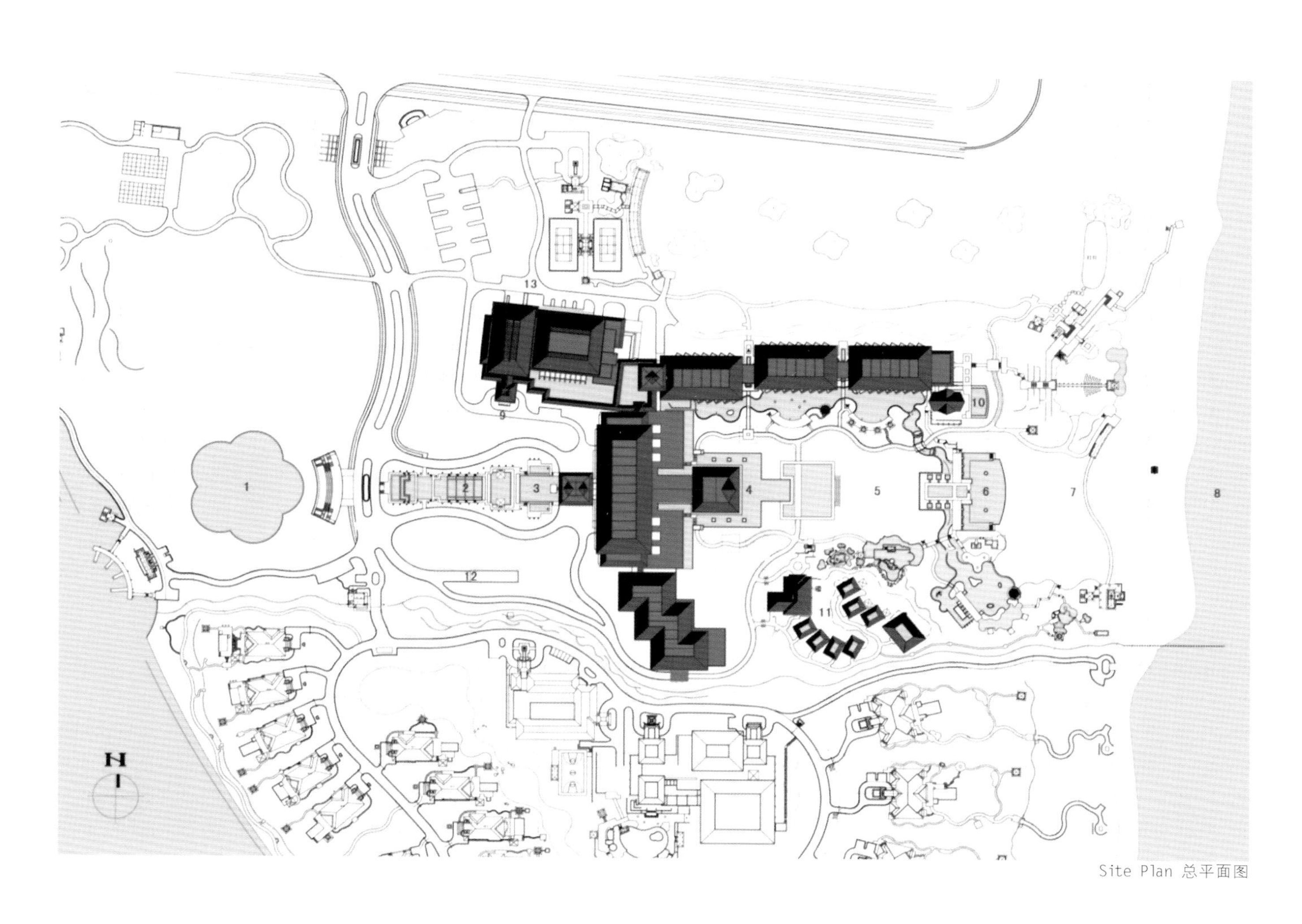

Site Plan 总平面图

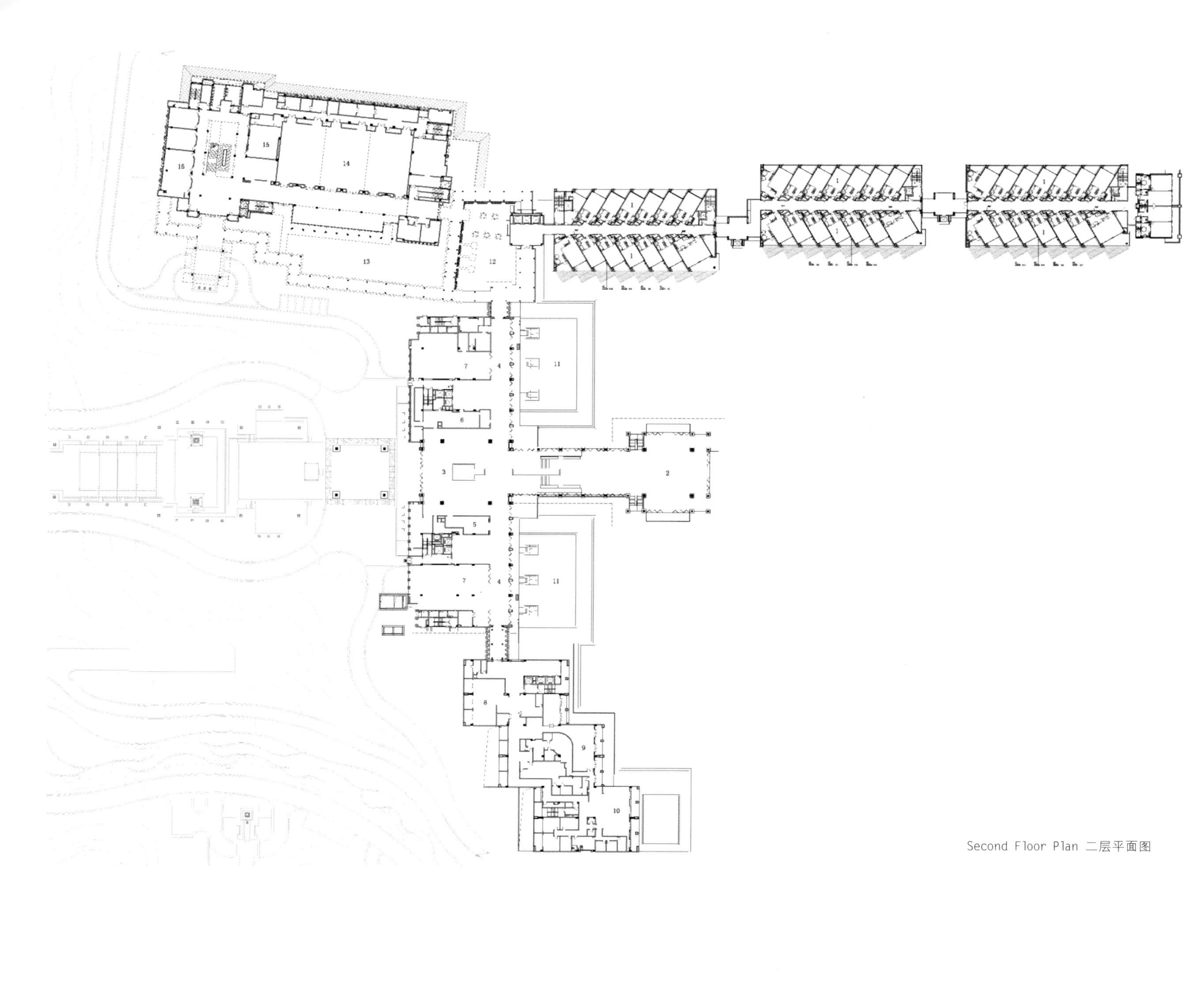

Second Floor Plan 二层平面图

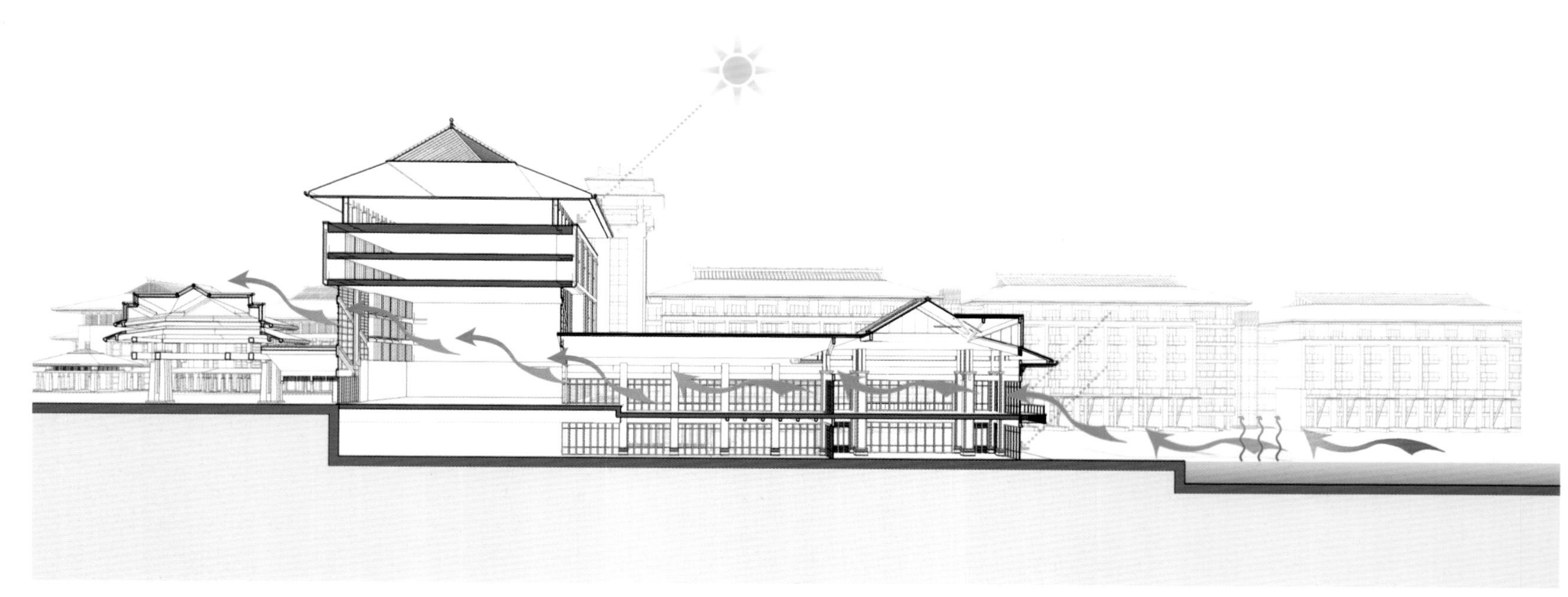